Aquatic Ecosystems: Assessment and Management

Aquatic Ecosystems: Assessment and Management

Edited by Ellie Ward

SYRAWOOD
PUBLISHING HOUSE
New York

Published by Syrawood Publishing House,
750 Third Avenue, 9th Floor,
New York, NY 10017, USA
www.syrawoodpublishinghouse.com

Aquatic Ecosystems: Assessment and Management
Edited by Ellie Ward

International Standard Book Number: 978-1-68286-699-3 (Hardback)

Cataloging-in-Publication Data

Aquatic ecosystems : assessment and management / edited by Ellie Ward.
p. cm.
Includes bibliographical references and index.
ISBN 978-1-68286-699-3
1. Aquatic ecology. 2. Aquatic biology. 3. Ecosystem management.
4. Ecological assessment (Biology). I. Ward, Ellie.

QH541.5.W3 A68 2019
577.6--dc23

Printed in China.

TABLE OF CONTENTS

PREFACE

The main aim of this book is to educate learners and enhance their research focus by presenting diverse topics covering this vast field. This is an advanced book which compiles significant studies by distinguished experts in the area of analysis. This book addresses successive solutions to the challenges arising in the area of application, along with it; the book provides scope for future developments.

Aquatic ecosystems are vast reserves of water biodiversity, essential for the maintenance of ecological balance in the environment. Aquatic ecosystems are primarily of two types- marine and freshwater. All water bodies are part of the aquatic ecosystem. They are essential for recycling of nutrients, purification of water, flood attenuation, recharge of groundwater and provision of habitat for different species of animals and plants. The health of an aquatic ecosystem is affected by the physical, chemical or biological changes in the environment. Overfishing, water pollution, climate change and coastal land encroachment have significant negative consequences on the aquatic ecosystem. This book is a compilation of chapters that discuss the most vital concepts and emerging trends in the field of aquatic ecosystem. Some of the diverse topics covered in this book address the varied branches that fall under this category. With state-of-the-art inputs by acclaimed experts in this field, this book targets marine biologists, oceanographers, limnologists, aquatic ecologists as well as students and researchers associated with this domain.

It was a great honour to edit this book, though there were challenges, as it involved a lot of communication and networking between me and the editorial team. However, the end result was this all-inclusive book covering diverse themes in the field.

Finally, it is important to acknowledge the efforts of the contributors for their excellent chapters, through which a wide variety of issues have been addressed. I would also like to thank my colleagues for their valuable feedback during the making of this book.

Editor

1

Scavenging along an ecological interface: utilization of amphibian and reptile carcasses around isolated wetlands

Erin F. Abernethy,[1,2,4,†] Kelsey L. Turner,[2,3] James C. Beasley,[2,3] and Olin E. Rhodes Jr.[2]

[1]*Odum School of Ecology, University of Georgia, Athens, Georgia 30602 USA*
[2]*Savannah River Ecology Lab, University of Georgia, Aiken, South Carolina 29802 USA*
[3]*Warnell School of Forestry and Natural Resources, University of Georgia, Athens, Georgia 30602 USA*

Abstract. Scavenging is an influential ecosystem process, providing many weak, but stabilizing food web links. Amphibians and reptiles (i.e., herpetofauna) utilize land/water interfaces and can have large biomasses concentrated here, making them important vectors of nutrients between aquatic and terrestrial systems and a valuable resource to scavengers. There are limited data on the fate of herpetofauna carrion, and thus, it is unknown how such carcasses are incorporated into food webs or how habitat and carcass characteristics influence which scavengers acquire these resources. We used camera traps to determine the proportion of experimental amphibian and reptile carcasses (lizards, salamanders, frogs, and toads) scavenged by vertebrates and invertebrates in two different habitats (wetland and upland) during the spring in the southeastern USA. Vertebrates removed 19.8% of all herpetofauna carcasses in both habitats, with the remaining carcasses removed by invertebrates (80.2%). We used log-linear models to evaluate the impact of habitat and carcass type on scavenging. We determined that habitat, but not carcass type, was influential in determining whether a vertebrate or invertebrate scavenged a carcass, with vertebrates able to scavenge more carcasses in wetland than in upland areas. In less than two days, 86.2% of all carcasses in both habitat types were completely scavenged, and carcasses in wetlands were removed more quickly, suggesting intense competition between vertebrates and invertebrates. Additionally, we observed low vertebrate scavenger richness, which we attribute to high invertebrate activity in the spring and the small size of the herpetofauna carcasses (<100 g). We identified the scavenging pathways along which herpetofauna are incorporated into the food web and the factors that influence those pathways, suggesting the large extent and spatial scale of this resource's influence on food web connections and consequently food web and ecosystem stability.

Key words: ants; beetle; carrion; frog; herpetofauna; invertebrate; lizard; opossum; raccoon; salamander; scavenger; toad.

† **E-mail:** efabernethy@gmail.com

Introduction

Carcasses located along ecological interfaces (e.g., riparian areas and beaches) have been recognized as important sources of nutrients within those ecosystems (Cederholm et al. 1999, Huijbers et al. 2013, 2016, Schlacher et al. 2013, Brown et al. 2015). This research accompanies growing recognition for the amount of necromass processed by scavengers and the prevalence of facultative scavenging across vertebrate species (Putnam 1983, DeVault et al. 2003, Wilmers et al. 2003, Wilson and Wolkovich 2011, Mateo-Tomás et al. 2015). Incorporating scavenging into food webs establishes a number of links that connect predators to the detrital pool, which in theory contribute to the overall stability of food webs (DeAngelis et al. 1989, McCann et al. 1998, Wilson and Wolkovich 2011, Moleón et al. 2014). Carcass characteristics, particularly size, and abiotic factors, such as habitat and season, are known to influence spatial and temporal engagement in scavenging behaviors and contribute to the diversity of scavenger communities (DeVault et al. 2004, Selva et al. 2005, Sharanowski et al. 2008, Parmenter and MacMahon 2009, Ruzicka and Conover 2012, Moleón et al. 2015, Turner et al. 2017).

While studies have examined the importance of fish carcasses to terrestrial ecosystems, how carcasses of other species at this land/water interface structure scavenger dynamics is unknown. Many amphibians and some reptiles have high densities and large biomasses concentrated at this interface, which makes them influential in ecosystem processes, such as nutrient cycling, in both terrestrial and aquatic ecosystems (Arnold and Wassersug 1978, Deutschman and Peterka 1988, Petranka and Murray 2001, Russell et al. 2002, Davic and Welsh 2004, Regester et al. 2006, Whiles et al. 2006, Beard et al. 2009). Salamanders in Hubbard Brook Experimental Forest, for example, can reach densities of 2950 individuals/ha, approximately equal to the biomass of mammals and twice that of avian species (Burton and Likens 1975). While there has been research on how herpetofauna (i.e., amphibians and reptiles) influence energy flow as consumers of resources (Stewart and Woolbright 1996, Kupferberg 1997, Beard et al. 2003), their contributions as carcass resources to other organisms are rarely studied. Previous research examining scavenging of herpetofauna focused on road-associated mortality (Antworth et al. 2005, Beckmann and Shine 2011, DeGregorio et al. 2011, Santos et al. 2011), which likely overestimates scavenging rates due to the high visibility of the carcasses and habituation of vertebrate scavengers to road-kill. But do herpetofauna expire in the land/water interfaces that they use throughout their life cycles? Two studies at the US Department of Energy's Savannah River Site (SRS) found low percentages of adults returned to wetlands to breed. Scott et al. (2007) found 1.1%–45.2% (average 21.1%) of salamander metamorphs (*Ambystoma opacum*) returned to breed over 13 cohorts, and Semlitsch et al. (1996) found only 5% of the metamorph cohort returned as breeding females for a range of amphibian species ($N = 13$) over 16 yr. Considering the low percentage of adults that return to breed, other sources of mortality (e.g., starvation, exposure, accidents, and disease) undoubtedly occur, producing carcasses at this ecological interface. Thus, adult herpetofauna present an abundant and potentially widespread carrion resource that could contribute substantially to food web connections and stability (Wilson and Wolkovich 2011).

Quantifying how amphibian and reptile carcasses are partitioned between vertebrate vs. invertebrate scavengers and within scavenger groups is critical for understanding the influence of herpetofauna carcasses on food web dynamics. Carcass size has been shown to influence these scavenger dynamics, but studies on carcass size of many herpetofauna (<100 g) are lacking (Moleón et al. 2015, Turner et al. 2017). Given current global herpetofauna declines (Gibbons et al. 2000, Houlahan et al. 2000), it is important to understand the extent to which vertebrate scavengers utilize these carcass resources. Quantifying vertebrate scavenging on small herpetofauna species (<100 g) within their native ranges where they reach high densities and biomass, such as in the southeastern USA, is critical to understanding the potential influence of these carcasses on ecosystem processes. Furthermore, these data can help predict how changing population sizes, considering both global declines (decrease in populations) and invasions (increase), may influence ecosystem stability.

Our overall objective was to evaluate consumption of herpetofauna (lizard, salamander, frog, and toad) carcass in a temperate ecosystem. Specifically, we had the following two objectives: (1) to quantify the percentage of herpetofauna carcasses scavenged by invertebrates vs. vertebrates and (2) to determine whether habitat and/or carcass type influences the proportion of carcasses removed and the elapsed time to removal by vertebrates vs. invertebrates and among species within these scavenger communities. We predicted that the majority of herpetofauna carcasses would be incorporated into the lower levels of food webs by invertebrate scavengers (DeVault et al. 2004, Abernethy et al. 2016). We also expected habitat to influence scavenger dynamics and time to removal, with vertebrates scavenging more carcasses in wetland habitats and swifter removal times by all scavengers in wetlands. We based this prediction on evidence that suggests mesocarnivores, which are highly efficient facultative scavengers (DeVault et al. 2011, Olson et al. 2012), focus foraging activity around wetland habitats during spring when herpetofauna are emerging and migrating from wetlands (Beasley et al. 2007, Beasley and Rhodes 2010). We anticipated time to removal to be swifter in wetlands, as a result of increased competition between vertebrates and invertebrates driven by the focused foraging of mesocarnivores. Furthermore, we predicted that carcass type would influence the partitioning of carcass resources, specifically that vertebrates would scavenge toads less frequently than other carcass types, as toads possess a bufotoxin that makes them distasteful to some mammalian predators but not invertebrates (Abernethy et al. 2016). We also predicted that salamanders and frogs with thin, poorly keratinized mucous skin would have a faster removal time than toads and lizards, that possess thicker, and in the case of toads, tuberculated skin (Cornaby 1974, Cooper 2012).

Methods

Study system

We conducted this research on the Department of Energy's SRS in South Carolina, USA. Eighty-four percentage of the ~800-km^2 SRS was forested (57% pine forest, 21% hardwood forest, 6% mixed forest, and 16% other, e.g., buildings) and managed by the USDA Forest Service, representing ecosystems and management practices prevalent throughout the southeastern USA. Camera traps were deployed around Carolina bays (Fig. 1), which are small, isolated wetlands found along the US Atlantic Coastal Plain that provide critical habitat for both amphibians and reptiles (Semlitsch and Bodie 1998, 2003). We characterized two habitat types around Carolina bays (i.e., wetland and upland), with wetland habitat occurring within 1–10 m of the water's edge and upland habitat being dry, forested areas located >100 m from the water's edge (101–2000 m).

We chose the following native herpetofauna common to the SRS as experimental carcasses: lizards (green anoles, *Anolis carolinensis*; 1–3 g),

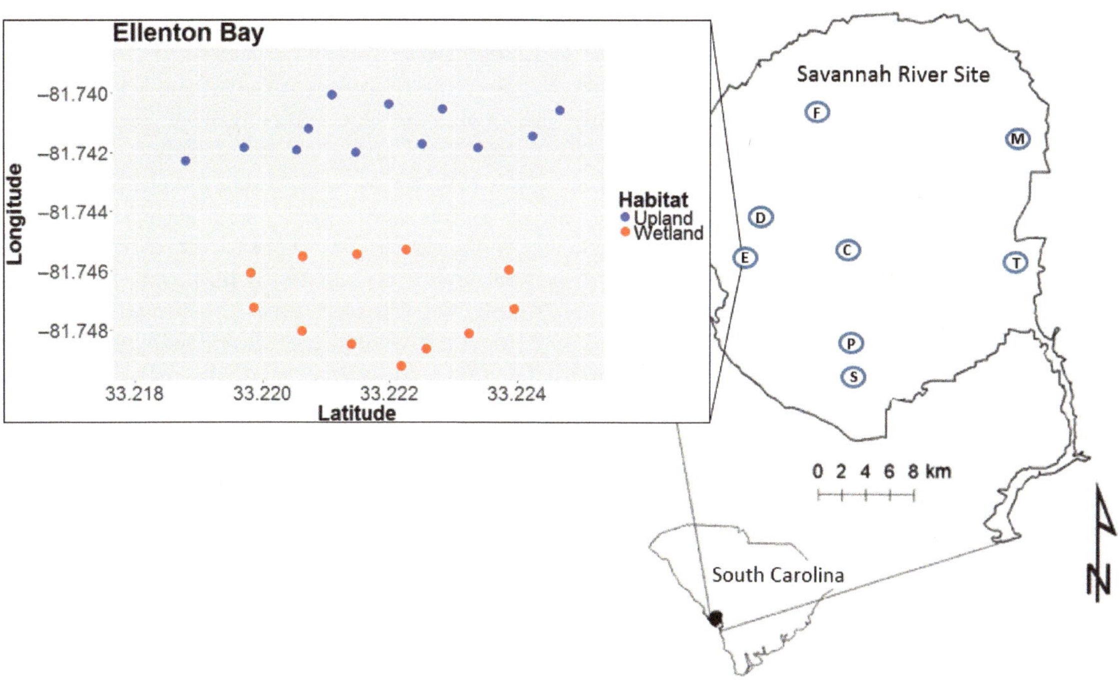

Fig. 1. Study area map showing the locations of Carolina bays around which scavenging trials were conducted on the Savannah River Site in Aiken, South Carolina. Inset image shows exact camera trap locations in wetland and upland habitats (Appendix S1 contains maps for each bay).

salamanders (mole salamanders, *Ambystoma talpoideum*; and marbled salamanders, *Ambystoma opacum*; 2–3 g), frogs (bullfrogs, *Rana catesbeiana*; and southern leopard frogs, *Rana sphenocephala*; 4–100 g), and toads (southern toads, *Anaxyrus terrestris*; 3–46 g). These species are abundant throughout the southeastern United States, and the amphibians, as well as the green anoles, depend heavily on isolated wetlands and the nearby upland habitat (Russell et al. 2002). We chose to perform the study in the summer to correspond with late emergences of herpetofauna metamorphs from Carolina bays and adult presence in nearby upland habitats. We obtained live animals either from a commercial supplier or collected them under SC (04-2014) and GA (29-WJH-14-93) collection permits. Animals were handled and euthanized following an Institutional Animal Care and Use Committee protocol approved through the University of Georgia (A2014 04-018-Y1-A0). We also used carcasses from previous experiments at the Savannah River Ecology Lab that had been frozen immediately following euthanasia or death. All carcasses were in a predominantly intact state, and no chemicals were used during euthanasia.

Experimental design

We chose eight Carolina bays of similar size and hydrology and paired each of these with nearby upland sites. Camera traps were located along pre-determined transects in wetland habitats, 1–10 m from water's edge, and upland habitats, 101–2000 m from water's edge, that were ≥50 m from major roads. Along transects, camera traps were located ≥200 m from all other active trials (Michaud et al. 2012). Camera trap trials, which consisted of one carcass placed in front of a camera (Reconyx HyperFire PC900; Reconyx, Holmen, Wisconsin, USA), were deployed from June to July 2014, each recording photographs for 7 d. Camera trap trials were started at different times from dawn to dusk each week, so as not to introduce systematic bias based on time of carcass placement (although all trials were started during daylight hours). We ran 24 camera trap trials with each carcass type (i.e., lizards, salamanders, frogs, and toads) in each habitat type (i.e., wetland and upland; $N = 192$). During a trial, a single carcass was placed on a pressure-sensitive, external triggering device located approximately 0.3–1 m in front of a camera that was attached to a tree a few centimeters to 0.5 m above the ground (DeVault and Rhodes 2002, Abernethy et al. 2016). Cameras were programmed to record three photographs when triggered by motion or by the pressure-sensitive, external trigger, and an additional time-lapse photograph was taken every 15 min. After 7 d, any remaining carcass material was removed, and camera traps were relocated 50–100 m away along transects and reset with fresh carcasses. (Appendix S1 shows camera locations for each Carolina bay and paired upland habitat.)

To confirm the independence of our carcasses, we conducted post hoc spatial clustering analyses using Moran's *I* statistic in R version 3.2.4 for all carcasses located within 2 km of each other (Moran 1950, R Development Core Team 2015). While we observed significant clustering of scavengers at two of the eight Carolina bays, further examination of the data showed that these results were driven by a few temporally isolated events rather than systematic, repetitive scavenging by a single vertebrate species, and thus, individual carcasses were considered independent throughout analyses.

Data analysis

For each carcass, we identified the primary scavenger to species or lowest possible taxonomic level and the time at which scavenging occurred. Since vertebrates quickly scavenged entire carcasses, they were assigned a single removal time (i.e., elapsed time from carcass deployment to removal); however, for invertebrates, we recorded time to removal, as well as detection and handling time. Detection time was measured from carcass deployment to when a carcass was initially detected by the invertebrate group that would be its ultimate scavenger, and handling time was the amount of time that invertebrate group needed to scavenge (time between detection and removal of a carcass). The one scavenger (or one group of scavengers, in the case of ants [Family: Formicidae] and beetles [Order: Coleoptera]) that consumed the majority of the flesh or removed the carcass from the field of view of the camera was considered the primary scavenger of that carcass for our analysis. While invertebrates consumed a portion of carcasses before vertebrates, we were unable to quantify this amount from the photographs. To be consistent, we chose to assign each carcass to the scavenger that clearly removed the majority of the resource. To better taxonomically describe the invertebrate scavenging community, we collected voucher specimens from additional herpetofauna carcasses over two days and nights in October 2014. These carcasses were not used in trials presented herein, and invertebrate specimens were identified and archived at the Georgia Museum of Natural History (Appendix S2).

To quantify the percentage of carcasses scavenged by vertebrates vs. invertebrates, objective 1, we calculated the percentage of each carcass type that was scavenged (i.e., removed from field of view or the majority of flesh was consumed) by vertebrates vs. invertebrates in each habitat. To evaluate factors influencing how carcasses were partitioned between vertebrates vs. invertebrates and within these two scavenger groups, objective 2, we used log-linear models to compare the number of carcasses (dependent variable) of each type (lizards, salamanders, frogs, and toads) scavenged by vertebrates and invertebrates in the two habitats (wetland and upland). Initially, we evaluated the three-way interaction between habitat, carcass type, and scavenger (independent variables); if non-significant, we further evaluated each of the pertinent two-way interactions.

We then constructed a log-linear model to compare the number of carcasses scavenged (dependent variable) by dominant members within the invertebrate scavenging community (ants and beetles) with habitat, carcass type, and scavenger as the independent variables. We evaluated three- and two-way interactions as appropriate. We constructed log-linear models using CATMOD procedure in SAS, version 9.3 (SAS Institute, Cary, North Carolina, USA). Due to low scavenging rates of individual vertebrate species, we could not conduct this analysis, nor the analysis described below for time to removal, within the vertebrate community.

To evaluate time to removal for herpetofauna carcasses, we used a three-way ANOVA to evaluate the time (dependent variable) from carcass deployment to when a carcass was scavenged (i.e., removal time) with habitat, carcass type, and scavenger as independent variables. Additionally, we conducted three-one-way ANOVAs to compare removal, detection, and handling time between dominant invertebrate scavengers.

Results

Proportion of carcasses removed by vertebrate vs. invertebrate scavengers

At the end of all experimental trials, the majority of carcasses were completely consumed or removed. Of the carcasses that remained, no flesh persisted, with only dry skin stretched over the skeleton ($N = 6$) or bones ($N = 18$); thus, all of the carcasses were considered scavenged. Invertebrates scavenged the majority of carcasses (80.2%), with vertebrates (seven species) removing 19.8% of total carcasses in all habitats (see Appendix S2 for taxa list and Appendices S3, S4 for raw data). Raccoon and opossum were the dominant vertebrate scavengers, together removing 14.6% of all carcasses (Fig. 2). Some species (bobcat, *Lynx rufus;* wild pig, *Sus scrofa;* and gray squirrel, *Sciurus carolinensis*) were observed at carcasses ($N = 5$) but did not scavenge, and a few (skunk, *Mephitis mephitis;* coyote, *Canis latrans;* wild pig; and gray squirrel) were recorded at the site of carcass placement after carcasses ($N = 22$) had already been removed. The following species are not widely considered scavengers but were observed scavenging carcasses entirely or in part: great-crested fly catcher (*Myiarchus crinitus*), Carolina wren (*Thryothorus ludovicianus*), and barred owl (*Strix varia*). With respect to invertebrates, ants scavenged 41.1% of total carcasses and beetles 22.9%. For the remaining 16.1% of carcasses ($N = 28$), it was unclear what type of invertebrate (ant, beetle, or other invertebrate) had scavenged the majority of the carcass. We categorized these carcasses as unknown invertebrates (Appendices S3, S4). For analyses that treated invertebrates as a collective group, these 28 trials were included, but excluded for analyses of the dominant invertebrate scavengers (ants and beetles).

Factors influencing proportions of carcasses removed

In our log-linear analysis to determine whether habitat or carcass type influenced the proportion of carcasses scavenged by vertebrates and invertebrates, the three-way interaction between scavenger, habitat, and carcass type was not significant, but the two-way interaction between scavenger and habitat was significant (scavenger × habitat; df = 1; $\chi^2 = 8.30$; $P = 0.0040$;

Fig. 2. Images from our scavenging research on the Savannah River Site (SRS) in Jackson, Carolina bays, around which scavenging trials were conducted on the SRS in Aiken, South Carolina, showing a red-shouldered hawk removing a toad carcass (a), a copperhead snake swallowing a frog carcass (b), a raccoon scavenging a salamander (c), an opossum removing a toad (d), and a black racer snake at the camera trap after the carcass has been removed (e).

Table 1). These results support our predictions concerning habitat that invertebrates would take more carcasses than vertebrates and that the strength of this relationship would vary by habitat (Fig. 3). In wetlands, invertebrates removed 71.9% of all carcasses, whereas vertebrates removed 28.1%. In upland habitats, invertebrates removed 88.5% of carcasses, whereas vertebrates removed 11.5%. Thus, vertebrate scavengers in wetland habitats consumed 2.5× more carcasses than vertebrates in upland habitats. No evidence was found that carcass type influenced this dynamic.

For the dominant invertebrate groups, we found that the proportion of scavenged carcasses was influenced by an interaction between scavenger (i.e., ant and beetle) and carcass (scavenger × carcass; df = 3; $\chi^2 = 19.39$; $P =$

Table 1. Results of log-linear models used to analyze herpetofauna data.

Scavenger	Model	df	χ^2	P
Vertebrates vs. invertebrates	Scavenger/carcass/habitat	3	2.52	0.4725
	Scavenger/carcass	3	2.20	0.5326
	Scavenger/habitat	1	8.30	0.0040*
Ant vs. beetle	Scavenger/carcass/habitat	3	0.43	0.9349
	Scavenger/carcass	3	19.39	0.0002*
	Scavenger/habitat	1	0.27	0.6059

Notes: Two-way interactions were analyzed if three-way interactions were non-significant. Interactions and main effects of scavenger, carcass, and habitat that significantly influenced the scavenging efficiency of vertebrates compared to invertebrates and within the invertebrate community (ant and beetles) are shown.

* Indicates significance ($P < 0.05$).

Fig. 3. Graphical representation of the significant effect of habitat on the proportions of carcasses scavenged by invertebrates vs. vertebrates. Invertebrates (represented by an ant) removed 71.9% and vertebrates (represented by a raccoon) removed 28.1% of all carcasses placed in wetlands (circle on left). In upland habitats, invertebrates removed 88.5% of carcasses and vertebrates removed 11.5% (rectangle on right).

0.0002; Table 1) but not habitat. Thus, of the carcasses removed by invertebrates, ants removed the most lizards, salamanders, and toads (71.8%, 62.5%, and 41.0%, respectively) and beetles removed the most frog carcasses (55.6%).

Factors influencing scavenging time

Only 186 of the 192 trials were used in analyses of time to removal, with six trials censored due to difficulty in determining exactly when carcasses were scavenged. The mean time to carcass removal was 26.65 h. Furthermore, 87.63% of the 186 carcasses were scavenged in <48 h, with vertebrates removing 94.74% and invertebrates 85.81% of the carcasses that each group would ultimately scavenge within each trial.

There was no statistically significant difference between mean vertebrate and invertebrate removal times of carcasses based on the two- or three-way interactions between scavenger, carcass type, or habitat or the main effects of scavenger or carcass type as determined by ANOVA. In alignment with our predictions, habitat did influence removal times, with faster rates observed in wetland habitats (ANOVA, $F_{1,170} = 3.715$, $P = 0.0556$; Fig. 4). Vertebrates and invertebrates removed carcasses in wetland habitats after an average of 23.16 h (range = 1.11–106.41), whereas in upland habitats it took on average 30.29 h (range = 1.55–130.19). No evidence was found that carcass type influenced this attribute of carcass removal.

For carcasses removed by dominant invertebrate scavengers, the total time from carcass placement to removal was not significantly different between the two groups (ants and beetles). However, one-way ANOVAs revealed significant differences between invertebrate groups relating to the time it took invertebrate scavengers to detect (ANOVA, $F_{1,120} = 99.78$, $P < 0.0001$) and fully remove carcasses, that is, handling time (ANOVA, $F_{1,120} = 30.31$, $P < 0.0001$). Ants took significantly less time to detect carcasses than beetles (2.81 h on average, compared to 19.69), but beetles handled carcasses for shorter amounts of time than ants (6.40 h on average, compared to 28.75 h).

Discussion

Invertebrate scavengers removed the majority of herpetofauna carcasses (80.2%), and most carcasses were removed in <2 d during this study. We found that habitat significantly influenced

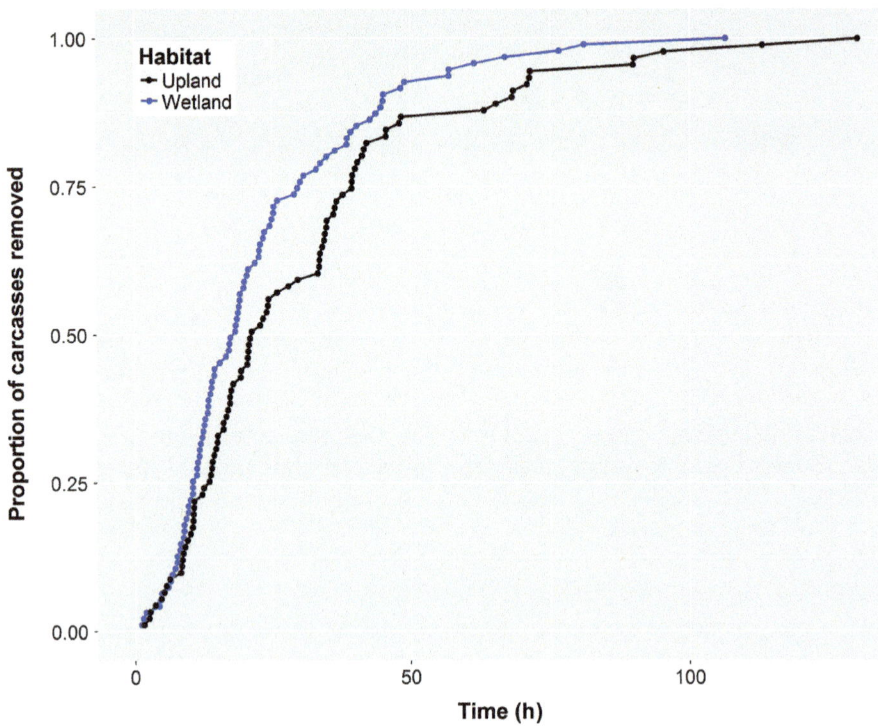

Fig. 4. The proportion of carcasses removed by all scavengers is shown as a function of time, with each line representing the two different habitats (wetland and upland). The elapsed time to carcass removal was unknown for six carcasses, which were deleted from this analysis.

the percentage of carcasses removed and the time to removal between vertebrates and invertebrates. Vertebrates removed more than twice as many carcasses in wetland (28.1%) than in upland habitats (11.5%), and carcasses were removed faster in wetland habitats, regardless of scavenger. Carcass type did not influence the proportion of carcasses scavenged or removal time of a carcass when comparing vertebrates vs. invertebrates, but it did influence the proportion of carcasses removed by individual invertebrate groups (i.e., ants vs. beetles).

While there are many abiotic and biotic factors known to influence scavenger dynamics, our study provides additional evidence for the importance of habitat, especially those occurring at the ecological interfaces, and carcass type on scavenging ecology. Habitat influenced both the percentage of carcasses scavenged by vertebrates (more around wetlands) and the overall speed at which carcasses were removed (faster around wetlands). The higher percentage of carcasses scavenged by vertebrates around wetlands was driven by raccoons and opossums, both of which are known to focus foraging efforts in these areas (Greenwood 1982, Hofmann 1991, Beasley et al. 2007, Beasley and Rhodes 2010). Other vertebrate scavengers (e.g., rodents, birds, and reptiles) scavenged an equal number of carcasses between wetland and upland habitats. While top scavengers (raccoons and opossums) observed in this research are also top scavengers in other scavenging studies conducted within our study area, overall scavenger species richness for herpetofauna carcasses was relatively low due to the summer season in which our study was conducted. We observed seven vertebrate species scavenging, whereas DeVault et al. (2004) and Turner et al. (2017) observed 17 and 12 species, respectively, scavenging rodent carcasses throughout the year. During the winter season, these two studies saw an increase in vertebrate scavenger species richness, likely because carcasses were available to be scavenged for a longer period of time due to decreased invertebrate activity.

While the only effect of carcass type we observed was on invertebrate community dynamics, it is likely that the small size (<100 g) of many herpetofauna carcasses influences which vertebrate scavengers will utilize these resources. Small mesocarnivores may find it energetically cost-effective and less risky to scavenge small carcasses, where they are less likely to be excluded and/or predated upon by large carnivores than at large carcasses (McKillup and McKillup 1994, Moleón et al. 2015, Turner et al. 2017). In South Carolina, small mesocarnivores (raccoons and opossums) are more likely than larger mesocarnivores (coyotes and bobcats) or obligate scavengers (vultures, *Coragyps atratus, Cathartes aura*) to utilize small carcasses (Turner et al. 2017). While small carcasses may be valuable to scavenge, mesocarnivores can experience enhanced competition with invertebrates when utilizing small carcasses. In our study, the percentage of carcasses consumed by invertebrates (80.2%) was high compared to other studies conducted in warm weather with small-sized mammal and avian carcasses (31% average, range = 6–75%, DeVault et al. 2003). The carcasses used in these studies were larger than our herpetofauna; thus, the size of our carcasses could have contributed to the success of invertebrate scavengers. Our study contributes to evidence suggesting intense competition between vertebrates and invertebrates for carcass resources in the summer, which could be mediated by carcass size, temperature, local abundance and/or foraging behavior of scavengers, and the predictability of the resource in the environment.

Our results also indicate that intense competition occurs within the invertebrate community, with ants scavenging significantly more carcasses than beetles, and that this dynamic is influenced by carcass type. The influence of carcass type on the proportion of carcasses scavenged by different invertebrate groups may have depended on size differences of a few 10s of grams and integument thickness. When considering the analysis on scavenging efficiency and time, we wish to point out the following aspects of our experimental design that influence this analysis. Carcasses were placed in the field during the day, which favors diurnal scavengers, and time to removal for beetle-scavenged carcasses occurred when a carcass was buried, rather than fully consumed as was the case for ants. Considering that the data are applicable to carcasses that expired during the day and that beetles are making carcasses less accessible to vertebrates and other invertebrates by burying them, we believe that analyzing these data is informative. It took ants significantly less time to detect a carcass of which they would be the primary scavengers, compared to beetles (3 h vs. 20 h). The average detection time for scavenging beetles, which are generally nocturnal, can be considered an average of the first two nights, since the majority of carcasses were scavenged in <2 d. Contributing to this competitive dynamic, beetles were able to remove carcasses on average only 6 h after initially detecting them, whereas ants required much longer (30 h) to fully consume the flesh from carcasses. Thus, beetles may be more efficient than recognized in this study, since some herpetofauna surely die at night.

As community dynamics and environmental factors mediate competition for carcass resources between and within vertebrate and invertebrate scavengers, the number and strength of those trophic links involving carcass resources influences ecosystem function and stability. This study suggests that herpetofauna carcasses in the southeastern USA are an important resource in food webs, particularly for invertebrates at lower trophic levels. With high local densities and biomass, herpetofauna carcasses likely influence ecosystem function and stability on a global scale to an extent that is greater than currently appreciated. How global declines in herpetofauna populations will affect the strength of these trophic connections is unknown but deserves further investigation.

Acknowledgments

We thank David Scott and Stacey Lance for assistance with site selection and study animal collection; Linda Lee, Robert Kennamer, and Donald Mosser for their help with permitting; and Wes Flynn for contributing code to create Fig. 4. Rick Hoebeke and Doug Booher of the Georgia Museum of Natural History were invaluable to us by identifying our representative invertebrate specimens. This research was supported by a contract from the US Department of Energy through Cooperative Agreement Number DE-FC09-07SR22506 with the University of Georgia Research Foundation. This report was prepared as an account of work sponsored by an agency of the United States Government. Neither the United States Government nor any agency

Literature Cited

Abernethy, E. F., K. L. Turner, J. C. Beasley, T. L. DeVault, W. C. Pitt, and O. E. Rhodes Jr. 2016. Carcasses of invasive species are predominantly utilized by invasive scavengers in an island ecosystem. Ecosphere 7:e01496. https://doi.org/10.1002/ecs2.1496

Antworth, R. L., D. A. Pike, and E. E. Stevens. 2005. Hit and run: effects of scavenging on estimates of roadkilled vertebrates. Southeastern Naturalist 4: 647–656.

Arnold, S. J., and R. J. Wassersug. 1978. Differential predation on metamorphic anurans by garter snakes (Thamnophis): social behavior as a possible defense. Ecology 59:1014–1022.

Beard, K. H., A. K. Eschtruth, K. A. Vogt, D. J. Vogt, and F. N. Scatena. 2003. The effects of the frog *Eleutherodactylus coqui* on invertebrates and ecosystem processes at two scales in the Luquillo Experimental Forest, Puerto Rico. Journal of Tropical Ecology 19:607–617.

Beard, K. H., W. C. Pitt, and E. A. Price. 2009. Biology and impacts of Pacific Island invasive species. 5. *Eleutherodactylus coqui,* the coqui frog (Anura: Leptodactylidae). Pacific Science 63:297–316.

Beasley, J. C., T. L. DeVault, M. I. Retamosa, and O. E. Rhodes Jr. 2007. A hierarchical analysis of habitat selection by raccoons in northern Indiana. Journal of Wildlife Management 71:1125–1133.

Beasley, J. C., and O. E. Rhodes Jr. 2010. Influence of landscape attributes on the spatial dynamics of raccoons in agriculturally fragmented landscapes. Canadian Journal of Zoology 88:161–169.

Beckmann, C., and R. Shine. 2011. Toad's tongue for breakfast: exploitation of a novel prey type, the invasive cane toad, by scavenging raptors in tropical Australia. Biological Invasions 13:1447–1455.

Brown, M. B., T. A. Schlacher, D. S. Schoeman, M. A. Weston, C. M. Huijbers, A. D. Olds, and R. M. Connolly. 2015. Invasive carnivores alter ecological function and enhance complementarity in scavenger assemblages on ocean beaches. Ecology 96: 2715–2725.

Burton, T. M., and G. E. Likens. 1975. Salamander populations and biomass in the Hubbard Brook Experimental Forest, New Hampshire. Copeia 3:541–546.

Cederholm, C. J., M. D. Kunze, T. Murota, and A. Sibatani. 1999. Pacific salmon carcasses: essential contributions of nutrients and energy for aquatic and terrestrial ecosystems. Fisheries 24:6–15.

Cooper, J. E. 2012. The estimation of post-mortem interval (PMI) in reptiles and amphibians: current knowledge and needs. Herpetological Journal 22: 91–96.

Cornaby, B. W. 1974. Carrion reduction by animals in contrasting tropical habitats. Biotropica 6:51–63.

Davic, R. D., and H. H. Welsh. 2004. On the ecological roles of salamanders. Annual Review of Ecology, Evolution, and Systematics 35:405–434.

DeAngelis, D. L., P. J. Mulholland, A. V. Palumbo, A. D. Steinman, M. A. Huston, and J. W. Elwood. 1989. Nutrient dynamics and food-web stability. Annual Review of Ecology and Systematics 20:71–95.

DeGregorio, B. A., T. E. Hancock, D. J. Kurz, and S. Yue. 2011. How quickly are road-killed snakes scavenged? Implications for underestimates of road mortality. Journal of the North Carolina Academy of Science 127:184–188.

Deutschman, M. R., and J. J. Peterka. 1988. Secondary production of tiger salamanders (*Ambystoma tigrinum*) in three North Dakota prairie lakes. Canadian Journal of Fisheries and Aquatic Sciences 45:691–697.

DeVault, T. L., I. L. Brisbin, and O. E. Rhodes Jr. 2004. Factors influencing the acquisition of rodent carrion by vertebrate scavengers and decomposers. Canadian Journal of Zoology 82:502–509.

DeVault, T. L., Z. H. Olson, J. C. Beasley, and O. E. Rhodes Jr. 2011. Mesopredators dominate competition for carrion in an agricultural landscape. Basic and Applied Ecology 12:268–274.

DeVault, T. L., and O. E. Rhodes Jr. 2002. Identification of vertebrate scavengers of small mammal carcasses in a forested landscape. Acta Theriologica 47:185–192.

DeVault, T. L., O. E. Rhodes Jr., and J. A. Shivik. 2003. Scavenging by vertebrates: behavioral, ecological, and evolutionary perspectives on an important energy transfer pathway in terrestrial ecosystems. Oikos 102:225–234.

Gibbons, J. W., et al. 2000. The global decline of reptiles, déjà vu amphibians. BioScience 50:653–666.

Greenwood, R. J. 1982. Nocturnal activity and foraging of prairie raccoons (*Procyon lotor*) in North Dakota. American Midland Naturalist 107:238–243.

Hofmann, J. E. 1991. Status and distribution of wetland mammals in Illinois. Pages 409–415 *in* L. M. Page and M. R. Jeffords, editors. Our living heritage: the biological resources of Illinois. Bulletin 34. Illinois Natural History Survey, Champaign, Illinois, USA.

Houlahan, J. E., C. S. Findlay, B. R. Schmidt, A. H. Meyer, and S. L. Kuzmin. 2000. Quantitative evidence for global amphibian population declines. Nature 404:752–755.

Huijbers, C. M., T. Schlacher, R. R. McVeigh, and R. M. Connolly. 2016. Functional replacement across species pools of vertebrate scavengers separated at a continental scale maintains an ecosystem function. Functional Ecology 30:998–1005.

Huijbers, C. M., T. A. Schlacher, D. S. Schoeman, M. A. Weston, and R. M. Connolly. 2013. Urbanisation alters processing of marine carrion on sandy beaches. Landscape and Urban Planning 119:1–8.

Kupferberg, S. 1997. Facilitation of periphyton production by tadpole grazing: functional differences between species. Freshwater Biology 37:427–439.

Mateo-Tomás, P., P. P. Olea, M. Moleón, J. Vicente, F. Botella, N. Selva, J. Vinuela, and J. A. Sánchez-Zapata. 2015. From regional to global patterns in vertebrate scavenger communities subsidized by big game hunting. Diversity and Distributions 21:913–924.

McCann, K., A. Hastings, and G. R. Huxel. 1998. Weak trophic interactions and the balance of nature. Nature 395:794–798.

McKillup, S. C., and R. V. McKillup. 1994. The decision to feed by a scavenger in relation to the risks of predation and starvation. Oecologia 97:41–48.

Michaud, J. P., K. G. Schoenly, and G. Moreau. 2012. Sampling flies or sampling flaws? Experimental design and inference strength in forensic entomology. Journal of Medical Entomology 49:1–10.

Moleón, M., J. A. Sánchez-Zapata, E. Sebastián-González, and N. Owen-Smith. 2015. Carcass size shapes the structure and functioning of an African scavenging assemblage. Oikos 124:1391–1403.

Moleón, M., J. A. Sánchez-Zapata, N. Selva, J. A. Donázar, and N. Owen-Smith. 2014. Inter-specific interactions linking predation and scavenging in terrestrial vertebrate assemblages. Biological Reviews 89:1042–1054.

Moran, P. A. P. 1950. Notes on continuous stochastic phenomena. Biometrika 37:17–23.

Olson, Z. H., J. C. Beasley, T. L. DeVault, and O. E. Rhodes Jr. 2012. Scavenging community response to the removal of a dominant scavenger. Oikos 121:77–84.

Parmenter, R. R., and J. A. MacMahon. 2009. Carrion decomposition and nutrient cycling in a semiarid shrub- steppe ecosystem. Ecological Monographs 79:637–661.

Petranka, J. W., and S. M. Murray. 2001. Effectiveness of removal sampling for determining salamander density and biomass: a case study in an Appalachian streamside community. Journal of Herpetology 35:36–44.

Putnam, R. J. 1983. Carrion and dung: the decomposition of animal wastes. Edward Arnold, London, UK.

R Development Core Team. 2015. R: a language and environment for statistical computing. R Foundation for Statistical Computing, Vienna, Austria.

Regester, K. J., K. R. Lips, and M. R. Whiles. 2006. Energy flow and subsidies associated with the complex life cycle of ambystomatid salamanders in ponds and adjacent forest in southern Illinois. Oecologia 147:303–314.

Russell, K. R., H. G. Hanlin, T. B. Wigley, and D. C. Guynn Jr. 2002. Responses of isolated wetland herpetofauna to upland forest management. Journal of Wildlife Management 66:603–617.

Ruzicka, R. E., and M. R. Conover. 2012. Does weather or site characteristics influence the ability of scavengers to locate food? Ethology 118:187–196.

Santos, S. M., F. Carvalho, and A. Mira. 2011. How long do the dead survive on the road? Carcass persistence probability and implications for road-kill monitoring surveys. PLoS ONE 6:e25383. https://doi.org/10.1371/journal.pone.0025383

Schlacher, T. A., S. Strydom, and R. Connolly. 2013. Multiple scavengers respond rapidly to pulsed carrion resources at the land-ocean interface. Acta Oecologica 48:7–12.

Scott, D. E., E. D. Casey, M. F. Donovan, and T. K. Lynch. 2007. Amphibian lipid levels at metamorphosis correlate to post-metamorphic terrestrial survival. Oecologia 153:521–532.

Selva, N., B. Jędrzejewska, W. Jędrzejewski, and A. Wajrak. 2005. Factors affecting carcass use by a guild of scavengers in European temperate woodland. Canadian Journal of Zoology 83:1590–1601.

Semlitsch, R. D., and J. R. Bodie. 1998. Are small, isolated wetlands expendable? Conservation Biology 12:1129–1133.

Semlitsch, R. D., and J. R. Bodie. 2003. Biological criteria for buffer zones around wetlands and riparian habitats for amphibians and reptiles. Conservation Biology 17:1219–1228.

Semlitsch, R. D., D. E. Scott, J. H. K. Pechmann, and J. W. Gibbons. 1996. Structure and dynamics of an amphibian community: evidence from a 16-year study of a natural pond. Pages 217–248 *in* M. L. Cody and J. Smallwood, editors. Long-term studies of vertebrate communities. Academic Press, San Diego, California, USA.

Sharanowski, B. J., E. G. Walker, and G. S. Anderson. 2008. Insect succession and decomposition patterns on shaded and sunlit carrion in Saskatchewan in three different seasons. Forensic Science International 179:219–240.

Stewart, M. M., and L. L. Woolbright. 1996. Amphibians. Pages 274–320 *in* D. P. Reagan and R. P. Waide, editors. The food web of a tropical rainforest. University of Chicago Press, Chicago, Illinois, USA.

Turner, K. L., E. F. Abernethy, L. M. Conner, O. E. Rhodes Jr., and J. C. Beasley. 2017. Abiotic and biotic factors modulate carrion fate and vertebrate scavenging communities. Ecology 98:2413–2424.

Whiles, M. R., et al. 2006. The effects of amphibian population declines on the structure and function of Neotropical stream ecosystems. Frontiers in Ecology and the Environment 4:27–34.

Wilmers, C. C., R. L. Crabtree, D. W. Smith, K. M. Murphy, and W. M. Getz. 2003. Trophic facilitation by introduced top predators: grey wolf subsidies to scavengers in Yellowstone National Park. Journal of Animal Ecology 72:909–916.

Wilson, E. E., and E. M. Wolkovich. 2011. Scavenging: how carnivores and carrion structure communities. Trends in Ecology and Evolution 26: 129–135.

2

Exploring benefits of spatial cooperative harvesting in a sea urchin fishery: an agent-based approach

Nicolas L. Gutierrez,[1,4,†] Peter Halmay,[2] Ray Hilborn,[1] Andre E. Punt,[1] and Stephen Schroeter[3]

[1]*School of Aquatic and Fishery Sciences, University of Washington, Box 355020, Seattle, Washington 98195-5020 USA*
[2]*San Diego Watermen's Association, 11103 Highway 67, Lakeside, California 92040 USA*
[3]*Marine Science Institute, University of California Santa Barbara, Isla Vista, California 93106 USA*

Abstract. Sedentary or low-mobility organisms show a high degree of dependency with their substrate, where its heterogeneity often determines small-scale spatial patterns of distribution, life history traits, and fishery yields. For sea urchins, this spatial structure is usually shaped by food availability, habitat structure, individual movement, and fishery dynamics. All of these have a significant impact on their physiological and reproductive status and in particular on their gonadal content. These patterns are of particular interest considering that the sea urchin fishery is a roe fishery where marketability depends on gonad yield and quality, which in turn is related to spatial and temporal variations in associated kelp beds. Thus, better gonad quality and yields generate higher profits for both fishers (divers) and processors. However, competition among divers within a non-cooperative system creates a "race for shellfish" precluding higher gonad yields per unit of effort. A spatially explicit agent-based model for the San Diego, California red sea urchin (*Strongylocentrotus franciscanus*) fishery was developed in order to assess the benefits of cooperative harvesting by depicting spatial and temporal variations in fishery yields. A cooperative harvesting scenario where divers consistently target those areas with higher yields avoiding low-quality sea urchins was compared against a non-cooperative situation where divers harvest at random or based only on densities of sea urchins. Sea urchin population at the end of the simulation period was 20% higher for the most cooperative scenario compared to the non-cooperative fishery. Further, for the most cooperative scenario where information sharing among divers is greatest and harvest is coordinated, sea urchin catches were at least 10% higher and gonad yield 35% higher than in the non-cooperative scenario. In this model, information sharing and organized harvesting typical of well-functioning cooperatives allowed fishers to optimize the use of the resource in terms of higher gonad yields per unit of effort while maintaining the productivity of the stock. This study also highlights the importance of community-based management (i.e., collaborative efforts in assessment, management, and governance of fisheries between fishers, scientists, and managers) toward improving fisheries sustainability.

Key words: community-based management; cooperative harvesting; sea urchins: individual-based models; small-scale fisheries; spatially explicit modeling.

† **E-mail:** nicolas.gutierrez@fao.org

Introduction

Spatial heterogeneity in the distribution and abundance of organisms deeply affects ecological processes at local and regional scales (Tilman and Kareiva 1997). These spatial patterns of abundance and life history traits may be determined by biological characteristics of a species (Gutierrez and Defeo 2003, 2005), the physical features of its environment (Fogarty and Botsford 2007, Gutierrez et al. 2008), and the fishing effort and fleet dynamics (Booth 2000, Branch et al. 2006). Management of marine resources has historically been characterized by relative uniformity of regulatory actions over space. This mismatch of ecological and management scales could lead to very fine-scale forms of serial depletion (Johnson et al. 2013). However, over the past decade, there have been improvements in fine-scale monitoring of species abundance and oceanographic processes that affect spatial abundance. As a result, fishery scientists have begun to incorporate spatial structure into both stock assessments (Cope and Punt 2011) and management plans (Wilen 2004, Prince 2010).

Spatial management is not only relevant to fisheries managers, but also has a great impact on users who fish in a specific area only (e.g., specialists; Hilborn and Ledbetter 1985) and on those exploiting sedentary or low-mobility resources (Caddy 1975, Orensanz and Jamieson 1998). Fishers have always used logbooks, information-sharing groups, folk knowledge, and other informal methods to track and monitor differences in spatial abundance and productivity of target species. Thus, both knowledge and technology have increased the searching efficiency of harvesters. However, the overall economic efficiency may be far from optimal when the flow and exchange of information is low (Wilson 1990, Little et al. 2004). Collection of local knowledge can be increased in quality and precision when fishers interact with each other and have access to spatially explicit information of the fishing grounds and their differences in productivity (Curtis and McConnell 2004). Yet in competitive fisheries there is no reason for fishers to share information freely with one another and there is usually a considerable level of secrecy and deceit (Gatewood 1984, Palmer 1990). This is more obvious in sedentary or low-mobility resources since the knowledge one obtains on these aggregations lasts longer and is more valuable than for highly mobile or migratory species (Wilson and Acheson 1980). Further, the need for fine-scale information also increases in sedentary resources with a high degree of spatial structure (Orensanz and Jamieson 1998, Gutierrez et al. 2011*a*).

Information gathering and sharing is usually enhanced in fishing communities with strong and well-organized local institutions such as cooperatives or committees. Community-based data collection programs improve the quality and quantity of relevant fishery information by enhancing spatial, temporal, and categorical resolution (Prince 2003, Schroeter et al. 2009), as well as significantly reduce the monetary costs of data collection (Johnson and van Densen 2007). Within cooperative agreements, the value of these data collection programs is maximized by a dynamic flow of spatiotemporal information among users, where cooperative members often have less or no incentive to compete with one another for sub-stocks with high fishery yields and economic value (Deacon et al. 2010). Thus, coordinated harvesting typical of well-structured fishing cooperatives has the potential to alleviate the "race for fish" often associated with other forms of traditional management and optimize the extraction of fishery resources improving yields and revenues (Dreyfus-Leon 2006, Costello and Deacon 2007, Ovando et al. 2012).

Evaluating the benefits of cooperative and coordinated harvesting requires a bionomic model that includes the biological dynamics of the resource, the dynamics of the harvesters and their choice of fishing times and locations, and the feedback between these two elements. Individual or agent-based models (ABMs) are of particularly suitable for this purpose since it allows for a natural link between processes at the individual level via bioenergetics (e.g., by relating the effect of food availability and diet on consumption, growth, and reproduction) and population characteristics and trajectories (Grimm et al. 2005, Beauchamp et al. 2007). ABMs are especially useful in spatially explicit settings because of their ease in simulating movement and in considering the spatial structure of exploited resources and their users (DeAngelis and Gross 1992). However, most studies designed to evaluate the effects of cooperative harvesting scenarios have focused on fishers as agents (e.g., Little et al. 2009, Cabral et al. 2010) and not based on the individuals or agents being targeted by the fishery (e.g., sea urchins).

Sea urchin populations, as other sedentary or low-mobility resources, are spatially structured and exhibit a tight ecological coupling with the rest of their ecosystem (Botsford et al. 2004). Their distribution is generally associated with kelp beds (*Laminaria* sp.) and their aggregations occur primarily because of habitat structure and food availability, generating highly patchy distributions and fine-scale spatial patterns in life history traits and/or physiological status. Some species of sea urchin (e.g., *Strongylocentrotus franciscanus*) store nutrients in their gonads year-round as a reserve for shortages in food supply (Scheibling and Hatcher 2001). Sea urchin gonads act as a reserve of energy necessary for survival, growth, and reproduction, but are also the target and final product of the sea urchin fishery (sold as *uni* in sushi restaurants). Thus, gonad yield (as percentage of the total body wet weight) and quality (i.e., texture and color), which are both affected by food availability, diet composition, reproduction periods, and individual movement, rather than the age/size of the mature sea urchins (McBride et al. 1999, 2004), will affect marketability of sea urchins and profits for divers and processors.

The San Diego sea urchin fishery concentrates the majority of its effort in the southern part of the Point Loma giant kelp (*Macrocystis pyrifera*) bed where 12 active divers fish about 168 d a year. Although the actual area that is fished for sea urchins (i.e., area covered by fishing effort which is not the entire area of distribution of the resource) is relatively small (~20 km^2), scientists and commercial divers recognize the great deal of spatial structure within this area with respect to the physical structure of the bottom, kelp coverage, sea urchin density, and the proportion of marketable urchins (i.e., good gonad quality or *uni*; Parnell et al. 2010*a*, 2010*b*). Thus, the high economic and societal value of red sea urchins and their fishery, the relevance of the gonad yield and quality for its marketability, the small-scale spatial patterns in growth, consumption, and reproduction, and the small-scale nature of this fishery make a spatially explicit bioenergetics model coupled with an ABM, an appropriate approach to tackle both ecological and fishery management research questions.

Here, an ABM is developed in order to evaluate alternative harvesting options for red sea urchins (*S. franciscanus*) off San Diego, California. The effects of spatially allocating effort to a particular area on both fishery yield and sea urchin population abundance are investigated for different competitive and cooperative harvesting scenarios. Model limitations and potential extensions are presented and the implications of well-structured cooperatives for successful fisheries management discussed.

Methods

Overview

The purpose of this ABM is to assess the benefits of cooperative harvest tactics in optimizing fishery yields and in contributing to the overall productivity of the population. The model was organized as follows: (1) an initialization step of kelp habitat and sea urchin distribution; and several modules updated by monthly steps and tracking (2) kelp dynamics; (3) sea urchin movement; (4) sea urchin bioenergetics; (5) survival; (6) recruitment; and (7) fishery. The sub-models were implemented in R (R Development Core Team 2010) following the order mentioned (Fig. 1). In general, the model structure and equations were designed to represent the red sea urchin dynamics, our own knowledge of the Point Loma kelp bed and the San Diego sea urchin fishery described in the literature (e.g., Parnell et al. 2010) and from the California Department of Fish and Game, from the processing plant, and from the San Diego community-based data collection program (SDCP; Schroeter et al. 2009; Appendix S1: Table S1).

Data available

Fishery-independent estimates of density, abundance, and size structure of sea urchin populations with high levels of spatial and temporal variability may require more resources than are typically available for agencies tasked with such management (Kalvass and Hendrix 1997, Parnell et al. 2006). For the San Diego sea urchin fishery, these limitations have been solved by enlisting fishery members in a community-based data collection program (SDCP; Schroeter et al. 2009). The

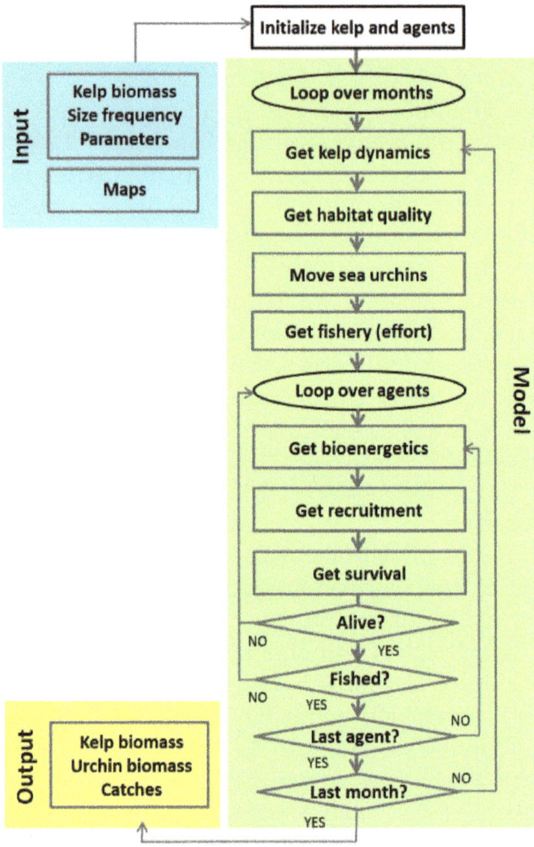

Fig. 1. Flow diagram illustrating model processes and general scheduling of the red sea urchin (*Strongylocentrotus franciscanus*) population model. Initial spatial arrays are based on user inputs including spatial extension and biomass of kelp bed (i.e., number of cells) and agents' characteristics.

San Diego Watermen's Association has developed a sampling protocol that allows working divers to collect random samples of sea urchin density and size distributions during the course of normal harvesting operations, capturing the relevant spatial variability typically of kelp bed areas. This sampling protocol has been calibrated by scientific surveys to validate their accuracy and precision (Schroeter et al. 2009). Divers collect fine-scale data of two kinds since 2004: (1) fishery-dependent data (catch, effort, catch size distributions, gonad yield and quality) and (2) fishery-independent data (density and size distributions independent of catch), as well as some environmental variables (temperature, bottom type, algae composition; Appendix S1: Table S2).

In addition to the SDCP, the San Diego sea urchin industry (i.e., processors) has also been contributing to the data collection and analysis by providing information on catches (in lbs and in US$) and gonad yield and quality by divers and regions from 2004 to 2010 (Appendix S1: Tables S1, S2). This information is available at relevant spatial and temporal scales, constituting a complete census of daily harvests by individual divers used to further corroborate and validate the information obtained from the community-based data collection program used in our operating model.

The environment

This study is based in the Point Loma kelp forest located along the western edge of Point Loma between the entrance to Mission Bay and San Diego Bay, California (32°42′ N; 117°16′ W). This kelp bed covers an area of ~10 km long by ~1 km wide, and a depth range between 15 and 45 m. The kelp forest occurs on a broad, gently sloping mudstone–sandstone terrace, with pockets of sand, cobbles, and boulders. The kelp forest is mainly composed of floating canopy algae (*Macrocystis*) and prostrate canopy algae (*Laminaria* sp.; Dayton 1985).

The ABM simulated the population dynamics of the red sea urchin (*Strongylocentrotus franciscanus*) in the southern tip of the Point Loma kelp bed (Fig. 2a, b), geographically matching the habitat distribution of kelp resources and sea urchins and the spatial dynamics of the fishery in a 9.5 km (latitude) by 5.5 km (longitude) area. The simulations were run on a monthly time step because shorter steps would require excessive computation time and longer time steps increase the likelihood of missing sea urchin–kelp transient dynamics and temporal patterns in fishery yields. A spatially explicit approach was used to capture the spatial structure of the resource and its fishery. The region was divided into 100 × 100 m (1 ha) cells to approximately cover the size of the harvested area by divers (95 cells in latitude and 55 cells in longitude) and to capture the movement of sea urchins at the temporal scale chosen (Fig. 2c). Cells were referenced by latitude (x-axis) and longitude (y-axis), and the key habitat-specific state variable was the amount of food (i.e., kelp) available.

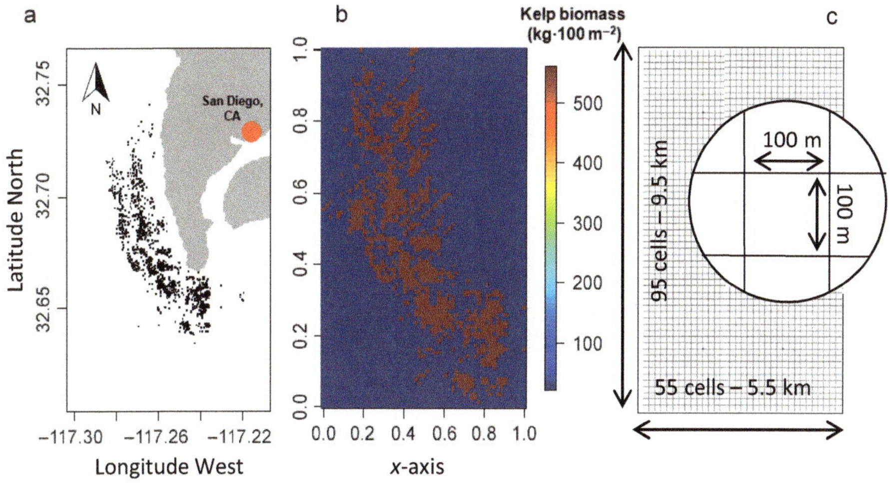

Fig. 2. Inputs used in the agent-based model of red sea urchin (*Strongylocentrotus franciscanus*) in Point Loma, San Diego. (a) Map of diving sites; (b) simulated kelp bed data matching distribution of diving sites and information from the San Diego community-based data collection program (2004–2009). Cells in red represent areas covered by kelp (A_1), and cells in blue represent areas outside the kelp bed (A_0); (c) grid representation of the study area (cell = 100 m^2 = 1 ha; total area 95 × cells × 55 cells = 5.22 km^2).

Initialization

Kelp bed (*Macrocystis pyrifera*) distribution was matched to historical data (2004–2010) on sea urchin harvesting sites from the SDCP (Fig. 2b). The whole region (A) was gridded into cells occupied by the kelp bed (A_1) and cells outside the kelp bed (A_0). Cells within the kelp bed were initialized with values of densities (in kg/ha) for the study area obtained from the literature and cells outside the kelp bed were assigned with a baseline value of drift *M. pyrifera* as well as other types of algae considered as available food for sea urchins (Harrold and Reed 1985). Each agent represented 100 individuals with identical characteristics, and the total number of agents for the study area was estimated from the SDCP data. Sea urchin locations (Fig. 2a) and initial size frequency (Fig. 3a) were initialized with historical information from the SDCP, and the model was then run for a period of 30 yr (burn-in) to stabilize size frequencies including recruits. Information on size and weight has been gathered from two sources: (a) measurements taken at the processing plant and corrected by a loss of water content (15%; D. Rudie, *unpublished data*) and (b) measurements taken in the field in Point Loma kelp bed (S. Teck, *unpublished data*). Parameter values and definitions are listed in Table 1, and state variables, bioenergetics parameters, and agent attributes in Appendix S1: Tables S3, S4.

Kelp biomass

Diet of *S. franciscanus* within kelp beds in southern California has been identified as primarily brown algae and specifically *M. pyrifera* (between 80 and 100% of total stomach contents; Dean et al. 1984, McBride et al. 1999). Although sea urchins may actively graze on *M. pyrifera* kelp stipes causing detachment and loss of canopy, most sea urchins within the kelp bed feed on drift algae. The abundance of drift algae has been identified as proportional to the attached algal biomass and composed by 87–98% of *Macrocystis* sp. (Harrold and Reed 1985). In this model, drift and attached algae were not differentiated and all available food $K_{j,t}$ (kg/ha) in cells j at time t within the kelp bed was calculated as the biomass of *M. pyrifera* using parameter values obtained in the literature for the study area (Senyk and Sanchez 2005, Rassweiler et al. 2008, Reed et al. 2008). A Schaefer-type production model with stochasticity was considered as an appropriate approach given the assumptions of our model, the available information, and the objectives of the study (Eq. 1):

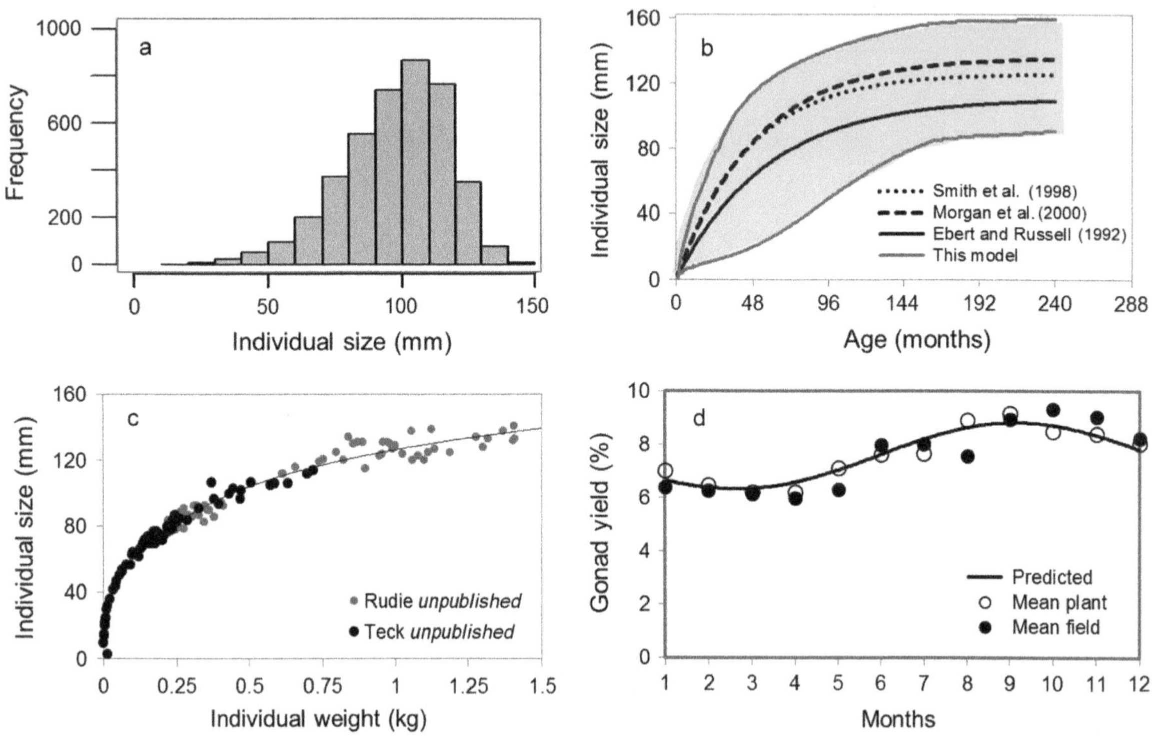

Fig. 3. Biometric information for red sea urchin in Point Loma, California. (a) Size frequency distribution of adult sea urchins from the San Diego community-based data collection program (SDCP; field data); (b) monthly sea urchin growth in size (mm) from the literature and for this model under different levels of kelp biomass (gray area; lower band: $K_{j,i}$ = 1 t/ha and upper band: $K_{j,i}$ = 60 t/ha); (c) size–weight relationship from field and fishery samples; and (d) monthly mean gonad yield (%) estimates from field and processor plant data for the period 2006–2009 and model fit.

$$K_{j,t+1} = \begin{cases} K_{j,t} + (r + \varepsilon_{j,t}) \cdot K_{j,t} \cdot \\ \quad (1 - K_{j,t}/k + \eta_{j,t}) - \sum C_{i\in j,t} & \text{if } j \in A_i \\ D - \sum C_{i\in j,t} & \text{if } j \in A_o \end{cases} \tag{1}$$

$$\varepsilon_{j,t} \sim N(0, \sigma_r^2) \quad \eta_{j,t} \sim N(0, \sigma_r^2),$$

where r is the average intrinsic growth rate of kelp (in month^{-1}), k the average kelp carrying capacity for each cell j (kg/ha), and $\sum C_{i\in j,t}$ the total consumption of kelp by all agents i in cell j (kg·kg^{-1}·month^{-1}) during time step t. Since kelp distribution and growth rates show small-scale spatial and temporal variabilities due to habitat heterogeneity and environmental conditions, spatial (over cells j) stochasticity and temporal (over time t) stochasticity in kelp growth rates ($\varepsilon_{j,t}$) and carrying capacity ($\eta_{j,t}$) were considered (Dayton et al. 1999, Senyk and Sanchez 2005). Both σ_r and σ_k were chosen to cover parameter values of kelp growth and carrying capacity for different kelp beds in California and for the period 1980–2009 obtained from the literature (Appendix S1: Table S1). In order to account for other food sources (e.g., other type of algae and drift or detached *M. pyrifera*), all cells outside the kelp bed (A_0) were assigned with a baseline value of food availability (D = 100 kg/ha, Harrold and Reed 1985, Dayton et al. 1992).

Individual growth

Sea urchin individual growth rates may vary considerably due to food quality and availability, where individuals in kelp bed areas grow faster than sea urchins in barrens or feeding fronts (Edwards and Ebert 1991, McBride et al. 1999; Fig. 3b). Thus, food availability has a major impact on red sea urchin *S. franciscanus* morphometrics

Table 1. Indexing and parameters used in the agent-based model of red sea urchin (*Strongylocentrotus franciscanus*) in Point Loma, San Diego, California.

Name	Description	Value	Unit	References
nsteps, t	Number of monthly steps	360	month	This study
nagents, i	Number of initial total agents	20,000	agents	This study
agents	Number of urchins per agent	100	individuals	This study
j	Cell	5225	cells	This study
d	Dive	80	dives/month	This study
A	Total area	52.25	km^2	This study
A_1	Kelp bed area of occurrence	6.9	km^2	This study based on Parnell et al. (2010*a*)
r	Kelp intrinsic growth rate	0.45	m^{-1}	Senyk and Sanchez (2005); Rassweiler et al. (2008); Reed et al. (2008)
k	Kelp carrying capacity	60	ha	Senyk and Sanchez (2005); Rassweiler et al. (2008); Reed et al. (2008)
σ_r	Standard deviation growth rate	0.01		Based on Harrold and Reed (1985); Reed et al. (2008)
σ_k	Standard deviation carrying capacity	0.10		Based on Harrold and Reed (1985); Reed et al. (2009)
D	Other algae as source of food	0.1	ha	Harrold and Reed (1985); Dayton et al. (1992)
c_{max}	**Maximum kelp consumption**	**1.4**	**kg/kg**	Based on McBride et al. (1997, 1998)
c_{50}	**Consumption half-saturation**	**200**	**kg**	Based on McBride et al. (1997, 1998)
ρ	**Metabolic cost / kg of urchin**	**80**	**kcal/kg**	Based on McBride et al. (1997, 1998, 1999)
e	Energetic content of kelp	580	kcal/kg	Paine and Vadas (1996)
m_b	**Urchin maturity parameter *b***	**0.05**		This study
m_{50}	**Urchin size at 50% maturity**	**50**	**mm**	This study
υ	**Energy-to-weight conversion**	**8×10^{-4}**	**kg/kcal**	Based on Farmanfarmaian and Phillips (1962); Webster and Giese (1975)
κ	**Proportion of energy to gonad**	**1.5×10^{-5}**		Based on Raymond et al. (2004); Siikavuopio et al. (2007)
ψ	**Proportion of gonad spawned**	**0.04**		This study
δ	**Recruits per kg eggs spawned**	**0.1**	**agents/kg_{egg}**	This study
α	Allometric growth parameter	24.7		This study; Smith et al. (1998); Morgan et al. (2000)
β	Allometric growth parameter	0.37		This study; Smith et al. (1998); Morgan et al. (2000)
s_{min}	**Minimum survival**	**0.6**	**yr^{-1}**	Based on Ebert and Russell (1992); McBride et al. (1999)
s_b	**Survival parameter *b***	**0.03**		Based on Ebert and Russell (1992); Ebert (1987)
s_{50}	**Survival parameter 50%**	**30**		Based on Ebert and Russell (1992); Ebert (1987)
o	Senescence parameter	30	yr	Ebert and Southon (2003)
p	Proportion of urchins accessible to divers	0.4		This study

Note: Rows in boldface represent final parameter values used in model calibration and based on annotated references.

and on both somatic and gonadal growth (Ebert 1980, McShane and Anderson 1997).

Monthly consumption $C_{i,t}$ (in kg) is related to kelp biomass in the cell in which agent i is found j (i) at time t, $K_{j(j),t}$ using a Holling type II functional response relationship (Holling 1959; Eq. 2) based on the assumption that sea urchins are limited by its capacity to process food (Breitburg 1996):

$$C_{i,t} = W_{i,t} \cdot c_{max} \cdot \left(\frac{K_{j(i),t}}{c_{50} + K_{j(i),t}} \right) \quad (2)$$

where W is the wet weight of sea urchins (kg), c_{max} is the maximum monthly consumption rate (kg·kg^{-1}·month^{-1}), and c_{50} is the half-saturation constant. Although sea urchin consumption was negatively correlated with sea water temperature in experimental conditions (from 10.1 to 16.8°C; McBride et al. 1999), red sea urchins in the field (i.e., southern California kelp bed) did not show significant changes in consumption with sea water temperature (McBride et al. 2004, Foster et al. 2015 and reference therein).

The total energy $E_{i,t}$ is given by the amount of kelp consumed by agent i in time t, $C_{i,t}$, and the energetic content of kelp, *M. pyrifera*, e (in kcal/kg; Eqs. 3–4; Zimmerman and Kremer 1986, Paine and Vadas 1996):

$$E_{i,t} = C_{i,t} \cdot e \tag{3}$$

The surplus energy $Z_{i,t}$ of the agent i at time t is then given by:

$$Z_{i,t} = E_{i,t} - (\rho \cdot W_{i,t}) \tag{4}$$

where ρ is the energy (in kcal/kg) used for bioenergetics processes including metabolism, egestion, excretion, and specific dynamic action (Thompson 1984, McBride et al. 1998).

Adult sea urchins store nutrients (i.e., surplus energy $Z_{i,t}$) in their gonads year-round prior to the production of gametes (Scheibling and Hatcher 2001). This allocation of nutrient to gonads may involve a consumption threshold (McBride et al. 2004, Lawrence 2007): (1) a minimal amount of food necessary for maintenance (i.e., metabolic cost ρ, below which no somatic growth or gonad development would occur) and (2) above this threshold, surplus energy Z is allocated to somatic E^{togrow} and gonadal growth E^{togonad} (in kcal) as a function of individual size and reproductive maturity (Eqs. 5, 6):

$$E_{i,t}^{\text{togrow}} = \begin{cases} Z_{i,t} \cdot \left(1 - \left(\frac{1}{1+\exp(-(m_b \cdot (L_{i,t} - m_{50})))}\right)\right) & \text{if } Z_{i,t} > 0 \\ 0 & \text{if } Z_{i,t} < 0 \end{cases} \tag{5}$$

and

$$E_{i,t}^{\text{togonad}} = \begin{cases} Z_{i,t} \cdot \left(1 - \left(\frac{1}{1+\exp(-(-m_b \cdot (L_{i,t} - m_{50})))}\right)\right) & \text{if } Z_{i,t} > 0 \\ 0 & \text{if } Z_{i,t} < 0 \end{cases} \tag{6}$$

(Note that $Z_{i,t} = E_{i,t}^{\text{togrow}} + E_{i,t}^{\text{togonad}}$),

where m_b determines the shape of the energy allocation function and m_{50} the sea urchin size L (in mm) at which 50% of individuals are mature. In these equations, the total energy obtained from food consumption and not spent on metabolic cost (Z) is inversely allocated to somatic growth or gonadal growth as a function of the energy allocation function. Thus, when $Z < 0$, no energy is allocated to either somatic or gonad growth. Moreover, sea urchins generally start to invest in gonadal fattening when reaching maturation ages/sizes (i.e., at sizes L of ca. 40 mm; Kato and Schroeter 1985).

Monthly mass for sea urchins is represented in our model as the sum of the somatic growth and gonadal increment (in kg). This way the model is able to account for differences in gonad index (gonad wet weight/total sea urchin wet weight) among agents (Eq. 7):

$$W_{i,t+1}^{\text{tot}} = W_{i,t} + G_{i,t} \tag{7}$$

Both somatic W and gonadal G weights were calculated by using the following general bioenergetics equations:

$$W_{i,t+1} = W_{i,t} + (E_{i,t}^{\text{togrow}} \cdot \nu) \quad \textit{and}$$
$$G_{i,t+1} = G_{i,t} + (E_{i,t}^{\text{togonad}} \cdot \kappa) - \text{Eggs}_{i,t} \tag{8}$$

where υ and κ are the energy to somatic and gonadal weight conversion, respectively (kg/kcal), and Eggs is the amount of gonad mass released as gametes (see *Recruitment* section). A υ of 8×10^{-6} and κ of 2×10^{-5} kg/kcal were assumed to match the range of parameter values obtained from the literature (McBride et al. 1997, Siikavuopio et al. 2007, Christiansen and Siikavuopio 2007). Gametogenic growth in sea urchins does not seem to involve de novo production, but a transfer of nutrients from gonadal reserves to gametogenesis (Pearse et al. 1986, Walker and Lesser 1998). Further, histological studies of gonads from laboratory and field experiments indicated that neither food nor temperature affected gametogenesis (McBride et al. 1997).

Finally, if the change in somatic weight (Eq. 3) is positive, size L (in mm) is updated by a size–weight regression of the form (Eq. 9):

$$L_{i,t} = \alpha \cdot W_{i,t}^{\beta} \tag{9}$$

where α and β are constants derived from a size–weight relationship of red sea urchins in Point Loma ($r^2 = 0.93$; $N = 208$; Fig. 3c).

Recruitment

Sea urchins are broadcast spawners where both females and males release similar mass of eggs and sperm, respectively (McBride et al. 1999, Pearce et al. 2004). Reproductive effort was calculated as the decrease in the mass of the gonads from spawning as a percentage of total gonad mass (Eq. 3; Raymond et al. 2004). The amount of gametes released by agent *i* at time *t*, $\text{Eggs}_{i,t}$ (in kg), is then calculated as follows (Eq. 10):

$$\text{Eggs}_{i,t} = G_{i,t} \cdot \left(\frac{1}{1+\exp(-(m_b \cdot (L_{i,t} - m_{50})))}\right) \cdot \psi \cdot (\sin[0.6 \cdot (t - 5.6)] + 1.2) \quad (10)$$

where G is the gonad weight (in kg) and ψ the proportion of gonad weight released as eggs as a function of the maturity ogive (see Eq. 4). Although sporadic spawning may occur year-round, peaks are evident in June/July and represented in this model by a seasonal component in Eq. 10 (McBride et al. 1999). Parameters of the seasonal component have been estimated from SDCP data (Fig. 3d). Even though adult densities may have an effect on fertilization success and larval production through Allee effects (Lundquist and Botsford 2011), our model did not account for such effect. After eggs are fertilized, planktonic larvae remain in the water column for ca. 30 d before settlement (Schroeter et al. 1996, Miller and Emlet 1997). Thus, considering the model monthly time step, an average time-lag of 1 between spawning and settlement was assumed. The number of recruits per spawner, R, was calculated as:

$$R_{i,t+1} = \text{Eggs}_{i,t} \cdot \delta \cdot \varepsilon_t \quad (11)$$

where δ is the number of agents produced by unit of spawn (released eggs; McBride et al. 1998) and ε_t is a stochastic component log-normally distributed $\sim \ln N(\mu, \sigma^2)$ to represent variability in environmental conditions affecting larvae survival. Total number of recruits at settlement per time step (month) were obtained by adding the number of recruits per spawner from the total population N and rounded to the nearest agent:

$$R_t = \text{round}\left(\sum_{i=1}^{N} R_{i,t}\right) \quad (12)$$

New recruits were allocated to random cells within the study region, initialized at size of 3 mm, and total and gonad weights were computed following the allometric relationships described earlier and obtained from the literature (Byrne et al. 1998). Food availability at settlement affects consumption and growth of new recruits as described in previous sections.

Survival

Analysis of transitions in the size distributions of red sea urchins (*S. franciscanus*) suggested that annual survival may be size specific: 0.6 for individuals of 11–40 mm test diameter, 0.82 for individuals of 41–70 mm, and 0.91 for those of 71–100 mm (Ebert and Russell 1992). Further, resource allocation and energetic budget are the main conditions in determining survival probability in *S. franciscanus* (Ebert 1987). While gut content is used for short-term nutrient storage, materials in the gonads especially become more important as starvation increases (McBride et al. 1999). Thus, all of the main body components may be used as nutrient reserves during starvation (Lares and Pomory 1998). In this study, the probability of survival per month was modeled as a logistic function of sea urchin size and a measure of starvation (Eq. 13; Ebert and Russell 1992, McBride et al. 1999):

$$S_{i,t} = s_{\min} + (1 - s_{\min}) \cdot \left(\frac{1}{1+\exp(-(-s_b \cdot (L_{i,t} - s_{50})))}\right) \cdot \lambda_{i,t} \quad (13)$$

where $s_{\min}$ is the minimum monthly survival rate, s_b is the slope parameter of the survival function, s_{50} is the 50% of the survival function, and $\lambda_{i,t}$ is a measure of condition. This condition is calculated as the ratio of actual total weight W^{tot} (body plus gonad) to expected somatic weight W calculated from current size L and the allometric parameters α and β (as suggested in Rashleigh and Grossman 2005; Eq. 14). In other words, well-fed sea urchins are able to store nutrients in the gonads that may be used during periods of starvation, thus increasing gonad weight, λ, and therefore survival. This formulation reflects the study area where predators are scarce and sea urchins have extremely high probability of survival (~1.0) when food is abundant (Ebert et al. 1999).

$$\lambda_{i,t} = \frac{W_{i,t}^{\text{tot}}}{\alpha \cdot L_{i,t}^{\beta}} \tag{14}$$

In order to stochastically simulate the mortality of agents, an i agent dies at time t if a randomly generated number that is uniformly distributed [0–1] is higher than $S_{i,t}$. Further, all individuals die once they reach age of o (parameter values for age senescence, o, were taken from Ebert and Southon 2003; Table 1). Mass mortality due to catastrophic events such as strong El Niño and pathogens has not being considered in our model. Lastly, compensatory density dependence processes in our model are linked to food limitations (i.e., kelp carrying capacity at each cell), affecting both somatic growth and gonadal weight, which in turn determines energetic condition linked to survival.

Individual movement

Red sea urchin movement has shown small-scale spatially explicit variations in speed and intensity between sites with kelp bed and barrens (i.e., areas where population growth of sea urchins is extreme due to lack of predators, causing destructive grazing of kelp beds; Mattison et al. 1977, Laur et al. 1986). In addition, Harrold and Reed (1985) and Lauzon-Guay and Scheibling (2007) found an inverse correlation between temporal patterns of algae biomass (especially *Macrocystis*) and sea urchin movement rates. Thus, movement rate and direction were modeled as a function of kelp availability and biomass of sea urchin grazing. On each time step (month), all agents examine neighboring cells and move if the quality Q (as a probability function of kelp biomass and sea urchin biomass in each cell) is higher in an adjacent cell (Eq. 15):

$$Q_{j,t} = \frac{K_{j,t} / \sum W_{i,t}^{\text{tot}}}{\sum_{j=1}^{9} \left(K_{j,t} / \sum_{i \in j} W_{i,t}^{\text{tot}} \right)} \tag{15}$$

where $K_{j,t}$ is the kelp biomass at cell j and time t, and $\sum_{i \in j} W_{i,t}^{\text{tot}}$ is the biomass of sea urchins in cell j and time t, that is, the sum of total weight for all agents i in cell j at time t, and 9 is the number of cells where sea urchins can possibly move to or remain with. Thus, the movement rule used in our model is as follows: An agent senses its neighboring cells, if the ratio of food availability to biomass of agents grazing is higher than the ratio at the current cell, the agent will move to the cell with highest ratio of kelp to sea urchin biomass during that time step. Movement occurs each month, where the spatial extent of each cell is consistent with the monthly movement rates documented for the red sea urchin (Mattison et al. 1977, Harrold and Reed 1985). Movement across the boundaries is reflective, that is, urchins exiting one edge of the model reentered at the same cell. This is a convenient means of approximating movement of urchins to and from the area beyond the modeled region. Previous work on movement of red sea urchins in southern California indicates that our model constraints (e.g., short movements, lack of predatory effect) are consistent with the processes occurring in the study area (Tegner and Dayton 1991, Foster and Schiel 2010). However, sea urchin or other species may show different movement behavior, and movement might depend on temperature in some cases (Dumon et al. 2009). In this model, given the relatively restricted spatial extent of the study region and the small-scale spatial resolution, it is unlikely that significant temperature differences exist among cells in a given time (Roughan et al. 2005).

Fishery

The present California sea urchin management scheme is based on size limits, restricted access, and seasonal closures. Size limits (i.e., 82.5 mm) were set at a level that allowed several years of spawning before harvest and appears to have helped prevent fishery collapse (Hilborn et al. 2007). The Point Loma red sea urchin fishery began in the early 1970s and currently involves about 12 divers operating 168 ± 11 d a year (mean ± SE; 2004–2009; D. Rudie, *unpublished data*). In 2007, the estimated fishing effort for the southern tip of Point Loma was 514 vessels days (Parnell et al. 2010). The fishery involves small boats (<10 m) and divers use mostly hookah or self-contained underwater breathing apparatus (SCUBA) equipment, where the unintended by-catch or direct disruptions to the habitat or other species are negligible. Average annual catches for the period 2003–2010 were 305 ± 21 t and the

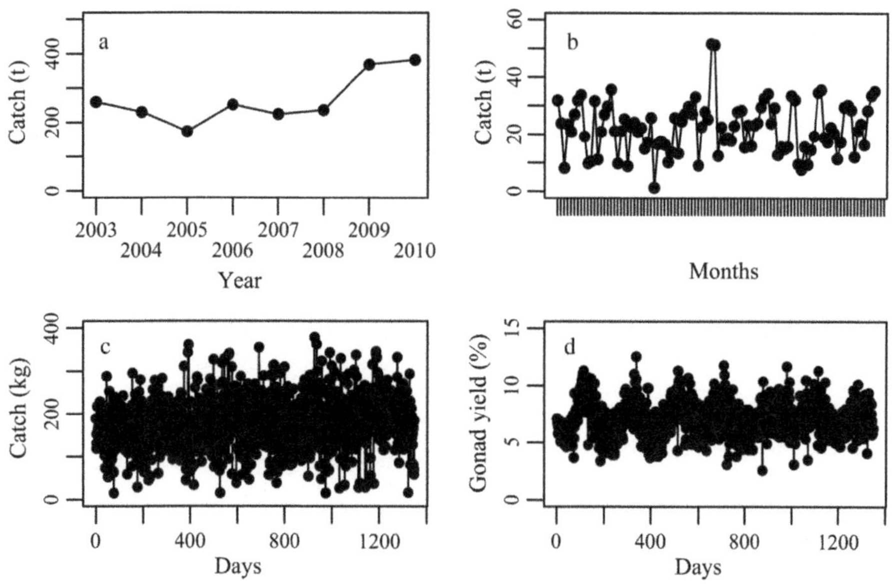

Fig. 4. Fishery yields for the study area obtained at the landing port, processing plant (D. Rudie, *unpublished data*) and from the San Diego community-based data collection program for the period 2003–2010: (a) annual catches (in t); (b) monthly catches (in tons); (c) mean catch per dive (in kg); and (d) mean gonad yield (gonad weight/total body weight) per dive.

average catch per dive was 212 ± 3.4 kg with an average gonad yield of 7.1 ± 2.1% (Fig. 4).

In this model, different levels of fishing effort were based on daily information from the SDCP, the local processor, and from Parnell et al. (2010*a*, 2010*b*). Individual dives (d) target one cell at each time step with a mean search area of 50 m^2 per dive (SDCP, *unpublished data*), thus representing 50% of the total area of a cell. In addition, considering sea urchin are often hidden in crevices and under rocks or reefs, and visibility conditions inside the kelp bed may reduce spotting sea urchins, only ca. 80% percent of the total legal agents in a cell are harvested (SDCP, *unpublished data*). Although the proportion of sea urchin hidden may be related to food availability and/or habitat complexity (e.g., rugosity), this proportion was assumed constant in space and time and independent of the size of sea urchins (i.e., constant vulnerability to harvesting). Thus, about 40% of the total legal sea urchins (>82.5 mm) are harvested in each cell and time step ($P = 0.5 \cdot 0.8 = 0.4$). After fishing effort is allocated, legal agents are removed from the population and added to the monthly total catch (Eq. 16).

$$H_t^s = \sum_1^d \sum_1^{p.i} W_{i,t}^{tot} \quad \text{and}$$

$$H_t^u = \sum_1^d \sum_1^{p.i} G_{i,t} \ldots i \in [L_{i,t}] > 82.5\,\text{mm} \qquad (16)$$

where H_t^s and H_t^u are the total sea urchin catch and total gonad (*uni*) catch per month in time t, respectively; i represents (legal size) sea urchin agents; p is the proportion of agents i accessible to the divers; and d is the total number of dives or cells (1 dive = 1 cell) per month (i.e., 80 dives/month).

Different fishing scenarios were simulated to capture an open access non-cooperative harvesting strategy (actually used by most San Diego sea urchin divers) and a cooperative and more selective fishery that optimizes sea urchin productivity and yields (used by few divers; Table 2):

1. Non-selective fishery (open access scenario).

   a. Divers decide on a monthly basis which area (i.e., cells c) within the fishing ground to target at random with replacement. This scenario assumes very limited or imperfect

Table 2. Model specifications under competitive and cooperative harvest strategies.

Scenario	Strategy	Information sharing	Direct fishing effort to cell (*c*) where	Harvest within each cell (*c*)
1a	Competitive	None	At random (replacement)	All accessible legal urchins
1b	Competitive	Indirect	Sum of total sea urchin weight is maximum	All accessible legal urchins
2a	Cooperative	Direct	Sum of total gonad yield is maximum	All accessible legal urchins
2b	Cooperative	Direct	Sum of total gonad yield is maximum	Sea urchins with highest gonad yield

Note: Information sharing: indirect (e.g., by visualizing sea urchin divers in particular harvesting areas and using that information as indirect measure of yields); direct (e.g., by sharing information on best sea urchin and/or gonad yield at the end of the day).

knowledge on best fishing areas from previous months and no exchange of information among divers (most frequent scenario for new sea urchin divers and for areas not previously explored).

b. Divers decide on a monthly basis where to fish based on historical knowledge on sea urchin concentrations and harvest occurs within that cell without selecting for gonad quality. No updated information on gonad yields or information exchange mechanisms are available.

2. Spatially directed harvesting (cooperative scenario). At each monthly time step, quality of fishing sites is evaluated based on sea urchin biomass and gonad yield assuming perfect knowledge through information sharing among divers (Table 2):

   a. Each month, fishing effort is directed to "best" cells where the total gonad content (total weight of gonads per cell in kg) is highest and exploiting sea urchins independently of their gonad content. Although gonad content is highest on average in these areas, some sea urchins within that area may have low gonad yields and divers do not discriminate their extraction by gonad yields.
   b. This scenario also assumes perfect knowledge on best yield areas and divers target cells where the gonad index is highest (total weight of gonads/total weight of urchins) and harvesting only sea urchins with best gonad yield (e.g., divers crack sea urchins in the ground to check gonad yield and quality and harvest only patches with highest gonad yield, moving to a different patch when encounter poorly fed sea urchins; Halmay pers.comm.). This highly selective strategy is already in use by two divers, which target highly marketable sea urchins and leaving the rest in the ground in order to improve their gonad yields and quality ("fattening").

Results

Evaluation of model performance

The calibration and evaluation criteria were as follows: (1) The model produced realistic individual growth patterns; (2) the observed and predicted values for mean sea urchin densities, proportion of legal individuals, and adult size and weight had to be in reasonable agreement with the data; (3) the population persisted through the simulation; and (4) fishery yields (in total sea urchins and gonad yield) were consistent with the information available for the study area. Since an individual-based model is stochastic by nature, multiple runs (i.e., 10) were necessary to be able to compare output values with field data. To calibrate the model, the selected parameter values (final values highlighted in boldface in Table 1) were adjusted until results were comparable with field observations for the sea urchin population in Point Loma, California, and/or sensible with values obtained from the literature for other species and/or other regions (Appendix S1: Table S1).

The intent of our model was not to perfectly reproduce growth rates of individual sea urchins in Point Loma but rather show that the parameters based on the literature for other sea urchin species and/or regions could effectively capture the general growth patterns of red sea urchins in

Point Loma, California. Thus, individual growth trajectories for the simulated sea urchin population under different scenarios of food availability (i.e., kelp abundance) were consistent with patterns observed in several field studies in California (Ebert and Russell 1992, Smith et al. 1998; Morgan et al. 2000; Fig. 3b). As expected, simulations under low kelp abundance (<1 t/ha) predicted both slower growth rates and smaller adult sizes (50% lower than for conditions of abundant food >30 t/ha).

Finally, all data on the sea urchin resource and its fishery in Point Loma available from the SDCP and the processing plant and from a literature review on the genus *Strongylocentrotus* and/or the region under study (i.e., San Diego, California; Appendix S1: Table S1) were used as reference values for both model initialization and calibration.

To further validate the growth sub-routine of the ABM, size distributions of red sea urchins generated from our model were compared with size distributions of individuals sampled from Point Loma during 2004–2010. Field distributions were derived from both biological samples and fishery-dependent information obtained by the SDCP divers.

There were no significant differences (Wilcoxon rank-sum test, $P < 0.05$) between means of observed and predicted proportion of legal size sea urchins, and mean adult sizes and weights (Fig. 5). However, the number of juveniles predicted by the calibrated model was significantly greater than the observed number. These differences were attributed to the fact that the model includes complete sampling of recruits and of juveniles, but the field method does not.

Sea urchin and kelp dynamics

The output of an initial simulation with no fishing effort showed a reasonable relationship between size (mm) and (1) age (months); (2) weight (kg of agent); (3) gonad yield (% of total urchin weight); and (4) frequency distribution of gonad yield based on the information available from the SDCP, the processing plant, and the literature (Fig. 3b for size-at-age; Fig. 3c for size–weight relationship). Thus, the final sea urchin population was sensibly represented by agents of different ages, sizes, and physiological status (i.e., with different gonad yield-at-age). Individual sea urchin agents were followed over 30 yr to highlight the effect of food availability and the fine-scale spatial structure of the kelp bed on the sea urchin population dynamics (i.e., growth, consumption, gonad mass, and reproduction). Sea urchin agents located in areas outside the kelp bed exhibited very slow growth rates and a decrease in gonad mass to basal levels with consequent low reproductive output. The calibrated model was able to replicate the kelp–urchin dynamics typical of these systems, where in absence of fishing and when natural predators (e.g., lobster) are being harvested intensively (Cowen 1983), sea urchin populations grow rapidly overgrazing kelp beds forming barrens areas of poorly fed urchins affecting their growth and reproduction (Harrold et al. 1991).

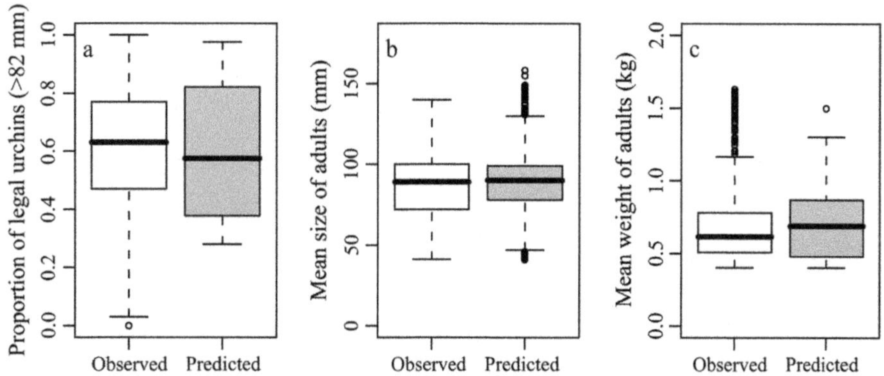

Fig. 5. Boxplots of observed and model predicted (a) proportion of legal sea urchins (>82 mm); (b) mean adult size (>40 mm); and (c) mean adult weight (kg) of red sea urchin *Strongylocentrotus franciscanus* in Point Loma, California (no significant differences at the 0.05 level).

Evaluating cooperative harvesting

Size frequency distribution of the sea urchin population at the end of the simulation period (year 30) for Scenario 1 was comparable with the size distribution of the sea urchin population in Point Loma, California, with exception of the juvenile size classes (Fig. 6). The latter was not represented in the field data due to limitations in the sampling design. For the competitive scenarios, size frequencies showed a more leptokurtic distribution compared to the cooperative scenarios, with lower numbers of legal size (>82.5 mm) sea urchins in the population and lower mean sea urchin sizes. This pattern may arise from the fact that divers under competitive scenarios harvest all legal sea urchins independently of their gonad content, while divers under Scenario 4 target only legal sea urchins with high gonad yields, leaving some large individuals with poor gonad content unharvested (Fig. 6).

For all scenarios, this relative rate of increase in sea urchin population diminished after the fishery started as expected and relative differences in sea urchin biomass among scenarios were established slowly in the model (Fig. 7a). The use of agents composed of 100 individuals each allowed the total number of individuals in the simulation to remain sufficiently large even with relatively high rates of fishing and natural mortality. Simulations of less than 200 time steps revealed few changes in sea urchin abundance, while by the end of the simulation period (30 yr), Scenario 1a (competitive fishery with random allocation of fishing effort) showed the highest biomass of sea urchin (91,540 t) and Scenario 1b (competitive harvesting targeting areas with maximum sea urchin density) the lowest (83,740 t; Table 3).

Catch of sea urchins showed a general increasing trend for the simulation period where Scenario 1a showed the lowest increase (5%), while Scenario 2b (cooperative harvest targeting sea urchins with highest gonad content) increased 32% in 30 yr (Fig. 7b). For this scenario, sea urchin catches were also 2% higher at the beginning of the fishery and 20% higher at the last year of the simulation compared to the most non-cooperative scenario (Scenario 1a). For Scenarios 1a and 2a, catches in year 30 were below the long-term average catch (251 t) of sea urchin from the Point Loma bed in 2004–2010, while

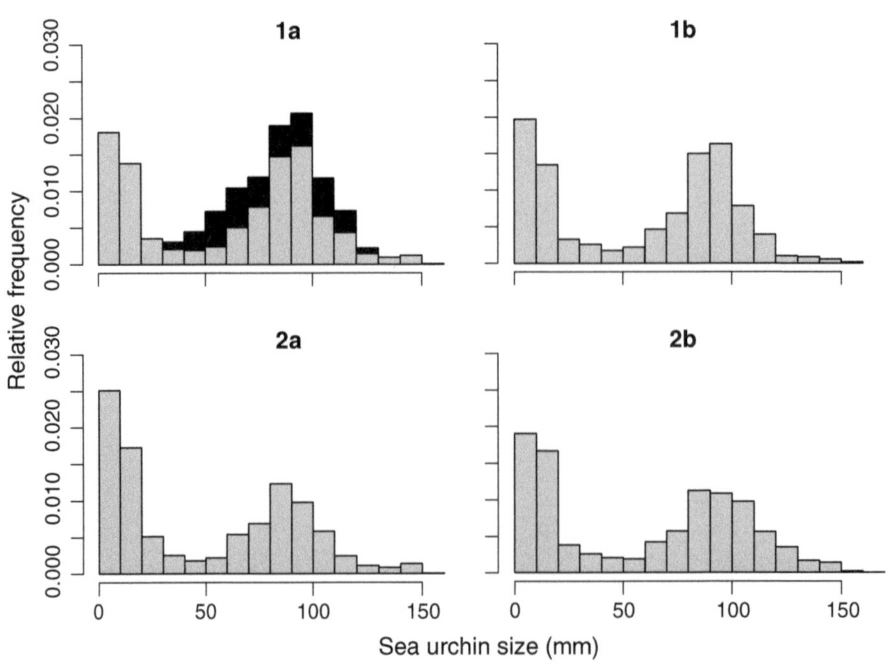

Fig. 6. Size frequency distributions of sea urchin *Strongylocentrotus franciscanus* from 10 simulations over 360 time steps (30 yr) and four different scenarios for a competitive (1a and 1b) and a cooperative (2a and 2b) fishery. Black bars correspond to field data obtained from San Diego community-based data collection program.

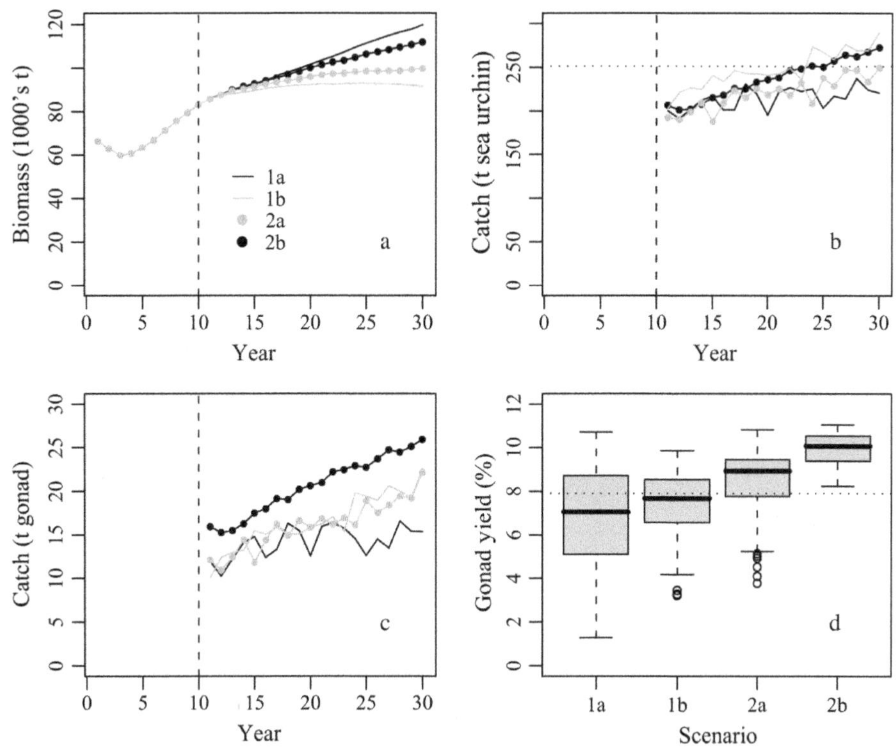

Fig. 7. Output of 10 simulations of 10,000 initial agents of *Strongylocentrotus franciscanus* over 30 yr and two cooperative and two non-cooperative scenarios as described in Table 2. (a) Sea urchin biomass (in 1000's t) aggregated by year; vertical lines denotes the time when fishery started; (b) catches in tons of total sea urchin; horizontal dotted lines show mean catches for the period 2003–2010; (c) catches in tones of sea urchin gonad or *uni*; and (d) boxplot (mean ± 95% CI) of gonad yields (as % of total body weight); horizontal line shows mean gonad yield (7.9%) of catches for the period 2003–2010 in Point Loma, California.

catches for Scenarios 1a and 2b were ca. 10% higher than the average catch (Fig. 7b; Table 3). Further, differences in catches of sea urchin in gonad weight among scenarios and particularly between competitive and cooperative scenarios became more evident since the beginning of the fishery, being ca. 90% higher for Scenario 2b than for Scenario 1a by the end of the simulation period. Lastly, Scenario 2b showed substantially less inter-annual variability in catches of gonad weight than the rest of the three scenarios (Fig. 7c; Table 3) given that divers harvest only sea urchins with high gonad yield and leave poorly fed urchins in the ground. Gonad yield (as % of gonad weight in total sea urchin weight) increased with the level of cooperation and

Table 3. Mean (SD) of kelp biomass and sea urchin biomass, abundance, catch in total sea urchins and in gonad, and gonad index (as % of gonad weight in total sea urchin weight).

	Kelp	Sea urchin				
Scenario	Biomass (1000's t)	Biomass (1000's t)	Abundance (million ind)	Catch (whole ind in t)	Catch (gonad in t)	Gonad Index (%)
1a	15,221 (1,902)	91.5 (16.7)	151 (2.53)	196.4 (43.6)	14.2 (6.5)	7.2 (2.7)
1b	16,858 (813)	83.7 (11.5)	126 (1.49)	210.9 (31.7)	16.5 (4.8)	7.8 (2.7)
2a	17,803 (498)	86.3 (13.5)	129 (1.72)	201.3 (41.8)	18.1 (5.8)	9.0 (1.7)
2b	16,212 (967)	89.7 (16.6)	147 (2.16)	217.4 (21.7)	22.8 (3.5)	11.0 (0.9)

coordination in harvesting, with both Scenario 2a and Scenario 2b showing higher gonad yield than the average for the study area from 2004 to 2010 (Fig. 7d).

Discussion

A spatially explicit, ABM was used to simulate the local dynamics of kelp and red sea urchin population in response to different urchin harvest strategies. Using this modeling approach, spatial variability in habitat quality and ecologically significant differences among individuals were simulated, factors that are difficult to incorporate in aggregated models. The model provided a reasonable agreement with field data for the Point Loma sea urchin population in terms of adult densities, size, and weight data as well as growth patterns, catches, and gonad yields. These results suggest that the model was consistent with the available biological data. However, it is not intended to represent an actual status of the Point Loma sea urchin stock nor an accurate representation of catches for this fishery.

To validate growth parameters, the model was initialized with size distributions of sea urchins in the Point Loma kelp bed. The SDCP sampling does not provide information on the numbers of young urchins (recruits and young-of-the-year; y-o-y), which are too small to be caught and landed by commercial divers. Further, the assessment of settlement is problematic because small-size juveniles are often hidden in crevices and because rapid mortality before observation leads to underestimation (Harrold et al. 1991, Ebert et al. 1994). Thus, sampling designs that capture intra- and inter-annual patterns of abundance and distribution of y-o-y sea urchins would be needed to improve the realism of the initial size frequency distributions used in our model. Beyond these limitations, our predicted distributions compared favorably with the field distributions of adult sea urchins.

A few caveats to our model should be noted. First, environmental conditions other than food availability were assumed spatially and temporally homogenous. Although the model focuses on a relatively small area (52 km^2) where environmental conditions such as water temperature, turbidity, salinity, and pollution are relatively homogenous, temporal variability and seasonality often characterize the study area (Parnell et al. 2006). Particularly, kelp resource can fluctuate significantly on annual-to-decadal time scales in response to oceanographic conditions associated with El Niño/La Niña and Pacific Decadal Oscillations (Parnell et al. 2010*a*). The combination of high kelp productivity in environments with the potential for high rates of disturbance can make kelp beds surprisingly transient. This is particularly important given the tight ecological coupling of sea urchins with their environment (Botsford et al. 2004) and particularly with the kelp bed (Steneck et al. 2002, Parnell et al. 2010*b*).

Second, habitat characteristics including bottom composition, relief, topographic features such as ledges and crevices and depth were not included in the model. Particularly, substrate complexity (e.g., rugosity) has been identified as an important aspect influencing sea urchin behavior, recruitment, and survival (Caddy 2007). Predator–prey interactions may also be affected by the availability of sea urchin refuge space through habitat cover (Grabowski et al. 2008).

Third, the model assumed a closed population where individuals recruit to the fishery only by reaching legal sizes (>82.5 mm). However, Hilborn et al. (2007) noted some discrepancies between the proportion of legal size urchins, growth patterns for the study area, and the observed fishing effort. Two alternative explanations to this discrepancy include (1) a considerable influx of legal size sea urchins from non-fished areas (i.e., deep areas non-accessible to diving) and (2) growth rates higher than the ones reported in the literature for the study area. To investigate these hypotheses, targeted surveys outside the fishing grounds and particularly in deep waters should be directed to assess the status of the resource in those areas, as well as additional growth studies performed at relevant spatial scales. Although our assumptions of "closed adult population" do not invalidate our results in terms of the observed differences in sea urchin catch and gonad yields under difference scenarios, including a proportion of adult sea urchins recruitment to the fishery from non-accessible areas would be essential to determine harvest strategies, catch quotas, and management plans in general.

Fourth, the model assumes that recruitment is directly proportional to the spawning stock (steepness of 1 in the stock–recruitment relationship).

Compensatory mechanisms in sea urchin recruitment may occur due to a greater availability of both food and space as shelter from predators (Tegner and Levin 1982, Caddy 2007). On the other hand, depensatory processes are also common in sea urchin and other broadcast spawners when low densities of spawners make fertilization less probable (Allee effect, Allee 1931). In addition, protection by adult spines canopy for conspecifics recruits and juveniles has also been suggested as a depensatory mechanism in purple and red sea urchins (Tegner and Dayton 1991, Botsford et al. 1994). In our model, compensatory density dependence processes occur through carrying capacity limitations in food availability (i.e., dense aggregations of sea urchins are more likely to overgraze the kelp bed, thus increasing sea urchin mortality through starvation and decreasing reproductive capacity due to weaker gonads).

Lastly, sea urchin divers were considered as homogeneous in terms of their catching strategies, capabilities, and efficiency in locating the most dense sea urchin patches and those urchins with higher gonad content. Diving times and areas covered by Point Loma sea urchin divers are variable, and the model did not consider this variability. In addition, fishing effort was assumed constant both intra- and inter-annually and restricted to a pre-specified number of dives per month obtained from the San Diego sea urchin fishery. A more comprehensive model including individual divers' characteristics, capabilities, and behavior would benefit the realism of the fishing process (Sanchirico and Wilen 1999).

Despite these limitations, our model was able to replicate a sea urchin population in a sufficiently realistic manner to be used as a base case scenario to project the Point Loma sea urchin population under different harvesting strategies. Our goal was to elucidate the effects of four different harvesting tactics in both the sea urchin population and its fishery. Two of these tactics were aimed at maximizing total sea urchin catch without considering gonad content (*fishing for spines*), and two scenarios directed to maximize total gonad yield (*fishing for uni*). Although each scenario corresponded to a particular harvest tactic, these were associated with different levels of cooperation and information sharing among divers, and levels of fishing mortality restrained by the current management regulations of limited entry and current number of divers exploiting this resource in Point Loma, California.

Within the non-cooperative cases, Scenario 1 corresponds to an open access competitive fishery where a "race for shellfish" and a lack of information transfer on best fishing areas preclude catch optimization. This scenario is still a common case for several sedentary fisheries worldwide and in particular for sea urchin resources (Lawrence 2007). Although this harvest tactic allowed the sea urchin population to increase 35% in 30 yr, catches in terms of total sea urchin and particularly in terms of gonad yield were minimum. Our second scenario assumes some degree of available information about the fishing ground, either directly by traditional ecological knowledge of experienced divers or indirectly by divers spotting fishing boats in areas thought to be the most productive. This case is more realistic to the sea urchin fishery in San Diego, where the relatively small area and the long-term experience of divers harvesting the Point Loma kelp bed allow them to efficiently target sea urchin aggregations. Although catches for this scenario were maximized, divers harvested a considerable proportion of poorly fed and low gonad content urchins that would likely decrease revenues and profits when sold to the local processor.

Both cooperative scenarios imply high degree of communication and information sharing among divers in order to support a coordinated harvesting strategy. Divers need to share daily or monthly information on what areas within the fishing ground show the highest densities of sea urchin with good gonad (*uni*) condition. This scenario is facilitated by the fact that (1) all San Diego sea urchin divers disembark their catches in two closely located ports (i.e., Mission Bay and Point Loma ports) and (2) most San Diego divers are affiliated to the California Sea Urchin Commission and/or the San Diego Fishermen Association, fostering periodic and real-time information exchange (P. Halmay, *personal communication*). However, even with coordinated harvesting aimed at targeting the best *uni* areas, individual divers may have different fishing strategies. Thus, we included two cooperative scenarios, one where divers harvest all legal sea urchins within the best *uni* areas in a particular month (Scenario 2a) and the other where divers move around different patches to find the best quality urchins within the best areas (Scenario 2b). For these two scenarios, total sea urchin catches showed similar trends and magnitude than the non-cooperative Scenario 2, but catches in gonad and mean gonad yield were significantly higher, especially for Scenario 2b with gonad yields 20% higher than the mean gonad yield for the San Diego fishery during 2004–2010. Although Scenario 2b could be considered an extreme case of *fishing for uni*, at least 10% of San Diego sea urchins divers are actually using this harvest tactic with 30% higher gonad yield than the rest of the fishers in Point Loma (D. Rudie, *unpublished data*). Further, the lower inter- and intra-annually variability in catches and gonad yield associated with Scenario 2b has been demonstrated to be beneficial providing a steady supply of fresh *uni* to local restaurants. Finally, size composition of the simulated population for both cooperative scenarios showed higher proportion of large sea urchins potentially contributing to the overall productivity of the population (Longhurst 2002, Berkeley et al. 2004).

Conclusions

Modeling results suggest that harvesting tactics aimed at maximizing gonad content instead of total sea urchin catch can be beneficial for the productivity of the stock (e.g., by increasing the proportion of large sea urchins) but especially in terms of maximizing revenues through better gonad yields (*uni*). However, assuming perfect knowledge of the fishing grounds and particularly of the sea urchin densities and gonad conditions required a high degree of daily or monthly information sharing by divers targeting a specific stock. It has been observed in several fisheries that fishers report their daily catches, location, composition, and other relevant information once they reach port or landing locations (Faunce 2011). Usually, this behavior involves strong and well-organized local institutions, with respected leaders, social cohesion, trust among fishers, and great sense of resource ownership and stewardship (Weber et al. 2004, Gutierrez et al. 2011*b*).

In addition to improving revenues by better gonad quality and steady supply, coordinated harvesting typical of well-structured fishing cooperatives diminishes the "race for fish" often associated with other forms of traditional management. The consequences for such a scenario may include overcapitalization, decreased fish quality, and unsafe fishing conditions. Although various forms of property rights such as individual or community quotas and territorial use rights for fisheries (TURFs) have been proposed to end this waste by giving incentives for stewardship and cost reduction (Castilla and Defeo 2001, Costello et al. 2008), this security over catch or space may not be sufficient to end the race to fish, due to a lack of incentives to coordinate harvest among quota holders. The biological and economic heterogeneity typical of sea urchin resources may induce a highly competitive scenario even when the right to harvest an allocated quota is secured (Scott 2000). Thus, in addition to ending the incentive to race, cooperative harvesting scenarios like the one used in our operating model can yield economic benefits to fishers by coordinating effort in order to avoid redundant searches, improving fish recovery rates and quality, and reducing regulatory bycatch, and biological benefits by allowing fish to grow and reproduce before being taken from the brood stock (Castilla et al. 1998, Ekerhovd 2008). Another area for cooperation is catch sharing or pooling (Deacon et al. 2010). Fishers compensate for losses by sharing catches in port, regardless of who caught more, effectively ensuring a minimum catch for each fisher for every fishing trip (Salas 2000). Processors are also likely to benefit from coordinated harvesting by having access to reliable supplies of top quality seafood and participating with fishers in developing new markets. Lastly, cooperative fisheries may be more likely to achieve sustainability standards, gaining access to seafood eco-labels with potential social, economic, and/or ecological benefits (Gutierrez et al. 2012, 2016, Pérez-Ramírez et al. 2016).

This modeling framework highlights the value of cooperation that is present in many fisheries, both real active cooperation and hidden and based on clues or visual sighting of other fishers (Dreyfus-Leon and Gaertner 2006, Brede and De Vries 2010). In real fisheries, groups of fishers exist based on friendship, background, ethnicity, or by belonging to the same company. Many aspects and dynamics of this cooperation have to be analyzed, especially to model and be able to compare with reality how these groups form and grow in real fisheries, their stability, composition, and optimality. Another important area of research is the joint importance of ecological, economic, and institutional criteria for shaping resource users' behavior and incentives. Fisheries whose social–ecological characteristics discourage voluntary approaches by resource users may call for greater investments in specific top-down incentives coupled with informal stakeholder engagement to maximize fisheries success and management objectives (Di Franco et al. 2016). On the other hand, more naturally amenable fishery resources may be better suited by bottom-up voluntary forms of management (e.g., co-management and self-governance; Gutierrez et al. 2011*b*, Cinner et al. 2012), in which case the best role for management and governance may be to remove legal and institutional impediments to such cooperation. Top-down management strategies, in developed countries, can be combined with informal stakeholder engagement in order to maximize fishery success. Equally important is for such institutions to contain mechanisms that consider the fine-scale ecological and social dynamics within the fishery and give fishers property rights over relevant scales (Johnson et al. 2012, Wilson et al. 2013). Lastly, as this study highlights, collaboration between scientists, managers, and fishers in data collection and research to inform management decisions is of utmost importance and should be encouraged.

Acknowledgments

We wish to express our gratitude to all San Diego sea urchin divers for their valuable collaboration on data collection and interpretation, as well as to David Rudie (Catalina Offshore Products Inc.) for guidance, data, and insights. D. Beauchamp (UW), L.W. Botsford (UC Davis), and J.T.T. Thorson (NOAA) provided useful suggestions to improve both models and manuscript. Financial support from Fulbright Scholarship, School of Aquatic and Fishery Sciences (University of Washington), and the California Sea Urchin Commission is greatly acknowledged.

Literature Cited

Allee, W. C. 1931. Animal aggregations: a study in general sociology. University of Chicago Press, Chicago, Illinois, USA.

Beauchamp, D. A., A. D. Cross, J. L. Armstrong, K. W. Myers, J. H. Moss, and L. J. Haldorson. 2007. Bioenergetic responses by Pacific salmon to climate and ecosystem variation. North Pacific Anadromous Fish Commission 4:257–269.

Berkeley, S. A., C. Chapman, and S. Sogard. 2004. Maternal age as a determinant of larval growth and survival in a marine fish, *Sebastes melanops*. Ecology 85:1258–1264.

Booth, A. J. 2000. Incorporating the spatial component of fisheries data into stock assessment models. ICES Journal of Marine Science 57:858–865.

Botsford, L. W., C. L. Moloney, A. Hastings, J. L. Largier, T. M. Powell, K. Higgins, and J. F. Quinn. 1994. The influence of spatially and temporally varying oceanographic conditions on meroplanktonic metapopulations. Deep-Sea Research II 41: 107–145.

Botsford, L. W., A. Campbell, and R. Miller. 2004. Biological reference points in the management of North American sea urchin fisheries. Canadian Journal of Aquatic Sciences 61:1325–1337.

Branch, T. A., et al. 2006. Fleet dynamics and fishermen behavior: lessons for fisheries managers. Canadian Journal of Aquatic Sciences 63:1647–1668.

Brede, M., and H. M. De Vries. 2010. Harvesting heterogeneous renewable resources: uncoordinated, selfish, team-, and community-oriented strategies. Environmental Modeling Software 25:117–128.

Breitburg, D. L. 1996. Consumer mobility and the relative importance of consumption and competition following physical disturbance. Marine Ecology Progress Series 138:83–92.

Byrne, M., N. L. Andrew, D. G. Worthington, and P. A. Brett. 1998. Reproduction in the diadematoid sea urchin *C. rodgersii* in contrasting habitats along the coast of Australia. Marine Biology 132:305–318.

Cabral, R. B., R. C. Geronimo, M. T. Lim, and P. M. Ali. 2010. Effect of variable fishing strategy on fisheries under changing effort and pressure: an agent-based model application. Ecological Modeling 221:362–369.

Caddy, J. F. 1975. Spatial model for an exploited shellfish population, and its application to the Georges Bank scallop fishery. Journal of the Fisheries Research Board of Canada 32:1305–1328.

Caddy, J. F. 2007. Marine habitat and cover: their importance for productive coastal fishery resources. Page 253. United Nations Educational Press, Paris, France.

Castilla, J. C., and O. Defeo. 2001. Latin America benthic shellfisheries: emphasis on co-management and experimental practices. Reviews of Fish Biology and Fisheries 11:1–30.

Castilla, J. C., et al. 1998. Artisanal "Caletas" as units of production and co-managers of benthic invertebrates in Chile. Canadian Special Publication in Fisheries and Aquatic Sciences 125:407–413.

Christiansen, J. S., and S. I. Siikavuopio. 2007. The relationship between feed intake and gonad growth of single and stocked green sea urchin (*Strongylocentrotus droebachiensis*) in a raceway culture. Aquaculture 262:163–167.

Cinner, J. E., T. R. McClanahan, A. MacNeik, N. A. J. Graham, and T. M. Daw. 2012. Co-management of coral reef social-ecological systems. Proceedings of the National Academy of Sciences of the United States of America 109:5219–5222.

Cope, J. M., and A. E. Punt. 2011. Reconciling stock assessment and management scales under conditions of spatially varying catch histories. Fisheries Research 107:22–38.

Costello, C., and R. Deacon. 2007. Efficiency gains from fully delineating rights in an ITQ fishery. Marine Resources Economics 22:347–361.

Costello, C., S. Gaines, and J. Lynham. 2008. Can catch shares prevent fisheries collapse? Science 321:1678–1681.

Cowen, R. K. 1983. The effects of sheephead (*Semicossyphus pulcher*) predation on red sea urchin (*Strongylocentrotus franciscanus*) populations: an experimental analysis. Oecologia 58:249–255.

Curtis, R. E., and K. E. McConnell. 2004. Incorporating information and expectations in fisherman's spatial decisions. Marine Resources Economics 19:131–143.

Dayton, P. K. 1985. Ecology of kelp communities. Annual Review in Ecology and Systematics 16: 215–245.

Dayton, P. K., M. J. Tegner, P. B. Edwards, and K. L. Riser. 1999. Temporal and spatial scales of kelp demography: the role of oceanographic climate. Ecological Monographs 69:219–250.

Deacon, R. T., D. P. Parker, and C. Costello. 2010. Overcoming the Common Pool Problem through Voluntary Cooperation: The Rise and Fall of a Fishery Cooperative. Working Paper No. 16339, NBER, Cambridge, Massachusetts, USA.

Dean, T. A., S. C. Schroeter, and J. D. Dixon. 1984. Effect of grazing of two species of sea urchins (*Strongylocenntrotus franciscanus* and *Lytechinus anamesus*) on recruitment and survival of two species of kelp (*Macroocystes pyriferia* and *Ptelodophora californica*). Marine Biology 70:301–313.

DeAngelis, D. L., and L. J. Gross. 1992. Individual-based models and approaches in ecology. Chapman and Hall, New York, New York, USA.

Di Franco, A., et al. 2016. Five key attributes can increase marine protected areas performance for small-scale fisheries management. Scientific Reports 6:38135.

Dreyfus-Leon, M. 2006. Modelling cooperation between fishermen with a cellular automaton. A framework for fishing effort spatial dynamics. Ecological Informatics 1:101–105.

Dreyfus-Leon, M., and D. Gaertner. 2006. Modeling performance and information exchange between fishing vessels with artificial neural networks. Ecological Modeling 195:30–36.

Dumon, C., J. H. Himmelman, and M. P. Russel. 2009. Size-specific movement of green sea urchins *Strongylocentrotus droebachiensis* on urchin barrens in eastern Canada. Marine Ecology Progress Series 276:93–101.

Ebert, T. A. 1987. Estimating growth and survival parameters by nonlinear regression using average size in catches. Pages 35–40 *in* D. Pauly and G. R. Morgan, editors. Theory and application of length-based methods in stock assessment. ICLARM, Manila, The Philippines.

Ebert, T. A., J. D. Dixon, S. C. Schroeter, P. E. Kalvass, N. T. Richmond, W. A. Bradbury, and D. A. Woodby. 1999. Growth and mortality of red sea urchins *Strongylocentrotus franciscanus* across a latitudinal gradient. Marine Ecology Progress Series 190:189–209.

Ebert, T. A. 1980. Relative growth of sea urchin jaws: an example of plastic resource allocation. Bulletin of Marine Science 30:467–474.

Ebert, T. A., and M. P. Russell. 1992. Growth and mortality estimates for red sea urchin *Strongylocentrotus franciscanus* from San Nicolas Island, California. Marine Ecology Progress Series 81:31–41.

Ebert, T. A., and J. R. Southon. 2003. Red sea urchins (*Strongylocentrotus franciscanus*) can live over 100 years: confirmation with A-bomb 14 carbon. Fishery Bulletin 101:915–922.

Ebert, T. A., S. C. Schroeter, J. D. Dixon, and P. Kalvass. 1994. Settlement patterns of red and purple urchins (*Strongylocentrotus franciscanus* and *S. purpuratus*) in California, USA. Marine Ecology Progress Series 111:41–52.

Edwards, P. B., and T. A. Ebert. 1991. Plastic response to limited food availability and spine damage in the sea urchin *Strongylocentrotus purpuratus* (Stimpson). Journal of Experimental Marine Biology and Ecology 145:205–220.

Ekerhovd, N. A. 2008. The effects of different strategic variables in non-cooperative fisheries games. Natural Resource Modeling 21:582–606.

Farmanfarmaian, A., and J. H. Phillips. 1962. Digestion, storage and translocation of nutrients in the purple sea urchin (*Strongylocentrotus purpuratus*). The Biological Bulletin 123:105–120.

Faunce, C. H. 2011. A comparison between industry and observer catch compositions within the Gulf of Alaska rockfish fishery. ICES Journal of Marine Science 68:1769–1777.

Fogarty, M. J., and L. W. Botsford. 2007. Population connectivity and spatial management of marine fisheries. Oceanography 20:112–123.

Foster, M. C., J. E. K. Byrnes, and D. C. Reed. 2015. Effects of five southern California macroalgal diets on consumption, growth, and gonad weight, in the purple sea urchin *Strongylocentrotus purpuratus*. PeerJ 3:e719. https://doi.org/10.7717/peerj.719

Foster, M. C., and D. R. Schiel. 2010. Loss of predators and the collapse of southern California kelp forests: alternatives, explanations and generalizations. Journal of Experimental Marine Biology and Ecology 393:59–70.

Gatewood, J. B. 1984. Cooperation, competition, and synergy: information-sharing groups among southeast Alaskan salmon seiners. American Ethnology 11:350–370.

Grabowski, J. H., H. A. Randall, and D. L. Kimbro. 2008. Habitat complexity influences cascading effects of multiple predators. Ecology 89:3413–3422.

Grimm, V., E. Revilla, U. Berger, F. Jeltsch, W. M. Mooil, S. F. Railsback, H. H. Thulke, J. Weiner, T. Wiegan, and D. L. DeAngelis. 2005. Pattern-oriented modeling of agent-based complex systems: lessons from ecology. Science 310:987–991.

Gutierrez, N., and O. Defeo. 2003. Development of a new scallop *Zygochlamys patagonica* fishery in Uruguay: latitudinal and bathymetric patterns in biomass and population structure. Fisheries Research 62:21–36.

Gutierrez, N. L., and O. Defeo. 2005. Spatial patterns in population dynamics of the scallop *Psychrochlamys patagonica* at the northern edge of its range. Journal of Shellfish Research 24:877–882.

Gutierrez, N. L., O. Defeo, S. R. Bush, D. S. Butterworth, C. A. Roheim, and A. E. Punt. 2016. The current situation and prospects of fisheries certification and ecolabelling. Fisheries Research 182:1–6.

Gutierrez, N. L., R. Hilborn, and O. Defeo. 2011*b*. Leadership, social capital and incentives promote successful fisheries. Nature 470:386–389.

Gutierrez, N. L., A. Martínez, and O. Defeo. 2008. Identifying environmental constraints at the edge of a species' range: scallop *Psychrochlamys patagonica* in the SW Atlantic Ocean. Marine Ecology Progress Series 353:147–156.

Gutierrez, N. L., A. Masello, G. Uscudun, and O. Defeo. 2011*a*. Spatial distribution patterns in biomass and population structure of the deep-sea red crab *Chaceon notialis* in Uruguayan waters of the Southwestern Atlantic Ocean. Fisheries Research 110:59–66.

Gutierrez, N. L., et al. 2012. Eco-label conveys reliable information on fish stock health to seafood consumers. PLoS ONE 7:e43765. https://doi.org/10.1371/journal.pone.0043765

Harrold, C., S. Lisin, K. Light, and S. Tudor. 1991. Isolating settlement from recruitment of sea urchins. Journal of Experimental Marine Biology and Ecology 147:81–94.

Harrold, C., and D. C. Reed. 1985. Food availability, sea urchin grazing, and kelp forest community structure. Ecology 66:1160–1169.

Hilborn, R., and M. Ledbetter. 1985. Determinants of catching power in the B.C. salmon purse seine fleet. Canadian Journal of Aquatic and Fishery Sciences 42:51–56.

Hilborn, R., S. Schroeter, and N. L. Gutierrez. 2007. Stock assessment of the red sea urchin (*Strongylocentrotus franciscanus*) in Point Loma, San Diego, CA. Page 57. California Sea Urchin Commission, San Diego, California, USA.

Holling, C. S. 1959. Some characteristics of simple types of predation and parasitism. Canadian Entomologist 91:385–398.

Johnson, T. R., and W. L. T. van Densen. 2007. The benefits and organization of cooperative research for fisheries management. ICES Journal of Marine Sciences 64:834–840.

Johnson, T. R., J. A. Wilson, C. Cleaver, C. Morehead, and R. L. Vadas. 2013. Modeling fine scale urchin and kelp dynamics: implications for management of the Maine sea urchin fishery. Fisheries Research 141:107–117.

Johnson, T. R., J. A. Wilson, C. Cleaver, and R. L. Vadas. 2012. Social-ecological scale mismatches and the collapse of the sea urchin fishery in Maine, USA. Ecology and Society 17:15. https://doi.org/10.5751/ES-04767-170215

Kalvass, P. E., and J. M. Hendrix. 1997. The California red sea urchin, *Stronglyocentrotus fransiscanus* fishery: catch, effort, and management trends. Marine Fisheries Review 59:1–17.

Kato, S., and S. C. Schroeter. 1985. Biology of the red sea urchin, *Strongylocentrotus franciscanus,* and its fishery in California. Marine Fisheries Reviews 47:1–20.

Lares, M. T., and C. M. Pomory. 1998. Use of body components during starvation in *Lytechinus variegatus* (Lamarck) (Echinodermata: Echinoidea). Journal of Experimental Marine Biology and Ecology 225: 99–106.

Laur, D. R., A. W. Ebeling, and D. C. Reed. 1986. Experimental evaluations of substrate types as barriers to sea urchin (*Strongylocentrotus* spp.) movement. Marine Biology 93:209–215.

Lauzon-Guay, J. S., andR. E. Scheibling. 2007. Behaviour of sea urchin *Strongylocentrotus droebachiensis* grazing fronts: food-mediated aggregation and density-dependent facilitation. Marine Ecology Progress Series 329: 191–204.

Lawrence, J. M. 2007. Edible sea urchins: biology and Ecology. Second edition. Page 529. Development in

Aquaculture and Fishery Sciences; Elsevier, Amsterdam, The Netherlands.
Little, L. R., S. Kuikkab, A. E. Punt, F. Pantus, C. R. Davies, and B. D. Mapstone. 2004. Information flow among fishing vessels modelled using a Bayesian network. Environmental Modeling Software 19:27–34.
Little, L. R., A. E. Punt, B. D. Mapstone, G. A. Begge, B. Goldman, and A. J. Williams. 2009. An agent-based model for simulating trading of multi-species fisheries quota. Ecological Modelling 220:3404–3412.
Longhurst, A. 2002. Murphy's Law revisited: longevity as a factor in recruitment to fish populations. Fisheries Research 56:125–131.
Lundquist, C. J., and L. W. Botsford. 2011. Estimating larval production of a broadcast spawner: the influence of density, aggregation, and the fertilization Allee effect. Canadian Journal of Aquatic and Fishery Sciences 68:30–42.
Mattison, J. E., J. D. Trent, A. L. Shanks, T. B. Akin, and J. S. Pearse. 1977. Movement and feeding activity of red sea urchins (*Strongylocentrotus franciscanus*) adjacent to a kelp forest. Marine Biology 39:25–30.
McBride, S. C., J. M. Lawrence, A. L. Lawrence, and T. J. Mulligan. 1998. The effect of protein concentration in prepared feeds on growth, feeding rate, total organic absorption, and gross assimilation efficiency of the sea urchin *Strongylocentrotus franciscanus*. Journal of Shellfish Research 17:1563–1570.
McBride, S. C., J. M. Lawrence, A. L. Lawrence, and T. M. Mulligan. 1999. Ingestion, absorption, and gonad production of adult *Strongylocentrotus franciscanus* fed different rations of a prepared diet. Journal of World Aquaculture Society 30:364–370.
McBride, S. C., R. J. Price, P. D. Tom, J. M. Lawrence, and A. L. Lawrence. 2004. Comparison of gonad quality factors: color, hardness and resilience, of *Strongylocentrotus franciscanus* between sea urchins fed prepared feed or algal diets and sea urchins harvested from the Northern California fishery. Aquaculture 233:405–422.
McShane, P. E., and O. F. Anderson. 1997. Resource allocation and growth rates in the sea urchin *Evechinus chloroticus* (Echinoidea: Echinometridae). Marine Biology 128:657–663.
Miller, B. A., and R. B. Emlet. 1997. Influence of nearshore hydrodynamics on larval abundance and settlement of sea urchins *Strongylocentrotus franciscanus* and *S. purpuratus* in the Oregon upwelling zone. Marine Ecology Progress Series 148:83–94.
Morgan, L. E., L. W. Botsford, S. R. Wing, and B. D. Smith. 2000. Spatial variability in growth and mortality of the red sea urchin, *Strongylocentrotus franciscanus*, (Agassiz) in northern California. Canadian Journal of Fisheries and Aquatic Sciences 57:980–992.
Orensanz, J. M., and G. S. Jamieson. 1998. The assessment and management of spatially structured stocks: an overview of the North Pacific Symposium on Invertebrate Stock Assessment and Management. Canadian Special Publication in Fisheries and Aquatic Sciences 125:441–460.
Ovando, D. A., et al. 2012. Conservation incentives and collective choices in cooperative fisheries. Marine Policy 37:132–140.
Paine, R., and R. L. Vadas. 1996. The effects of grazing by sea urchins, *Strongylocentrotus* spp., on benthic algal populations. Limnology and Oceanography 14:710–719.
Palmer, C. T. 1990. Kin-selection, reciprocal altruism, and information sharing among Maine lobster lobstermen. Ethology and Sociobiology 12:221–235.
Parnell, P. E., P. K. Dayton, R. A. Fisher, C. C. Loarie, and R. D. Darrow. 2010*b*. Spatial patterns of fishing effort off San Diego: implications for zonal management and ecosystem function. Ecological Applications 20:2203–2222.
Parnell, P. E., P. H. Dayton, C. E. Lennert-Cody, L. L. Rasmussen, and J. J. Leichter. 2006. Marine reserve design: optimal size, habitats, species affinities, diversity, and ocean microclimate. Ecological Applications 16:945–962.
Parnell, P. E., E. F. Miller, C. E. Lennert-Cody, P. K. Dayton, M. L. Carter, and T. D. Stebbins. 2010*a*. The response of giant kelp (*Macrocystis pyrifera*) in southern California to low-frequency climate forcing. Limnology and Oceanography 55: 2686–2702.
Pearce, C. M., T. L. Dagget, and S. M. C. Robinson. 2004. Effect of urchin size and diet on gonad yield and quality in the green sea urchin (*Strongylocentrotus droebachiensis*). Aquaculture 233:337–367.
Pearse, J. S., V. B. Pearse, and K. K. Davis. 1986. Photoperiodic regulation of gametogenesis and growth in the sea urchin *Strongylocentrotus purpuratus*. Journal of Experimental Biology 237:107–118.
Pérez-Ramírez, M., M. Castrejon, N. L. Gutierrez, and O. Defeo. 2016. The Marine Stewardship Council certification in Latin America and the Caribbean: a review of experiences, potentials and pitfalls. Fisheries Research 182:50–58.
Prince, J. D. 2003. The barefoot ecologist goes fishing. Fish and Fisheries 4:359–371.
Prince, J. D. 2010. Rescaling fisheries assessment and management: a generic approach, access rights, change agents, and toolboxes. Bulleting of Marine Science 86:197–219.
R Core Development Team. 2012. R: A language and environment for statistical computing. R Foundation for Statistical Computing, Vienna, Austria. http://www.R-project.org/

Rashleigh, B., and G. D. Grossman. 2005. An individual-based simulation model for mottled scuplin (*Cottus bairdi*) in a Southern Appalachian stream. Ecological Modelling 187:247–258.

Rassweiler, A., K. K. Arkema, D. C. Reed, R. C. Zimmerman, and M. A. Brzezinski. 2008. Net primary production, growth, and standing crop of *Macrocystis pyrifera* in southern California. Ecology 89:2068.

Raymond, J. F., J. H. Himmelman, and H. E. Guderley. 2004. Sex differences in biochemical composition, energy content and allocation to reproductive effort in the brooding sea star *Leptasterias polaris*. Marine Ecology Progress Series 283:179–190.

Reed, D. C., A. Rassweiler, and K. K. Arkema. 2008. Biomass rather than growth rate determines variation in net primary production by giant kelp. Ecology 89:2493–2505.

Roughan, M., E. J. Terrill, J. L. Largier, and M. P. Otero. 2005. Observations of divergence and upwelling around Point Loma, California. Journal of Geophysical Research 110:1–11.

Salas, S. 2000. Fishing strategies of small-scale fishers and implications for fisheries management Dissertation. Page 154.Resource Management and Environmental Studies, The University of British Columbia, Vancouver, British Columbia, Canada.

Sanchirico, J. N., and J. E. Wilen. 1999. Bioeconomics of spatial exploitation in a patchy environment. Journal of Environmental Economics Management 37:129–150.

Scheibling, R. E., and B. G. Hatcher. 2001. The ecology of *Strongylocentrotus droebachiensis*. Pages 271–306 *in* J. M. Lawrence, editor. Edible sea urchins: biology and ecology. Developments in Aquaculture and Fisheries Science No. 32, Elsevier Science Press, Amsterdam, The Netherlands; London, UK; New York, New York, USA.

Schroeter, S. C., J. D. Dixon, T. A. Ebert, and J. V. Rankin. 1996. Effects of kelp forests *Macrocystis pyrifera* on the larval distribution and settlement of red and purple sea urchins *Strongylocentrotus franciscanus* and *S. purpuratus*. Marine Ecology Progress Series 133:125–134.

Schroeter, S. C., N. L. Gutierrez, M. Robinson, R. Hilborn, and P. Halmay. 2009. Moving from data poor to data rich: a case study of community-based data collection for the San Diego red sea urchin (*Strongylocentrotus franciscanus*) fishery. Marine and Coastal Fisheries 1:230–243.

Scott, A. 2000. Moving through the narrows: from open access to ITQs and self-government, in use of property rights in fisheries management. Page 115 *in* R. Shotton, editor. Technical Paper 404/1, FAO, Rome, Italy.

Senyk, N. A., and J. Sanchez. 2005. Development of a dynamic optimal habitat model to describe the spatial and temporal habitat distributions of giant kelp, *Macrocystis pyrifera*. Page 47. Final Report, CEQA, Santa Barbara, California, USA.

Siikavuopio, S. I., J. S. Christiansen, S. Bjørn-Steinar, and D. Trine. 2007. Seasonal variation in feed intake under constant temperature and natural photoperiod in the green sea urchin (*Strongylocentrotus droebachiensis*). Aquaculture 272:328–334.

Smith, B. D., L. W. Botsford, and S. R. Wing. 1998. Estimation of growth and mortality parameters from size frequency distributions lacking age patterns: the red sea urchin (*Strongylocentrotus franciscanus*) as an example. Canadian Journal of Fisheries and Aquatic Sciences 55:1236–1247.

Steneck, R. S., M. H. Graham, B. J. Bourgue, D. Corbett, J. M. Erlandson, J. A. Estes, and M. J. Tegner. 2002. Kelp forest ecosystems: biodiversity, stability, resilience and future. Environmental Conservation 29: 436–459.

Tegner, M. J., and P. K. Dayton. 1991. Sea urchins, El Niños, and the long-term stability of southern California kelp forest communities. Marine Ecology Progress Series 77:49–63.

Tegner, M. J., and S. A. Levin. 1982. Do sea urchins and abalones compete in California kelp forest communities?. Pages 265–271 *in* J. M. Lawrence, editor. International Echinoderm Conference. A. A. Balkema, Rotterdam, The Netherlands.

Thompson, R. J. 1984. Partitioning of energy between growth and reproduction in three populations of the green sea urchin (*Strongylocentrotus droebachiensis*). Advances in Invertebrates Reproduction 3: 50–57.

Tilman, D., and P. Kareiva, editors. 1997. Spatial ecology: the role of space in population dynamics and interspecific interactions. Princeton Monographs in Population Biology, No. 30. Princeton University Press, Princeton, New Jersey, USA.

Walker, C. W., and M. P. Lesser. 1998. Manipulation of food and photoperiod promotes out-of-season gametogenesis in the green sea urchin, *Strongylocentrotus droebachiensis*: implications for aquaculture. Marine Biology 132:663–676.

Weber, J. M., S. Kopelman, and D. M. Messick. 2004. A conceptual review of decision making in social dilemmas: applying a logic of appropriateness. Personality and Social Psychology Review 8:281–307.

Webster, S. K., and A. C. Giese. 1975. Oxygen consumption of the purple sea urchin with special reference to the reproductive cycle. Biological Bulletin 148:165–180.

Wilen, J. E. 2004. Spatial management of fisheries. Marine Resources Economics 19:7–19.

Wilson, J. A. 1990. Fishing for knowledge. Land Economics 66:12–29.

Wilson, J. A., and J. Acheson. 1980. A model of adaptive behavior in the New England Industry. National Science Foundation, USA, Providence, Rhode Island, USA.

Wilson, J. A., J. M. Acheson, and T. R. Johnson. 2013. The cost of useful knowledge and collective action in three fisheries. Ecological Economics 96:165–172.

Zimmerman, R. C., and J. N. Kremer. 1986. In situ growth and chemical composition of the giant kelp, *Macrocystis pyrifera*: response to temporal changes in ambient nutrient availability. Marine Ecology Progress Series 27:277–285.

3

A budget of algal production and consumption by herbivorous fish in an herbivore fisheries management area, Maui, Hawaii

EMILY L. A. KELLY,[1,]† YOAN EYNAUD,[1] IVOR D. WILLIAMS,[2] RUSSELL T. SPARKS,[3] MEGHAN L. DAILER,[4] STUART A. SANDIN,[1] AND JENNIFER E. SMITH[1]

[1]*Center for Marine Biodiversity and Conservation, Scripps Institution of Oceanography, University of California San Diego, La Jolla, California 92093 USA*
[2]*Ecosystem Science Division (ESD), Pacific Islands Fisheries Science Center (PIFSC), National Marine Fisheries Service, NOAA IRC, Honolulu, Hawaii 96818 USA*
[3]*Division of Aquatic Resources, Maui Office, Department of Land and Natural Resources, Wailuku, Hawaii 96768 USA*
[4]*Department of Botany, University of Hawai'i, 3190 Maile Way, Honolulu, Hawaii 96822 USA*

Abstract. The relative rates of carbon fixed by primary producers vs. consumption by primary consumers shape the community of organisms in ecosystems. On coral reefs, it is important to understand the demography of algae, given known competitive dynamics with reef-building corals. Numerous studies have shown that fleshy algal abundance is enhanced in the absence of top-down control by herbivores on coral reefs. However, fewer studies have quantified and compared rates of production by the benthic fleshy algal community and consumption by herbivores. Here, we estimate a budget for fleshy algal growth and herbivorous fish consumption on a Hawaiian coral reef based upon integration of field-measured and taxonomically specific variables. Data were collected at Kahekili Herbivore Fisheries Management Area, Maui, established in 2009, in which taking of herbivorous fish and urchins is prohibited. Daily algal production was determined by quantifying benthic community composition, standing stock of algal biomass, and growth rates of common algal components. Consumption was determined using distributions of biomass and size classes of herbivorous fish species, consumption rates of herbivores on different algal species, and herbivore bite sizes. Our results show that throughout the first five years of herbivore protection, algal production consistently exceeded the grazing capacity of the herbivorous fish assemblage, but by a diminishing margin since 2010 (consumption 20.8% of production) to the end of the study in 2014 (consumption 67.0% of production). Further, larger size classes of herbivorous fishes in the scraper/excavator herbivore guild contributed more to consumption in later years, which could have additional feedbacks that promote reef-building taxa. Projecting the budget using data from a neighboring decades-long protected herbivorous fish assemblage indicated that the production and consumption budget for Kahekili could become balanced in future with continued management.

Key words: algae; coral reef; grazing; herbivory; net primary production; phase shift; productivity.

† **E-mail:** elkelly@ucsd.edu

INTRODUCTION

Primary production in ecosystems, be it by grasses and woody plants in temperate landscapes, phytoplankton in freshwater lakes, or benthic algae on corals reefs, provides the basis of food webs across the planet. Consumers of this primary production, including ungulates in forests and savannahs, zooplankton in lakes, and fishes and urchins on coral reefs, shape these ecosystems via top-down influence of grazing (Carpenter 1986, Belsky 1992, Shurin et al. 2002). Importantly, changes to top-down control can result in shifts in ecosystem dynamics and to phase shifts. For example, increased top-down influence can shift plant community structure (Ripple and Beschta 2003) or result in desertification of former grasslands (Andrew 1988). In lakes, the absence of top-down control can result in exponential growth of unpalatable phytoplankton (Benndorf et al. 2002). Finally, on coral reefs, removal of herbivores via over-extraction (Friedlander et al. 2007) or disease (Hughes 1994) can lead to phase shifts to fleshy algal-dominated reefs.

To understand the influence of top-down control in an ecosystem or how changes to top-down control may alter systems, it is important to understand net primary production in ecosystems. In temperate forests, using direct measurements of growth of individual plants, identifying allometric relationships, and scaling up these relationships, grazers are estimated to consume 10–50% of primary production (as reviewed in Gower et al. 1999). In African savannas, consumption of grasses by large ungulates, determined by net foliage consumption and ungulate energetic expenditure, was estimated to be 57% of primary production (Coe et al. 1976, Frank et al. 1998). Alteration of grazing populations in these ecosystems, for example through livestock grazing, can alter the portion of primary production that is consumed (Andrew 1988).

On coral reefs, upwards of 100% of the total benthic algal production may be consumed (Wanders 1977, Hatcher 1981, Carpenter 1986, van Rooij et al. 1998), which is believed to be important in allowing corals to maintain a competitive advantage for space over faster growing algae (as reviewed in McCook et al. 2001). Generally, estimates of total herbivory on reefs have been based on extrapolation from detailed observations of a few abundant herbivores (Klumpp and Polunin 1990, Paddack et al. 2006), calculated through estimates of metabolic demands of fishes (van Rooij et al. 1998), or based on consumption of algae on settlement tiles (Wanders 1976, Russ 2003). Production of benthic algal biomass on reefs has been estimated through laboratory measurements of oxygen production (Wanders 1976, Klumpp and Polunin 1990, van Rooij et al. 1998) and growth of algae in situ in the absence of herbivores (Carpenter 1986, Russ 2003, Paddack et al. 2006).

The proportion of benthic algal production consumed by herbivores varies according to the benthic composition of a given reef. On reefs with high coral-cover (>50%) herbivores were estimated to consume upwards of 100% or more of daily algal production (Wanders 1977, Hatcher 1981, Carpenter 1986, van Rooij et al. 1998). On naturally algal-dominated reef flats, site-attached damselfish and blennies combined with roving acanthurids and scarids were estimated to remove between a quarter and over double the daily production (Klumpp and Polunin 1990). Finally, following fleshy (non-calcareous turf and macroalgae) algal phase shifts on low coral-cover reefs (<10%), herbivores were estimated to consume 31–77% of algal production (Paddack et al. 2006). Thus, some of the variability in these estimates likely arises from differences in habitat and the herbivore community living within that habitat.

The balance of production and consumption on a coral reef based upon the herbivore assemblage and algal community could be used to project future trajectories for a given reef. On a reef in which estimates of production and consumption are balanced, in theory benthic community composition would be expected to remain fairly stable. However, if production grossly exceeds consumption there is potential for algal biomass to increase through time. The balance between production and consumption could be indicative of how fleshy algal standing stock on a reef may be increasing, decreasing, or remaining stable, although importantly changes in algal cover on reefs also depend on competition dynamics (McCook et al. 2001). To date, there has not been an examination of how this balance of production and consumption changes through time at a given site. Detailed field-derived rates of algal growth and herbivore assemblage consumption, while inherently variable, are needed to develop reasonable estimates of production–consumption budgets for coral reefs.

Directly increasing herbivore biomass to drive the reversal of phase shifts has been discussed as a management option for coral reefs (Green and Bellwood 2009, Rasher et al. 2013, Jackson et al. 2014). Several studies have shown a negative relationship between herbivore biomass and macroalgal cover across reefs in the Caribbean and the Pacific (Williams and Polunin 2001, Friedlander et al. 2007, Jouffray et al. 2015). Results from Fiji in shallow-water reefs lend promising support for herbivore management as a tool for coral reef recovery, with higher growth rates of corals and lower macroalgal biomass in protected reefs (Rasher et al. 2012). While there are few examples of such management being put into practice (Jackson et al. 2014), the decline in coral cover and increase in fleshy algal cover as well as fleshy algal blooms on the reef at Kahekili Beach Park on West Maui, Hawaii (Rodgers et al. 2015), led state managers to designate this reef an Herbivore Fisheries Management Area (HFMA) in 2009. This HFMA is the first of its kind in Hawaii. The goal of the Kahekili HFMA (KHFMA) was to promote coral cover by increasing herbivore biomass and consequent grazing pressure. Here, we use a calculated budget of benthic algal production and consumption by herbivorous fish to better understand the ability of the fish assemblage to reduce algal cover at the KHFMA.

The major components of this study were fourfold. First, we quantified herbivore consumption and algal production based on observational surveys and experimentation. Second, we tracked these rates through time from before the inception of an herbivore-specific fisheries management area to five years into herbivore protection in order to quantify community-level changes in algal growth and consumption over the duration of protected status. Next, through modeling production and consumption through time, we quantified the proportion of algal production consumed by the KHFMA herbivore assemblage to assess the amount of production consumed by the herbivorous fish assemblage. Finally, we calculated production and consumption using herbivore biomass from a nearby marine protected area that has been fully closed for 40 yr to understand how a future herbivore assemblage may impact the KHFMA benthic community.

Methods

Study area and overview

All data were collected at the KHFMA on West Maui, Hawaii, in relatively shallow (2–10 m deep) fringing reef habitat (shallow aggregated reef in Williams et al. 2016). Algal growth and consumption by herbivores were determined through analysis of benthic community composition, standing stock of algae, growth rates of dominant algal species, abundance and biomass of herbivorous fish species, the species-specific consumption rates of herbivores on different algal species, and the bite size of herbivores. These data were collected through a series of field and laboratory experiments as well as semi-annual monitoring surveys at KHFMA as described below.

Benthic community composition

Benthic cover was determined from surveys conducted twice a year from 2008 to 2014 by the NOAA Coral Reef Ecosystem Program and Hawaii Division of Aquatic Resources (DAR) (Williams et al. 2016). These surveys were conducted across the ~1 km^2 area of KHFMA and included ~20–50 transects per survey. Photoquadrats (~1 m^2) were taken every meter along each 25-m transect. Photographs were analyzed using the image analysis software PhotoGrid 1.0 in which 15–100 points were superimposed on the image in a stratified random fashion and benthic cover was identified to genus for corals and macroalgae or functional group for turf algae, crustose coralline algae, and cyanobacteria.

Algal standing biomass

Standing stock of macroalgae was determined by removing all fleshy algae inside eight 0.6 × 0.85 m quadrats placed haphazardly on the shallow aggregate reef at 5–8 m depth. Algae were collected by hand using blunt metal tools and razor blades, separated by species in the laboratory, wet-weighed, dried in the oven at 60°C for 24 h, and dry-weighed. Prior to the removal of macroalgae, photoquadrats were taken of the area and canopy heights were measured using a ruler. Benthic cover was then related to wet and dry standing stock biomass estimates within each quadrat.

To determine standing stock of mixed-assemblage turf algae, areas of turf covering dead massive *Porites* colonies were scraped off the calcium carbonate in situ and vacuumed into a bag using a marine battery and a bilge pump (Appendix S1: Fig. S1). A known area of turf was scraped using a razor blade, and tubing on the intake end of the pump was held adjacent to this area to capture all material removed. Tubing on the outgoing end of the pump was connected to a polyvinyl chloride (PVC) canister capped with Nitex mesh to catch all scraped material. Samples were transferred from Nitex mesh to filter paper in the laboratory and then decalcified for 24 h with 5% HCl to remove carbonate material. Samples were then filtered on pre-weighed Whatman Grade 1 qualitative cellulose ash-free filters, dried at 60°C for 24 h, and their dry weight recorded and normalized to the size of the scraped area.

Algal growth rates

To determine in situ growth rates of common macroalgal species, samples were first collected from the KHFMA and transported to a small field station in Lahaina in plastic bags of seawater. Samples were spun gently 20 times in a salad spinner, blotted with a paper towel, and weighed to obtain an initial wet weight (3–5 g). All samples were then placed in mesh pockets and stored overnight in ambient seawater aerated with bubblers. Samples were deployed on the reef at KHFMA the following morning by placing them in small cylindrical cages constructed out of 1 cm diameter clear plastic mesh and galvanized wire frame. Algae were deployed at 10-m intervals along a 150-m transect at depths of 2 and 8 m. Cages were attached to ropes, which were staked into the limestone substrate, and small floats were used to elevate samples off the benthos. After 4–7 d, algae were removed from cages and transported via labeled plastic bags to the laboratory to obtain final wet weight. Changes in weight were used to determine proportional algal growth rates over the period of deployment (change in weight × initial weight^{-1} × time^{-1}). Macroalgal species assessed here included the most common species found on the reef as follows: *Acanthophora spicifera* (red, non-native blooming species; $n = 9$), *Ulva lactuca* (green, native blooming species; $n = 10$), *Amansia glomerata* (red, native species; $n = 10$), and *Tricleocarpa fragilis* (red, native species; $n = 10$).

To determine growth rates of turf algae, plots that were cleared to quantify standing stock were caged with standard hardware 2.54 cm diameter vinyl-coated chicken wire preventing urchins with tests larger than 2.54 cm and fishes >5 cm from grazing inside. Turf samples were allowed to grow for 7–60 d ($n = 22$) and then resampled using the vacuum method as mentioned above, decalcified, and dried using the methods described above.

Herbivore abundance and biomass

Surveys to quantify herbivorous fish abundance and biomass were conducted twice per year by the NOAA Coral Reef Ecosystem Program and Hawaii Division of Aquatic Resources (Williams et al. 2016). In this study, ~20–50 25-m transects were sampled haphazardly in the shallow aggregate reef habitat inside the KHFMA. Fish densities of all observed species were estimated by visual strip transect along each transect line. On the outward-bound leg, fish >15 cm, within a 4 m wide belt centered on the diver, were recorded. On the return leg, fishes <15 cm were recorded within a 2 m wide belt. Fish sizes were estimated in 5-cm bins. Biomass was calculated using established species-specific conversion factors taken largely from FishBase (Froese and Pauly 2016).

Herbivores were categorized as grazers, scrapers, or browsers (Green and Bellwood 2009, Edwards et al. 2014). Detritivores were not included in the consumption calculation as their impact in removal of algal biomass is considered minimal or not well quantified (Choat et al. 2002, Tebbett et al. 2017), and by excluding them, estimates of consumption err on the side of conservative.

Herbivore grazing rates

Grazing rates were quantified for all observed species of protected herbivores at the KHFMA (acanthurids, scarids, and kyphosids, the last of which were not recorded on surveys at the reef during the course of this study). Grazing rates were determined using timed behavioral observations (Bellwood and Choat 1990, Hamilton et al. 2014, Kelly et al. 2016). Briefly, divers followed at least 10 and up to 60 individuals of each herbivorous fish species. Observations were made between 2 and 10 m depth in the shallow aggregate reef habitat. Divers recorded number of bites and type of substrate consumed by each fish during 3- to 5-min timed swims, allowing for multiple grazing forays. Divers terminated observations if fish behavior was altered by diver presence. Substratum type of bites was recorded to the finest taxonomic classification possible through field identification. Macroalgae were identified to genus while turf algae, crustose coralline algae, and other groups were identified to functional group.

Herbivore bite size

Laboratory experiments were conducted to calculate the weight of algae removed per bite by representative acanthurids and scarids of varying sizes. Bite size data collection was conducted at the Maui Ocean Center using fish on display at the aquarium. *Ulva lactuca* was used in these experiments because of low error in calculating wet weight for that species and its high degree of palatability. *Ulva lactuca* fragments were weighed initially and lowered into aquariums housing target fish species using fishing line. The number of bites taken by a single fish on a piece of algae was recorded. After grazing, the remaining algal tissue was removed from the tank and wet-weighed as described above. Bite size per species and size class were calculated from the change in weight of the algal specimen divided by the number of bites taken by the herbivore. These bite rate data were only used in this calculation to estimate biomass of algae removed per bite and not as bite rate data in the overall consumption calculation.

Literature values for bite size of herbivore species in this study were also used when available for various acanthurid and scarid species. Data were either taken directly from previous studies as measurements of algal biomass removed per bite (Klumpp and Polunin 1990) or were extrapolated from recorded values of scarid grazing scar area (Ong and Holland 2010) and combined with experimental turf standing stock data in this study to determine algal biomass removed per bite of a given species and size class.

Herbivore total length (TL; cm) was plotted against bite size (BS; g of algae removed per bite), and a power function was used to describe the relationship between these two factors, roughly following the allometric increase in body size by length of fishes (Huxley and Tesissier 1936, Zgliczynski and Sandin 2017). Because bite-sized data were not available for all species and size classes of fishes, the relationship between herbivore size and BS was estimated at family level (acanthurid or scarid) given gross mouth morphological similarities. This estimate further assumes that the same biomass of algae was removed for a bite on turf algae or macroalgae.

Overall fish consumption

In calculating total fish consumption for a given year on the reef, we used a bootstrapping approach to combine the multiple sources of uncertainty that contribute to the total algal consumption estimate. For a given type of fish (e.g., trophic group, size class), the mean algal consumption is a function of the fish biomass density sample (D: g_F/m^2) and the fish grazing capability (K: $g_A \cdot g_F^{-1} \cdot day^{-1}$). K is a function of the estimated fish bite rates (R: bites/day), the corresponding BS (S: g_A/bite), and species biomass (F: g_F). As explained above, the BS and fish biomass were both calculated using the fish TL associated with the BS. Thus, we can write K as:

$$K = \frac{R \times S}{F}.$$

The bootstrapping procedure involved generating series of length β (=10^4) of values of D and K, randomly resampled with replacement from survey and experimental data, and noted as $D^*$and K^*, respectively. Hence, the bth sampled mean, a_b, can be written as:

$$a_b = \frac{1}{n}\left(\sum_{i=1}^{n} D_i^*\right) \times \frac{1}{m}\left(\sum_{j=1}^{m} K_j^*\right),$$

where $b \in [1, \beta]$, n is the length of the fish biomass density sample, D, and m the length of the fish grazing capability sample, K. By repeating this process β times, we obtained a distribution of estimated means from which a global estimated mean and 95% confidence interval were calculated. Following this procedure, we calculated mean annual fish consumption at KHFMA for each year between 2008 and 2014.

Overall algal production

We calculated overall algal production in a similar manner to consumption using a bootstrapping approach. For a given type of algae (e.g., turf or macroalgae), the overall algal growth is a function of the algal growth rate (G: g/day) and the algal benthic cover (C: %). Because we used a bootstrapping procedure, we did not directly use our sample set, but rather a series, of length β, of data randomly resampled with replacement, and, respectively, noted $G^*$and C^*. Hence, the bth sampled mean, c_b, can be written as:

$$c_b = \frac{1}{n}\left(\sum_{i=1}^{n} G_i^*\right) \times \frac{1}{m}\left(\sum_{j=1}^{m} C_j^*\right),$$

where $b \in [1, \beta]$, n the number of algal growth rate samples, and m the number of algal benthic cover samples. By repeating this process β times, we obtained a distribution of estimated means from which a global estimated mean, noted A, and its 95% confidence interval were calculated. Following this procedure, we calculated mean annual algal production at KHFMA from 2008 to 2014.

Finally, using the bootstrapped data of β samples for both consumption and production we created a new distribution of net production (production remaining after herbivorous fish consumption). As with the above distributions, we obtained a global estimated mean of net production from this new distribution and determined its 95% confidence interval via quantiles generated by the bootstrap. Following this procedure, we calculated mean annual net production at KHFMA from 2008 to 2014.

To project potential future herbivore assemblage scenarios after many more years of herbivore management at the KHFMA, we calculated the net production at the KHFMA using KHFMA benthic data from 2014 and herbivore assemblage data in 2014 from Molokini Marine Life Conservation District (MLCD) just offshore of Maui that was closed to all fishing in 1977. Regular benthic and fish biomass data are collected as part DAR's long-term monitoring program (Sparks et al. 2015). Using the herbivorous fish assemblage data from Molokini, we compared the consumption estimates to the current benthic algal biomass production estimates from KHFMA to determine whether such a long-term protected herbivore population has the capacity to consume the algal production at the KHFMA.

Results

Benthic community composition

Over the course of this study (2008–2014), the reef was comprised primarily of turf algae (41.2–53.8%) and hard coral (34.0–37.3%), as well as crustose coralline algae (CCA) (0.3–9.2%), macroalgae (1.3–4.7%), cyanobacteria (0.01–1.4%), and other substrata including sand and other invertebrates (<10%; Fig. 1). CCA increased between 2008 and 2014 from 0.3% to 9.2% (t test, $p < 0.001$) and turf decreased between 2008 and 2014 from 53.8% to 41.2% (t test, $p < 0.001$). Coral cover and macroalgal cover did not significantly change over this time period (t test, $p = 0.840$ and $p = 0.467$, respectively).

Algal standing biomass

For ease of calculation with differing percent cover of algae across the reef, algal standing biomass was reported assuming 100% coverage of a given area by these algal groups. Therefore, macroalgal standing biomass was 51.5 g dry wt/m^2 (SE = 17.9), and turf algal standing biomass was 79.0 g dry wt/m^2 (SE = 11.8; Table 1). The majority of standing biomass of macroalgae was *Amansia glomerata*, which accounted for 38.0 g dry wt/m^2 (SE = 12.4).

Algal growth rates

Growth rates (g [g^{-1} algae]·day^{-1}, and mean [±SE]) for turf algae and four species of macroalgae were measured in situ at the KHFMA (Fig. 2). Turf algae had the fastest growth rate at 0.0708 (±0.0163), which was variable (Appendix S1: Fig. S2). The growth rates of macroalgal species were as follows: *Ulva lactuca* at 0.0341 (±0.0051),

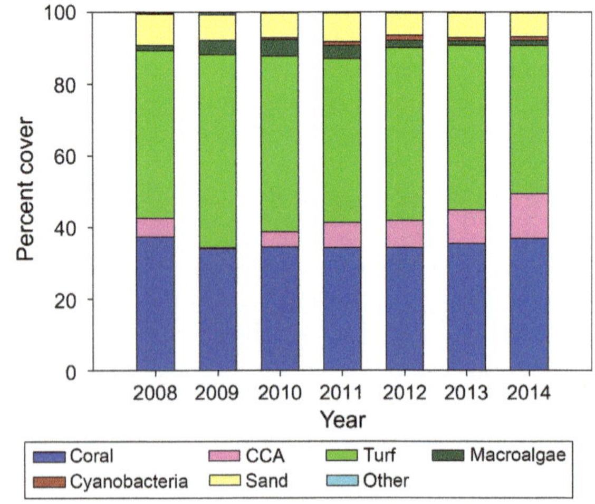

Fig. 1. Benthic community composition through time across all survey years. Turf and macroalgae cover for each season and year are used in the calculation of algal production. The 2009 surveys were completed a month after the Kahekili Herbivore Fisheries Management Area was legally established, but before signage or enforcement. n = 42 (2008), 23 (2009), 21 (2010), 54 (2011), 40 (2012), 39 (2013), 40 (2014) per year.

Table 1. Growth rate, standing biomass, and average percent cover of algae measured at the Kahekili Herbivore Fisheries Management Area.

Species	Growth rate (g dry wt·m^{-2}·day^{-1}) (±SE)	Standing biomass† (g dry wt/m^2) (±SE)	Average percent cover (across all years)
Turf algae			
Mixed	0.0708 (±0.0164)	78.98 (±11.83)	46.9
Macroalgae			
All	...	51.49 (±17.92)	4.0
Amansia glomerata	0.0093 (±0.0021)	38.02 (±12.36)	0.9
Tricleocarpa fragilis	0.0003 (±0.0046)	<0.1	<0.1
Acanthophora spicifera	0.0229 (±0.0099)	<0.1	<0.1
Ulva lactuca	0.0341 (±0.0051)	<0.1	<0.1

Note: Data are averages for each measurement across all years and seasons of the study (2008–2014).
† Standing biomass assumes 100% coverage of given alga in a square meter.

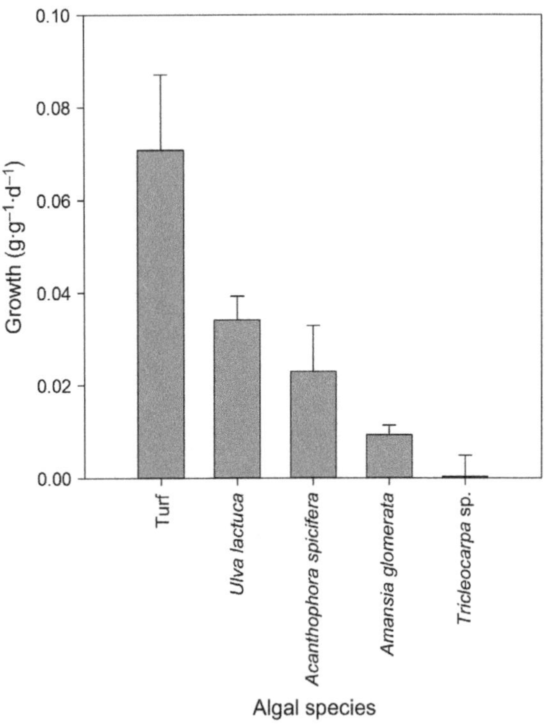

Fig. 2. Growth rates of turf algae and four species of macroalgae (g dry wt·g^{-1} algae·day^{-1}) at the Kahekili Herbivore Fisheries Management Area. Bars are SE. Turf algae, n = 22; *Ulva lactuca*, n = 10; *Acanthophora spicifera*, n = 9; *Amansia glomerata*, n = 10; *Tricleocarpa* sp., n = 10).

Acanthophora spicifera at 0.0229 (±0.0099), *Amansia glomerata* at 0.0093 (±0.0021), and *Tricleocarpa fragilis* at 0.0003 (±0.0046). Raw data from algal growth experiments, as opposed to means, are used in the calculation of algal growth below.

Herbivore biomass

Total herbivore biomass fluctuated throughout the duration of the study (Fig. 3) with the highest biomass occurring in 2014 (19.7 g/m^2 ± 2.1 SE). Total herbivore biomass declined between 2009 (16.4 g/m^2 ± 2.4 SE; when the HFMA was designated) and 2010 (9.4 g/m^2 ± 1.1 SE). However, when considering the feeding guilds separately (Appendix S1: Table S2), scraper and grazer biomass increased from 2010 to 2014 while browser biomass fluctuated but remained lower than both scrapers and grazers through all years. While grazer biomass showed an increasing trend from 2010, there were more smaller-bodied fishes that make up this biomass than in 2008 and 2009 (Fig. 4). For scrapers, the increase in biomass from 2008 to 2014 was the result of more biomass in the larger size classes. Browser biomass remained low throughout the study years but with all size classes present in the last three years of the study.

Herbivore grazing rates

Herbivore grazing rates varied across species of herbivores as well as for turf algae and macroalgae (Fig. 5). Grazer species had an overall higher grazing rate (with the exception of *Zebrasoma flavescens*) than scraper species, and both had higher average bite rates than all browser species.

Herbivore bite size

Experimental and literature values for acanthurid and scarid bite sizes resulted in the following relationships between fish TL (cm) and BS (g of algae removed per bite). Acanthurid BS was defined (Appendix S1: Fig. S4A) as follows:

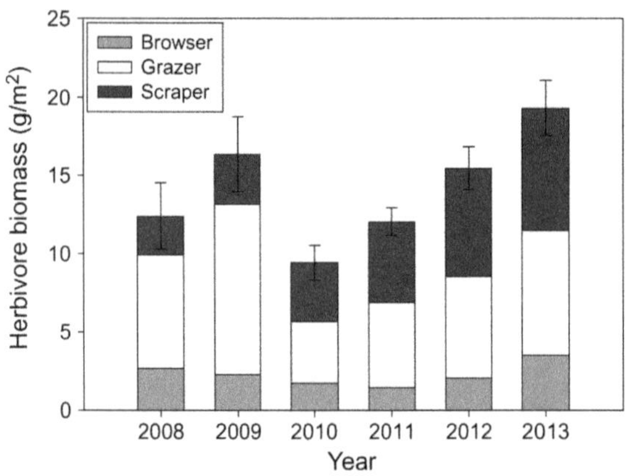

Fig. 3. Fish biomass through time across all survey years. Bars are SE for total herbivore biomass. The 2009 surveys were completed a month after the Kahekili Herbivore Fisheries Management Area was legally established, but before signage or enforcement. $n = 42$ (2008), 23 (2009), 21 (2010), 54 (2011), 40 (2012), 39 (2013), 40 (2014) per year.

$$BS = 4 \times 10^{-8} \times (TL)^{3.3307} \quad (R^2 = 0.641).$$

Scarid BS was defined (Appendix S1: Fig. S4B) as follows:

$$BS = 5 \times 10^{-10} \times (TL)^{4.3744} \quad (R^2 = 0.831).$$

These equations defined bite size on both turf and macroalgae for these groups.

Overall algal growth and fish consumption budget

Based upon our modeled estimates, benthic algal biomass production (turf and macroalgae) was consistently greater than consumption of algal biomass by herbivorous fishes throughout the length of the study (Fig. 6A). Production ranged from the lowest mean value of 1.9 g dry $wt{\cdot}m^{-2}{\cdot}day^{-1}$ in 2008 to the highest mean value of 2.8 g dry $wt{\cdot}m^{-2}{\cdot}day^{-1}$ in 2012. Consumption ranged from a low mean value of 0.5 g dry $wt{\cdot}m^{-2}{\cdot}day^{-1}$ in 2010 to a high mean value of 1.7 g dry $wt{\cdot}m^{-2}{\cdot}day^{-1}$ in 2014. At any given time point, the lowest amount of production consumed by the herbivore assemblage was 20.8% in 2010 and the greatest amount of mean production consumed was 67.0% in 2014. In comparison with other studies with calculated production and consumption, these values lie within the breadth of values found on other reefs (Table 2; Appendix S1: Fig. S3). Throughout the years of analysis at the KHFMA, mean net production (gross production—consumption) was >0, although not significantly so in 2008 and 2014 (95% CI included 0; Fig. 6B). However, if we used the Molokini Shoal herbivorous fish assemblage biomass data (including species-specific biomass and size classes of fishes) and compare grazing rates of this assemblage to algal biomass production at KHFMA in 2014 (Appendix S1: Table S1), mean net algal production is <0 (Fig. 6B). This exercise showed that it is possible with existing Hawaiian herbivore assemblages to balance the consumption/production budget on KHFMA reefs.

Estimated total consumption rates by different size classes within each herbivore guild varied across years for both turf (Appendix S1: Fig. S5) and macroalgae (Appendix S1: Fig. S6) as a result of changes in abundance of size classes. There was a shift in the size class distribution for the scrapers such that the upper three size classes removed the most turf biomass in 2012 and 2013 (Appendix S1: Fig. S5). Mid-sized grazers removed the most turf biomass in early years, but over the last three years, the smaller size classes removed the greatest daily biomass of turf algae. Based upon our calculations, browsers removed very little turf biomass overall. In terms of macroalgal consumption, our data suggest that browser size classes fairly evenly removed macroalgal biomass (Appendix S1: Fig. S6). Small grazers also appeared important in macroalgal removal. Scrapers removed very little macroalgal biomass overall, and all size classes contributed equally.

Discussion

Our study assessed the balance of fleshy algal production and herbivorous fish consumption of algae through time at an HFMA. We used a novel approach to assess the potential changes in reef function as a result of herbivore protection by asking whether herbivorous fish consumption has increased in the HFMA through time. Over the first five years of the HFMA designation, we found that mean algal production continued to exceed herbivorous fish consumption; however, the magnitude of difference decreased from 2010 until 2014. When the fish assemblage data from Molokini MLCD were used in the model, mean net production was less than zero. These findings

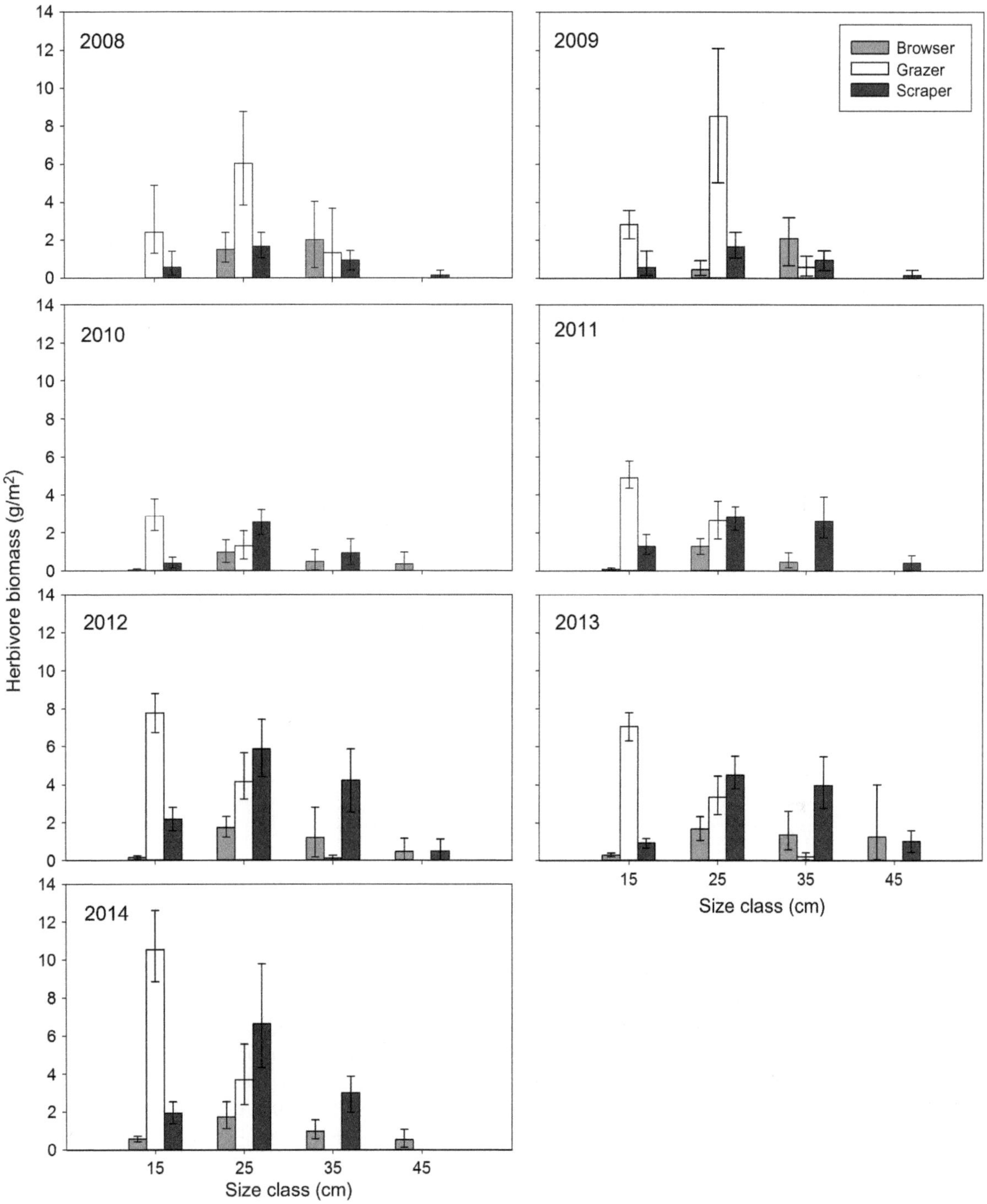

Fig. 4. Herbivore biomass per year broken into guilds and size classes. Bars are 95% confidence intervals. 2009 surveys were completed a month after the Kahekili Herbivore Fisheries Management Area was legally established, but before signage or enforcement.

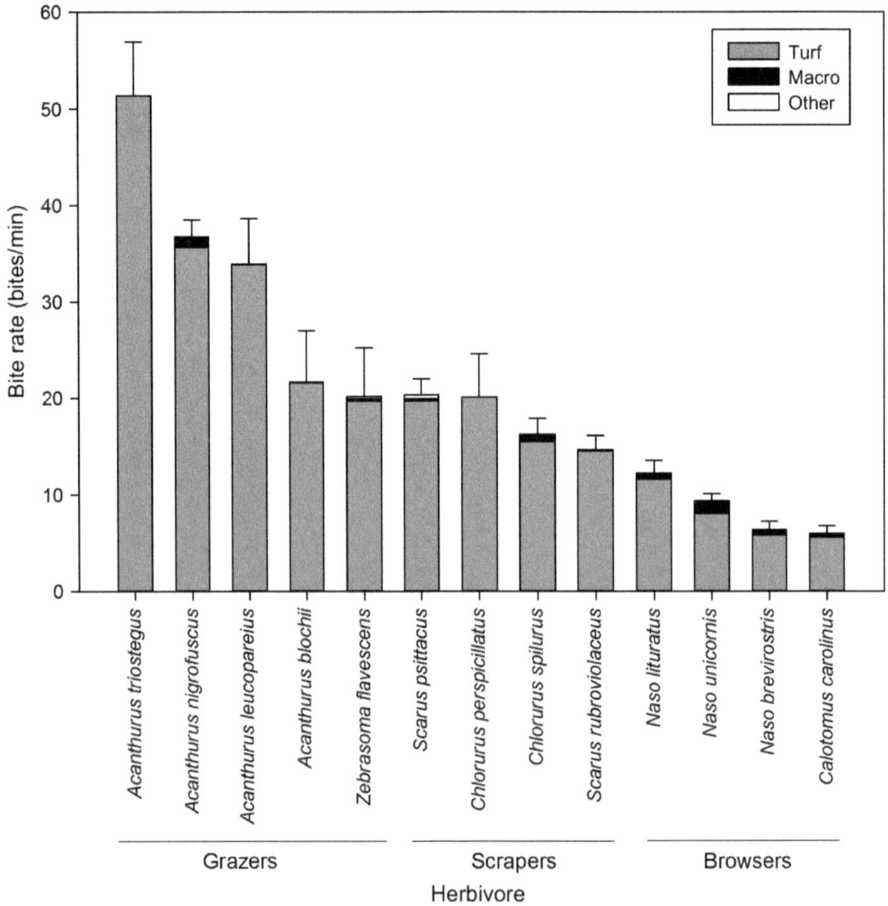

Fig. 5. Herbivore species mean bite rates on turf algae, macroalgae, and other substrates. Bars are SE. n = 10–64.

suggest that with further increases in fish abundance and biomass within the KHFMA to a level observed on long-term well-protected nearby reefs in Hawaii, consumption by the herbivorous fish assemblage could balance production at the reef.

In addition to small increases in consumption by the herbivorous fish assemblage through time, the composition of the major contributors to consumption shifted. Importantly, in the later years of the study, larger fishes from the scraper/excavator guild contributed more to consumption, particularly of turf algae, as a result of their increase in density in the fish assemblage (Appendix S1: Fig. S5). By scraping turf off the reef and exposing bare limestone, these fishes are known to promote crustose coralline algae (Belliveau and Paul 2002, Smith et al. 2010), an important reef builder (Smith et al. 2016) and substrate for coral larvae settlement (Price 2010, O'Leary et al. 2012). Bonaldo and Bellwood (2009) have further demonstrated differences in bite impact across species within this guild and the subsequent importance of these differences in maintaining algae and sediment-free space on the reef. Therefore, the contributions of scrapers to grazing are not only important for the consumed biomass of fleshy benthic algae, but also for the broader "ecological fate" of each bite (Kelly et al. 2016) in opening space for recruitment by reef builders.

In contrast to the scrapers, browsers played a larger role in consumption of macroalgae than turf algae, though they removed more turf biomass than macroalgal biomass (Appendix S1: Figs. S5, S6) per observations of their feeding (Fig. 5). Grazers of the two smallest size classes played important roles in turf and macroalgal removal, likely a result of their high bite rates,

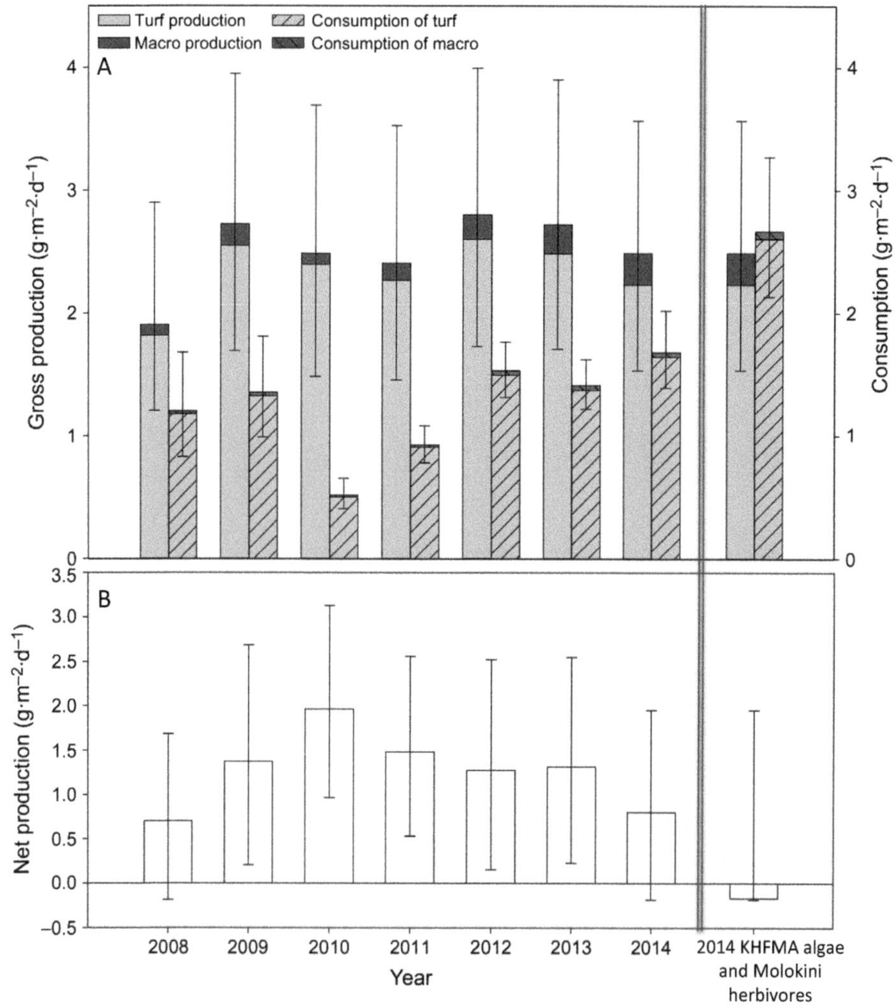

Fig. 6. (A) Estimated growth (solid bars) of macroalgae and turf algae (g dry wt·m^{-2}·day^{-1}) and consumption (hashed bars) of macroalgae and turf algae (g dry wt·m^{-2}·day^{-1}) by the herbivore assemblage from 2008 to 2014 and, after the black vertical line, using the herbivorous fish assemblage of Molokini Shoal reef. (B) Estimated net production of the algal community after accounting for consumption by the herbivores. Protection of Kahekili as an Herbivore Fisheries Management Area occurred in summer 2009. (C) Molokini's herbivorous fish assemblage can be used as a proxy for Kahekili in potential future years. Bars are 95% confidence intervals.

which exceeded those of both browsers and scrapers (Fig. 5). While the two smallest grazer size classes removed the most turf biomass of all fishes, their bites remove algal filaments without exposing bare limestone as those of scrapers do. The "ecological fate" of a bite from a small grazer is therefore to maintain productive turf algae. Thus, while our model focused on removal of algal biomass, there are additional dynamics to consider in how the herbivore assemblage shapes not just removal of fleshy algae but also what will occupy the space from where a bite was taken.

Integrating the impact of all herbivores, we saw a higher proportion of algae consumed by the herbivorous fish assemblage in 2014 (67.0%) as compared to just after designated an HFMA in 2010 (20.8%; Fig. 6B). There were two important benthic responses to this increased grazing. First, there was an increase in CCA from 2008 to 2014 from 0.3% to 9.2% (see Williams et al. 2016 for further details). Secondly, turf algae cover

Table 2. Comparison across studies that have measured benthic fleshy algal production and consumption by herbivores.

Study	Location	Mean production rate (g c·m^{-2}·day^{-1})	Mean consumption rate (g c·m^{-2}·day^{-1})	Percentage of production consumed
Hawaii				
This study	Maui	0.63–0.68	0.34–0.42	20.8–67.0
Caribbean				
Paddack et al. (2006)	Florida Keys	1.05	0.34–0.76	31–77
van Rooij et al. (1998)	Bonaire	1.46–1.86	1.74	101
Carpenter (1986)	St. Croix	1.58–3.11	2.5	97
Kopp et al. (2010)	Guadeloupe	0.35–0.95	0.1–0.7	27–200
Great Barrier Reef				
Russ (2003)	Myrmidon	0.2–1.8	0.2–2.1	38–100
Klumpp and Polunin (1990)	Davies	1.11–2.51	0.57–1.13	42–72
Hatcher (1981)	One Tree Reef	0.61–0.97	0.40–0.79	61–290

Note: For comparison to other studies, dry weight calculations in this study were converted to g C by assuming g C = 0.25 × g dry wt algae.

declined from 53.8% to 41.2% over this same period. That said, the magnitude of these two changes did not match the magnitude of the change in proportion of production consumed. We believe two factors were at play here. First, increasing grazing intensity on the reef may have resulted in increased productivity of fleshy algae, particularly turf algae (on reefs: Carpenter 1986, in grasslands: Westlake 1963, McNaughton 1979), which could have delayed a corresponding change in turf algae cover. Secondly, the functional group "turf algae" is a very broad category and a relatively small change in cover might have under-represented a potentially larger change in turf biomass if turfs were more heavily cropped turfs on average. Turf height was not examined in this study but could serve as a helpful metric in future work (as in Harris et al. 2015). A dynamic model in future studies utilizing the in situ rates provided in this study would further illuminate benthic response to increased grazing pressure.

Importantly, some of the production not consumed in our model would undoubtedly have been converted to detritus (Hatcher 1990), exported off the reef (Hamner and Hauri 1981), or consumed by other organisms. The sea urchin assemblage is known to be major a contributor to algal grazing on reefs (Ogden and Lobel 1978, Mumby 2006, Sandin and McNamara 2012), but urchin grazing was not assessed here. Sea urchins have also been shown to exhibit species-specific preferences for different algal species and functional groups (Stimson et al. 2007, Lewis et al., *unpublished data*), though their influence relative to herbivorous fishes can certainly vary (McClanahan et al. 1996). Further, our grazing budget considers only herbivores observed on the reef within the fish families protected by the KHFMA. The impact of some species from these families (e.g., *Ctenochaetus strigosus* and *Acanthurus olivaceus*) is still debated (Marshell and Mumby 2012, Tebbett et al. 2017), as these species are often classified as detritivores but are known to also remove some algal filaments (Choat et al. 2002, Kelly et al. 2016). These species likely have interesting non-linear impacts on fleshy algal dynamics via reduced algal shading by removal of detritus, which may increase algal productivity, as well as contributing to turf algal grazing via incidental consumption of algal filaments. In the interest of being conservative in our approach to estimating consumption, since a "balanced budget" of production and consumption would signal success of the HFMA management, and given the uncertainty in the impact of *C. strigosus* and *A. olivaceus*, we have excluded these species from the analysis. We also did not include territorial damselfishes or omnivorous fishes, including triggerfish such as *Melichthys niger*, one of the more abundant triggerfishes within the KHFMA (Williams et al. 2016) and known to consume benthic algae in some contexts. While these species were not within the goal of this study (given that the management measures do not provide explicit protection for these taxa), further research is needed to better understand their ecological role on reefs including the KHFMA.

Other studies that have calculated production and consumption on reefs through a variety of methods and on a variety of reef types have seen similar results to those reported here in proportion to benthic cover and herbivorous fish biomass (Table 2; Appendix S1: Table S3). On reefs with <10% coral and high algal cover in the Florida Keys, mean production was two times our estimates from KHFMA (1.05 vs. 0.48–0.70; Paddack et al. 2006), while KHFMA has three times the amount of coral cover and proportionally less fleshy algae. Meanwhile, herbivore fish biomass in the Florida Keys sites was 3–10 times that of the KHFMA and mean consumption rates were equivalent to 1.5 times the KHFMA rates, perhaps a result of different species and size class distributions within the herbivore communities. On high coral-cover reefs in the Caribbean (Wanders 1976, Carpenter 1986), production was 3–5 times that of the KHFMA despite having a smaller proportion of the benthos being covered by algae. These patterns are likely the result of higher growth rates of algae in the Caribbean (Roff and Mumby 2012) as well as the potential for high growth rates of algae subjected to intense grazing (Westlake 1963).

Consumption patterns on other reefs intuitively also vary with herbivore biomass, though herbivore community composition invariably can and does alter this relationship. Our estimates of total consumption by the herbivore community of benthic fleshy algal production were on the low end of the spectrum when comparing to other studies (Table 2; Appendix S1: Table S3). However, total fish biomass at the KHFMA was also on the low end of biomass estimates of these other studies, which in many cases were no-take areas or remote reefs, and consumption in this study was proportional to biomass as compared to other studies (Appendix S1: Fig. S3). Placing the KHFMA in a global context, both total herbivore biomass and biomass of individual herbivore guilds in 2014 were similar to the global mean for each of these metrics at fisheries accessible coral reefs (global mean = 20.5 g/m^2; scraper = 9.5 g/m^2; browser = 2.0 g/m^2; grazer = 17.5 g/m^2; Edwards et al. 2014). This level of herbivore biomass remains less than half the biomass found on reefs not accessible to fishing in the Main Hawaiian Islands (MHI; ~40 g/m^2; Edwards et al. 2014). On low coral-cover reefs (<10%), even a return of herbivore biomass to historically equivalent populations may not be enough to reverse phase shifts given the high algal standing stock now present on many reefs (Paddack et al. 2006). With >30% coral cover, after 5 yr of herbivore protection KHFMA was already showing signs of declining consistent with herbivore control on net algal production (Fig. 6B). These patterns are corroborated by recent findings showing an increasing cover of CCA and decreasing cover of turf algae (Williams et al. 2016). Thus, if herbivorous fish biomass becomes more similar to other Hawaiian protected areas such as the Molokini MLCD, consumption will likely become sufficient to control algal production and ultimately reduce algal standing stock (Fig. 6C).

There are important assumptions and caveats in using data from Molokini MLCD in this study. Importantly, this offshore reef is not exposed to the same nutrient and sediment loading that the nearshore KHFMA reef experiences (Dailer et al. 2010, Swarzenski et al. 2012) and benthic community composition differs (Rodgers et al. 2015). However, the use of Molokini herbivorous fish biomass is intended to provide an example of potential herbivore biomass for the KHFMA using a nearby reef with long-term protection. While fish biomass at the KHFMA in recent years has appeared to asymptote in the last two years (Williams et al. 2016), this is largely understood to be a result of poaching within the KHFMA, rather than indicating that the carrying capacity of the reef has been reached. Therefore, from an ecological standpoint, parrotfishes and surgeonfishes at KHFMA are probably not limited by the system itself, but rather by ongoing harvesting. That being said, it likely will not be possible for the KHFMA herbivore biomass to substantially increase beyond current levels without a meaningful reduction in the levels of poaching.

Given that herbivores consumed 67% of algal production in 2014 within the KHFMA (Fig. 6B), we suggest that herbivorous fish biomass will need to continue to increase to reduce algal biomass and ultimately reduce algal cover. Reefs in the MHI that are not accessible to fishing with herbivore biomass of 40 g/m^2 (Edwards et al. 2014) may be reasonable medium-term goals for herbivore biomass at KHFMA with continued herbivore management into the future. The results of the production–consumption budget for the KHFMA calculated with the 2014 herbivore assemblage data from the Molokini MLCD support this, providing a potential future scenario for the KHFMA using an ecosystem that has been protected for decades.

Our approach here incorporates functional roles of benthic cover and the fish assemblage, and by projecting fish biomass into the future, provides support for the potential of herbivore protection in balancing production and consumption within the KHFMA. By using these rates, we may be able to begin estimating how many and what type of herbivores are needed to elicit strong top-down control on algal communities. While herbivore management is currently promoted as a strategy to reduce algal overgrowth of corals, we need more tools to be able to predict management and set targets for the herbivore assemblages. These targets will be directly related to the benthic community structure of a given reef community and the production of the algal community.

Across terrestrial and marine ecosystems, a mass balance calculation for primary production and consumption by grazers can inform management efforts as to the deficit or excess in grazing in a system and provide a target window for ecosystem restoration. For coral reefs, the KHFMA and other herbivore protected areas around the world provide a unique new opportunity to manage a key ecosystem service, grazing, without completely eliminating fishing and have shown exciting positive results thus far. It is our hope that the framework provided in this study is used in other locations to begin to understand and estimate how many and what types of herbivorous fishes are needed to control algal abundance on a given reef. Herbivore management is an approach that can be used for both maintaining healthy reef ecosystems as well as for restoration and should be used more broadly for reef conservation.

Acknowledgments

We thank Darla White for ideas and field support and Penny Dockry for logistical support. Thanks to Jim Luecke and John Gorman and the other staff at Maui Ocean Center for access to facilities to measure fish BSs. We also thank Mark Miller, Donna Brown, Susan Kram, Samantha Clements, Jill Harris, Levi Lewis, Elizabeth Keenan, Aaaron Hartmann, Lauren Freeman, Simon Freeman, and Melissa Garren for help in the field. Funding was provided by NSF IGERT (grant no. 0903551), Hawaii Coral Reef Initiative, Mia Tegner Fellowship, Women Divers Hall of Fame, Explorers Club Exploration Fund, the Sussman Fellowship, and the Oceanids Memorial Fellowship.

Literature Cited

Andrew, M. H. 1988. Grazing impact in relation to livestock watering points. Trends in Ecology and Evolution 3:336–339.

Belliveau, S. A., and V. J. Paul. 2002. Effects of herbivory and nutrients on the early colonization of crustose coralline and fleshy algae. Marine Ecology Progress Series 232:105–114.

Bellwood, D. R., and J. H. Choat. 1990. A functional analysis of grazing in parrotfishes (family Scaridae): the ecological implications. Environmental Biology of Fishes 28:189–214.

Belsky, A. J. 1992. Effects of grazing, competition, disturbance and fire on species composition and diversity in grassland communities. Journal of Vegetation Science 3:187–200.

Benndorf, J., W. Böing, J. Koop, and I. Neubauer. 2002. Top-down control of phytoplankton: the role of time scale, lake depth and trophic state. Freshwater Biology 47:2282–2295.

Bonaldo, R. M., and D. R. Bellwood. 2009. Dynamics of parrotfish grazing scars. Marine Biology 156: 771–777.

Carpenter, R. C. 1986. Partitioning herbivory and its effects on coral reef algal communities. Ecological Monographs 56:345–364.

Choat, J. H., K. D. Clements, and W. D. Robbins. 2002. The trophic status of herbivorous fishes on coral reefs 1: dietary analyses. Marine Biology 140:613–623.

Coe, M. D., D. H. Cummings, and J. Phillipson. 1976. Biomass and production of large herbivores in relation to rainfall and primary production. Oecologia 22:341–354.

Dailer, M. L., R. S. Knox, J. E. Smith, M. Napier, and C. M. Smith. 2010. Using ∂15N values in algal tissue to map locations and potential sources of anthropogenic nutrient inputs on the island of Maui, Hawai'i, USA. Marine Pollution Bulletin 60: 655–671.

Edwards, C. B., A. M. Friedlander, A. G. Green, M. J. Hardt, E. Sala, H. P. Sweatman, I. D. Williams, B. Zgliczynski, S. A. Sandin, and J. E. Smith. 2014.

Global assessment of the status of coral reef herbivorous fishes: evidence for fishing effects. Proceedings of the Royal Society B 281:20131835.

Frank, D. A., S. J. McNaughton, and B. F. Tracy. 1998. The ecology of the Earth's grazing ecosystems. BioScience 48:513–521.

Froese, R., and D. Pauly, editors. 2016. FishBase. World Wide Web electronic publication. http://www.fishbase.org

Friedlander, A. M., E. Brown, and M. E. Monaco. 2007. Defining reef fish habitat utilization patterns in Hawaii: comparisons between marine protected areas and areas open to fishing. Marine Ecology Progress Series 351:221–233.

Gower, S. T., C. J. Kucharik, and J. M. Norman. 1999. Direct and indirect estimation of leaf area index, f(APAR), and net primary production of terrestrial ecosystems. Remote Sensing of Environment 70: 29–51.

Green, A. L., and D. R. Bellwood. 2009. Monitoring functional groups of herbivorous reef fishes as indicators of coral reef resilience: a practical guide for coral reef managers in the Asia Pacific Region. IUCN, Gland, Switzerland.

Hamilton, S. L., J. E. Smith, N. N. Price, and S. A. Sandin. 2014. Quantifying patterns of fish herbivory on Palmyra Atoll (USA), an uninhabited predator-dominated central Pacific coral reef. Marine Ecology Progress Series 501:141–155.

Hamner, W. M., and I. R. Hauri. 1981. Effects of island mass: water flow and plankton pattern around a reef in the Great Barrier Reef Lagoon, Australia. Limnology and Oceanography 26:1084–1102.

Harris, J. L., L. S. Lewis, and J. E. Smith. 2015. Quantifying scales of spatial variability in algal turf assemblages on coral reefs. Marine Ecology Progress Series 532:41–57.

Hatcher, B. 1981. The interaction between grazing organisms and the epilithic algal community of a coral reef: a quantitative assessment. Pages 515–524 *in* Proceedings of the Fourth International Coral Reef Symposium. Marine Science Center, University of the Philippines, Quezon City, Manila, Philippines.

Hatcher, B. G. 1990. Coral reef primary productivity. A hierarchy of pattern and process. Trends in Ecology & Evolution 5:149–155.

Hughes, T. P. 1994. Catastrophes, phase shifts, and large-scale degradation of a Caribbean coral reef. Science 265:1547–1551.

Huxley, J. S., and G. Tesissier. 1936. Terminology of relative growth. Nature 137:780–781.

Jackson, J. B. C., M. K. Donovan, K. L. Cramer, and W. Lam. 2014. Status and trends of Caribbean coral reefs: 1970-2012. Global Coral Reef Monitoring Network, IUCN, Gland, Switzerland.

Jouffray, J.-B., M. Nyström, A. V. Norström, I. D. Williams, L. M. Wedding, J. N. Kittinger, and G. J. Williams. 2014. Identifying multiple coral reef regimes and their drivers across the Hawaiian archipelago. Philosophical Transactions of the Royal Society B: Biological Sciences 370:20130268–20130268.

Kelly, E. L. A., Y. Eynaud, S. M. Clements, M. Gleason, R. T. Sparks, I. D. Williams, and J. E. Smith. 2016. Investigating functional redundancy versus complementarity in Hawaiian herbivorous coral reef fishes. Oecologia 182:1151–1163.

Klumpp, D. W., and N. V. C. Polunin. 1990. Algal production, grazers, and habitat partitioning on a coral reef: positive correlation between grazing rate and food availability. Pages 372–388 *in* M. Barnes and R. N. Gibson, editors. Trophic relationship in the marine environment. Proceedings of the 24th European Marine Biology Symposium. Aberdeen University Press, Aberdeen, UK.

Kopp, D., Y. Bouchon-Navaro, M. Louis, D. Mouillot, and C. Bouchon. 2010. Herbivorous fishes and the potential of caribbean marine reserves to preserve coral reef ecosystems. Aquatic Conservation: Marine and Freshwater Ecosystems 20:516–524.

Marshell, A., and P. J. Mumby. 2012. Revisiting the functional roles of the surgeonfish *Acanthurus nigrofuscus* and *Ctenochaetus striatus*. Coral Reefs 31:1093–1101.

McClanahan, T. R., A. T. Kamukuru, N. A. Muthiga, M. Gilagabher-Yebio, and D. Obura. 1996. Effect of sea urchin reductions on algae, coral, and fish populations. Conservation Biology 10:136–154.

McCook, L. J., J. Jompa, and G. Diaz-Pulido. 2001. Competition between corals and algae on coral reefs: a review of evidence and mechanisms. Coral Reefs 19:400–417.

McNaughton, S. J. 1979. Grazing as an optimization process: grass-ungulate relationships in the Serengeti. American Naturalist 113:691–703.

Mumby, P. J. 2006. The impact of exploiting grazers (Scaridae) on the dynamics of Caribbean coral reefs. Ecological Applications 16:747–769.

Ogden, J. C., and P. S. Lobel. 1978. The role of herbivorous fishes and urchins in coral reef communities. Environmental Biology of Fishes 3:49–63.

O'Leary, J. K., D. C. Potts, J. C. Braga, and T. R. McClanahan. 2012. Indirect consequences of fishing: Reduction of coralline algae suppresses juvenile coral abundance. Coral Reefs 31:547–559.

Ong, L., and K. N. Holland. 2010. Bioerosion of coral reefs by two Hawaiian parrotfishes: species, size differences and fishery implications. Marine Biology 157:1313–1323.

Paddack, M. J., R. K. Cowen, and S. Sponaugle. 2006. Grazing pressure of herbivorous coral reef fishes on low coral-cover reefs. Coral Reefs 25: 461–472.

Price, N. 2010. Habitat selection, facilitation, and biotic settlement cues affect distribution and performance of coral recruits in French Polynesia. Oecologia 163:747–758.

Rasher, D. B., S. Engel, V. Bonito, G. J. Fraser, J. P. Montoya, and M. E. Hay. 2012. Effects of herbivory, nutrients, and reef protection on algal proliferation and coral growth on a tropical reef. Oecologia 169: 187–198.

Rasher, D. B., A. S. Hoey, and M. E. Hay. 2013. Consumer diversity interacts with prey defenses to drive ecosystem function. Ecology 94:1347–1358.

Ripple, W. J., and R. L. Beschta. 2003. Wolf reintroduction, predation risk, and cottonwood recovery in Yellowstone National Park. Forest Ecology and Management 184:299–313.

Rodgers, K. S., P. L. Jokiel, E. K. Brown, S. Hau, and R. Sparks. 2015. Over a decade of change in spatial and temporal dynamics of Hawaiian coral reef communities. Pacific Science 69:1–13.

Roff, G., and P. J. Mumby. 2012. Global disparity in the resilience of coral reefs. Trends in Ecology and Evolution 27:404–413.

Russ, G. R. 2003. Grazer biomass correlates more strongly with production than with biomass of algal turfs on a coral reef. Coral Reefs 22:63–67.

Sandin, S. A., and D. E. McNamara. 2012. Spatial dynamics of benthic competition on coral reefs. Oecologia 168:1079–1090.

Shurin, J. B., E. T. Borer, E. W. Seabloom, K. Anderson, C. A. Blanchette, B. Broitman, S. D. Cooper, and B. S. Halpern. 2002. A cross-ecosystem comparison of the strength of trophic cascades. Ecology Letters 5:785–791.

Smith, J. E., C. L. Hunter, and C. M. Smith. 2010. The effects of top-down versus bottom-up control on benthic coral reef community structure. Oecologia 163:497–507.

Smith, J. E., et al. 2016. Re-evaluating the health of coral reef communities: baselines and evidence for human impacts across the central Pacific. Proceedings of the Royal Society B 283:20151985.

Sparks, R., K. Stone, D. White, and M. Ross. 2015. Maui and Lanai Monitoring Report. Hawaii Division of Aquatic Resources, Maui Office, Wailuku, Hawaii, USA.

Stimson, J., T. Cunha, and J. Philippoff. 2007. Food preferences and related behavior of the browsing sea urchin *Tripneustes gratilla* (Linnaeus) and its potential for use as a biological control agent. Marine Biology 151:1761–1772.

Swarzenski, P. W., C. D. Storlazzi, M. K. Presto, A. E. Gibbs, C. G. Smith, N. T. Dimova, M. L. Dailer, and J. B. Logan. 2012. Nearshore morphology, benthic structure, hydrodynamics, and coastal groundwater discharge near Kahekili Beach Park, Maui, Hawaii. USGS Report 2012-1166. U.S. Geological Survey, Reston, Virginia, USA.

Tebbett, S. B., C. H. R. Goatley, and D. R. Bellwood. 2017. Clarifying functional roles: algal removal by the surgeonfishes *Ctenochaetus striatus* and *Acanthurus nigrofuscus*. Coral Reefs 36:803–813.

van Rooij, J. M., J. J. Videler, and J. H. Bruggemann. 1998. High biomass and production but low energy transfer efficiency of Caribbean parrotfish: implications for trophic models of coral reefs. Journal of Fish Biology 53:154–178.

Wanders, J. B. W. 1976. The role of benthic algae in the shallow reef of Curacao (Netherlands Antilles). I: primary productivity in the coral reef. Aquatic Botany 2:235–270.

Wanders, J. B. W. 1977. The role of benthic algae in the shallow reef of Curaçao (Netherlands Antilles) III: the significance of grazing. Aquatic Botany 3: 357–390.

Westlake, D. F. 1963. Comparisons of plant productivity. Biological Reviews 38:385–425.

Williams, I. D., and N. V. C. Polunin. 2001. Large-scale associations between macroalgal cover and grazer biomass on mid-depth reefs in the Caribbean. Coral Reefs 19:358–366.

Williams, I. D., D. J. White, R. T. Sparks, K. C. Lino, J. P. Zamzow, E. L. A. Kelly, and H. L. Ramey. 2016. Responses of herbivorous fishes and benthos to 6 years of protection at the Kahekili herbivore fisheries management area, Maui. PLoS ONE 11: e0159100.

Zgliczynski, B. J., and S. A. Sandin. 2017. Size-structural shifts reveal intensity of exploitation in coral reef fisheries. Ecological Indicators 73:411–421.

4

Local and regional stressors interact to drive a salinization-induced outbreak of predators on oyster reefs

David L. Kimbro,[1,†] J. Wilson White,[2] Hanna Tillotson,[3] Nikkie Cox,[3] Megan Christopher,[3] Owen Stokes-Cawley,[1] Samantha Yuan,[3] Timothy J. Pusack,[4] and Christopher D. Stallings[4]

[1]*Department of Marine and Environmental Sciences, Northeastern University, Marine Science Center, Nahant, Massachusetts 01908 USA*
[2]*Department of Fisheries and Wildlife, Oregon State University, Corvallis, Oregon 97331 USA*
[3]*Apalachicola National Estuarine Research Reserve, East Point, Florida 32328 USA*
[4]*College of Marine Science, University of South Florida, St. Petersburg, Florida 33701 USA*

Abstract. Predator outbreaks are predicted to increasingly decimate economically and ecologically important prey populations because global climate change and food-web modifications frequently facilitate predators and stress prey. Natural systems are organized hierarchically, with processes operating at multiple scales giving rise to patterns of biodiversity, so predicting and managing outbreaks requires a framework that accounts for the effects of both local and regional stressors. Here, we used the comparative experimental approach to investigate whether the collapse of a nationally important oyster fishery in the Gulf of Mexico (Apalachicola Bay, Florida) could have been (1) caused proximally by a predator outbreak and (2) whether this outbreak was mediated by local- and/or regional-scale forces. During the fishery collapse, we paired experiments with monitoring in Apalachicola Bay and found elevated water salinity, high abundance of predatory snails, and intense oyster mortality due to predation. By repeating these experiments over 4 yr, we found that periods of reduced water salinity inhibited predation on oysters. To partition the influence of local-versus-regional factors on this predator outbreak, we simultaneously replicated the paired experiments and monitoring in a nearby bay (Ochlockonee Bay) that shares the same regional-scale rainfall conditions. Increasing freshwater withdrawals from the watershed that drains into Apalachicola Bay have increased salinities in that bay, but there have not been similar withdrawals in the Ochlockonee Bay watershed. Therefore, Apalachicola Bay experienced a localized anthropogenic stress, while both bays experienced regional stress from drought. In Ochlockonee Bay, our experiments demonstrated that the river maintained sufficiently low salinity to provide ~50% of oyster reefs with a refuge from predation. In contrast, salinity-dependent predation in Apalachicola Bay extended up to the river mouth. Given the stark differences in upstream water withdrawals between these watersheds, it is reasonable to surmise that these withdrawals exacerbated the stress of regional drought, created the difference in predation between the two bays, and thus may have precipitated the oyster fishery collapse. Our study provides empirical support for recent theory about the hierarchical organization of ecosystems, which predicts that stressors will interact across scales to cause localized predator outbreaks.

Key words: Apalachicola Bay; climate change; consumer front; drought; environmental stress; meta-ecosystem; oyster fishery; predation refuge; predator–prey; scale; water withdrawals.

† **E-mail:** d.kimbro@northeastern.edu

Introduction

Predators can maintain the community structure and ecosystem functioning of natural systems (Paine 1966, Schmitz 2008). However, under certain conditions, outbreaks and fronts of predators (or herbivorous consumers) can form, resulting in runaway consumption of important resources (Silliman et al. 2013). For instance, a prolonged drought associated with climate change promoted an outbreak of pine beetles that caused a regional loss of forest trees in Canada (Kurz et al. 2008). Likewise, nutrient-rich coastal runoff events triggered outbreaks of crown-of-thorns seastars on the Great Barrier Reef, which decimated living corals (Fabricius et al. 2010). Because the effects of predation on community structure can quickly shift from beneficial to detrimental, understanding the underlying cause(s) of such outbreaks is fundamental to conserving key resources such as hardwood forests and coral reefs.

Recently, the initiation of predator outbreaks was linked to environmental stressors that exceed a certain threshold and in turn either increase prey susceptibility to predation, increase the local density of the predator, or both (Silliman et al. 2013). But to understand how ecosystems will respond to environmental stressors, we must consider the relative influence of local- and regional-scale processes (Gouhier et al. 2010, Menge et al. 2015). For instance, the American Dustbowl of the 1930s involved a broad drought and catastrophic loss of topsoil that was associated with—yet could not be completely explained by—La Niña conditions of warmer sea surface temperatures in the tropical eastern Pacific (Cook et al. 2009). To accurately explain the location and magnitude of drought, models had to include the effect of local-scale devegetation and enhancement of dust aerosols, which were caused by the replacement of drought-tolerant prairie grasses with drought-sensitive wheat (Cook et al. 2009). If predator outbreaks result from an interaction between regional-scale (e.g., La Niña, drought) and local-scale stressors (e.g., devegetation), then our ability to mitigate outbreaks depends on understanding the relative roles of multi-scale processes in exceeding stress beyond critical thresholds (Scheffer and Carpenter 2003).

One important resource vulnerable to predator outbreaks is the oyster. Oysters form reefs that provide habitat for juveniles of commercially important invertebrates and finfishes, stabilize shorelines, filter coastal water, and remove excess nitrogen (Grabowski et al. 2012). Unfortunately, habitat degradation, overharvesting, and eutrophication have decreased the global abundance of this habitat by 85%, with most of the remaining reefs concentrated in six ecoregions, five of which are located in the United States (Beck et al. 2011). These remaining reefs continue to face a variety of threats, including environmentally triggered predator outbreaks. For example, eastern oyster (*Crassostrea virginica*) reefs began to decline in the southern reach of the Matanzas River estuary (northern Florida, USA) in 2008 (Garland and Kimbro 2015). The underlying cause of this decline was increased water salinity, due in part to a regional drought, which facilitated the reproductive success of the crown conch (*Melongena corona*), a gastropod predator responsible for the proximal cause of oyster loss (Garland and Kimbro 2015).

Regional drought alone, however, could not entirely explain the conch outbreak in the southern Matanzas River estuary. Oyster reefs in the northern reach of the same estuary lacked conchs and remained healthy, despite experiencing the same regional stress. To reconcile this contradiction, Garland and Kimbro (2015) observed that the southern reach of the estuary was flushed only by the Matanzas inlet (Sheng et al. 2008), while the northern reach was influenced by flushing of both the Matanzas inlet and the St. Augustine inlet. Because the Matanzas inlet has constricted by 63% since 1995 (Garland and Kimbro 2015), the southern reach now experiences less tidal flushing than the waters of the northern reach. Thus, the combination of regional drought and a local stress (reduced tidal flushing) was necessary to explain why southern reefs—but not northern reefs—were affected by a salinity-induced conch outbreak.

More recently, oysters in Apalachicola Bay, Florida (Fig. 1), experienced catastrophic mortality. The Apalachicola oyster fishery was one of the healthiest in the nation, consistently providing 10% of U.S. oyster landings (Beck et al. 2011). But sharp oyster declines between the summer of 2012 and the spring of 2013 resulted in the declaration of a federal fishery disaster (FFWCC 2013). The southern oyster drill (*Stramonita haemastoma*, hereafter "drill") is known to cause high oyster

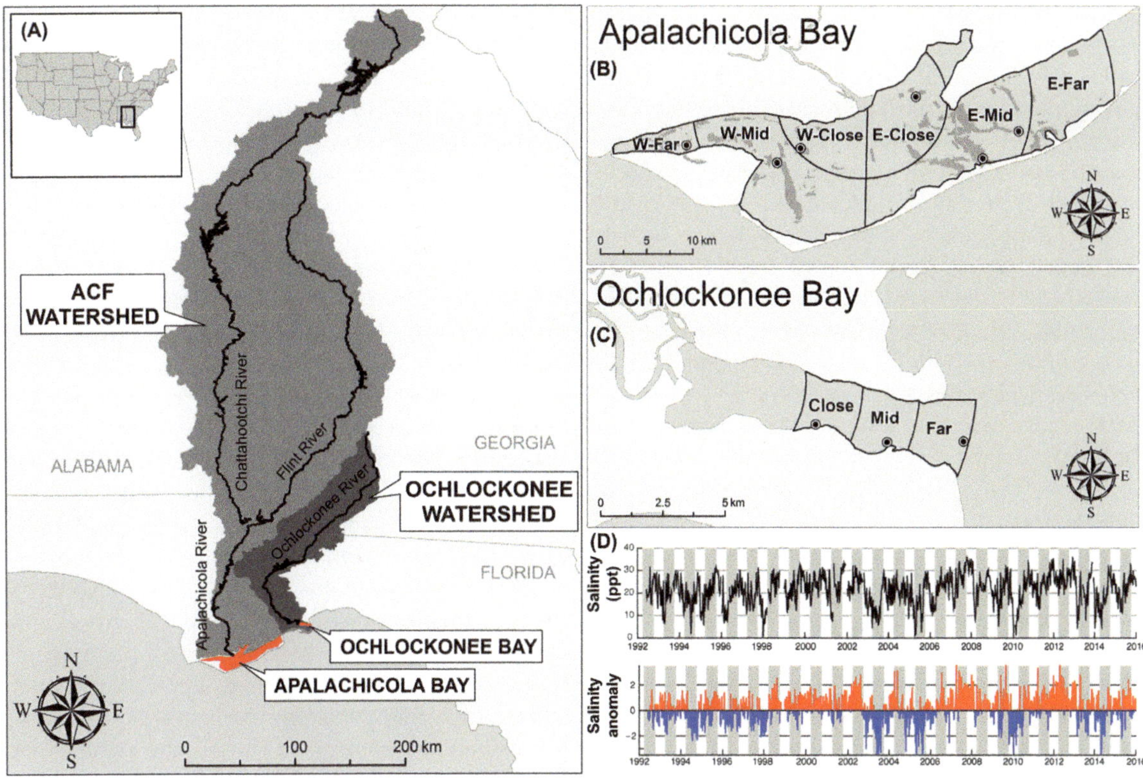

Fig. 1. (A) Map of study sites along the Florida panhandle. Apalachicola Bay and Ochlockonee Bay are highlighted in red. Shading distinguishes the watersheds of the Apalachicola River (dark gray) and the Ochlockonee River (darker gray). Map of Apalachicola Bay (B) and Ochlockonee Bay (C). In Apalachicola, dark shading illustrates distribution of oyster reefs. In both estuaries, concentric circles illustrate proportional distances (close, mid, far) of oyster reefs from river discharge. In Apalachicola, proportional distances extend west (W) and east (E) of the river. (D) Time series of weekly mean salinity and weekly salinity anomaly at Cat Point oyster bar in Apalachicola Bay from 1992 to 2016. Anomaly was calculated based on the climatological mean from 1992 to 2002; data obtained from Apalachicola National Estuarine Research Reserve (http://cdmo.baruch.sc.edu).

mortality, especially under high-salinity conditions (Menzel et al. 1966), and could have contributed to the oyster population collapse that occurred during anomalously high-salinity conditions of 2011–2013 (Fig. 1D). However, there is debate as to the contribution of local-versus-regional factors that produce high-salinity conditions favoring drill outbreaks in Apalachicola Bay. A regional drought in 2011–2013 may have increased the salinity of estuaries throughout northern Florida (i.e., a regional factor), but freshwater input into Apalachicola Bay also depends on flow from the Apalachicola-Chattahoochee-Flint (ACF) watershed, which lies mostly within the state of Georgia (Fig. 1A). Georgia's withdrawals from the ACF river system have increased over time (Marella and Fanning 2011), which may have also increased salinity in Apalachicola Bay (i.e., a factor local to the bay). Understandably, legal disputes have developed between Florida and Georgia over these freshwater withdrawals (Florida v. Georgia, 2014).

We used the comparative experimental approach (Menge et al. 2003) to address two questions about the cause(s) of drill outbreaks in Apalachicola Bay, Florida, that may have contributed to the oyster fishery collapse. First, are drill abundance and predation on oysters controlled by water salinity? Second, is the spatial distribution of drill outbreaks, and the ensuing predation, similar between two bays sharing the same regional drought stressor but with different local stressors via freshwater withdrawal rates? If upstream freshwater withdrawals enhance the intensity and spatial distribution of drill predation, then future management could reduce freshwater withdrawals from the ACF watershed. This is a challenging resource management scenario, because we lack data on drill predation in Apalachicola Bay prior to the collapse, and therefore cannot conduct a traditional Before-After-Control-Impact (BACI) assessment (Underwood 1993). Instead, we used manipulative field experiments in two estuaries to separate the contribution of local-versus-regional factors on drill outbreaks.

Methods

Study system

Apalachicola Bay is a large (400 km^2) and shallow (1.9 m average depth) estuary located in the Florida panhandle at the terminus of the ACF River system (Fig. 1A). The ACF watershed (50,000 km^2) has headwaters in northeastern Georgia and flows through west-central Georgia before entering Florida. The primary source of freshwater and nutrients to this bay is the Apalachicola River (Mortazavi et al. 2000, Putland et al. 2013). Consequently, it is the primary cause of salinity variation throughout the bay (Livingston et al. 2000). Maximum river flows occur during late winter months and reflect rainfall amounts in the upper basin; low flows occur during the late summer months (Morey et al. 2009). In Apalachicola Bay, oyster reefs cover approximately 5–13% of the bottom (16–24 km^2) and are distributed from areas close to the river mouth (lower salinity) to those far from it (higher salinity; Fig. 1B; FDEP and CAMA 2013, Zu Ermgassen et al. 2012).

Ochlockonee Bay is a shallow (1.0 m average depth) estuary located 30 km east of Apalachicola Bay (Kaul and Froelich 1984; Fig. 1C). Unlike the large ACF basin, the smaller watershed (6500 km^2) of this bay begins in southwest Georgia (Fig. 1C). Water salinity in Ochlockonee Bay (25 km^2 of water) is primarily influenced by flow from the Ochlockonee River, with maximum flows in late winter months and minimum flows in the late summer months (Kaul and Froelich 1984). Shellfish harvest area maps maintained by the state of Florida illustrate that this bay also has subtidal reefs both close to and far from the river (FDACS 2012*a*, *b*).

Hypothesis 1: Variation in salinity controls predation on Apalachicola Bay oysters

Experiments.—We conducted seven replicate rounds of a field experiment on oyster survival at multiple sites in Apalachicola Bay from 2013 to 2017, capturing both spatial and temporal variabilities in water salinity. These experiments are hereafter referenced as A1–A7 ("A" for Apalachicola) and are summarized in Table 1. Prior to the first experiment, we partitioned the subtidal oyster reefs of Apalachicola Bay into six zones that reflected spatial gradients in salinity. Specifically, reefs identified in a geophysical benthic habitat mapping program (Twichell et al. 2007) were assigned to a zone based on their relative distance from the river (close, mid, far) as well as their relative position to the river (east vs. west; Fig. 1B). In each zone, we randomly selected one reef to receive nine protective frames (1.2 m × 0.9 m × 0.6 m) constructed of steel rebar (13 cm thickness). On each reef, these relatively open rebar frames were deployed at 3.0-m linear increments from each other (9 frames × 6 zones = 54 total frames) and were used to protect the experiment from harvesting and boating activities (see Appendix S1 for further details).

One side of each rebar frame contained three posts separated by 0.4 m. Three experimental units were attached to these posts and randomly assigned among three treatments: (1) control, (2) cage, and (3) cage-control treatments. Thus, each experiment (A1–A7) consisted of 162 experimental units that were constructed from 0.2 m × 0.2 m squares of vinyl-coated wire mesh (5 mm × 5 mm mesh opening). The control treatment consisted of a single mesh panel to which adult oysters were attached; the panel was attached to a post and placed flat on the reef with oysters facing upwards. The cage treatment had the same orientation, but it was enclosed by additional mesh panels to form a 0.2 m × 0.2 m × 0.2 m cage. For the cage-control treatment, two mesh walls were removed from the full cage setup to maintain caging material effects, while also allowing access by predators. Adult oysters (mean ± SD length = 67.77 ± 17.21 mm) were collected from the east mid-zone of the bay (Fig. 1B). While final survival of these oysters in the cage treatment reflected the influence of the physiochemical environment and disease on oysters (e.g., Dermo, a disease caused by the protist *Perkinsus marinus*; Petes et al. 2012), survival in the

Table 1. Summary of repeated experiments conducted in Apalachicola Bay (AB) and Ochlockonee Bay (OB). For each experimental round (Apalachicola Bay, A1–A7; Ochlockonee Bay O1–O3), we provide the starting date, duration of the experiment (weeks), levels of oyster density tested, as well as the source and sample interval of the associated water salinity data.

Round	Start date	Duration (weeks)	Initial oyster density	Water salinity source	Interval of salinity samples
A1	May 2013	3	4	YSI Pro2030 Hydrodynamic model	Biweekly Daily
A2	August 2013	1	4	YSI Pro2030 Hydrodynamic model	Biweekly Daily
A3 O1	September 2014	12	1,3,5	Sea-Bird Electronics (A3) Hydrodynamic model (A3) Onset logger (O1)	Monthly Daily 15 min
A4	February 2015	14	1,3,5	Sea-Bird Electronics Hydrodynamic model	Monthly Daily
A5 O2	April 2015	14	1,3,5	Sea-Bird Electronics (A5) Hydrodynamic model (A5) Onset logger (O2)	Monthly Daily 15 min
A6 O3	October 2015	13	1,3,5	Sea-Bird Electronics (A6) Hydrodynamic model (A6) YSI Pro and Sea-Bird Electronics (O3)	Monthly Daily 8–10 d
A7	February 2016	11	1,3,5	Sea-Bird Electronics Hydrodynamic model	Monthly Daily

control treatment reflected the influences of both the environment and predators. Predation effects were calculated as the difference in survival between the cage treatment and the control treatment, standardized by survival in the cage treatment: ([cage − control]/cage; Trussell et al. 2006). We also quantified whether oysters in the control treatment died because of a destructive crushing predator (e.g., crab) or a non-destructive predator that attacks between the valves (e.g., drill).

The first two rounds of this experiment (A1 and A2) used an initial density of four adult oysters for each experimental unit, because average density per unit area of our cages (0.09 m^2) on the primary commercial reefs was 3.09 oysters in 2012 (FFWCC 2013). A1 lasted three weeks during the high-salinity conditions associated with the oyster fishery collapse (Fig. 1D). A2 followed the same procedure as A1, but occurred after two large precipitation events with lower-salinity conditions (Fig. 2C). A2 lasted one week due to logistical constraints. During each experiment, we quantified predator abundance within each rebar frame as well as the ambient salinity and temperature of each zone at the surface (1.0 m depth) and subsurface (1.0 m above benthos) with a handheld meter (YSI Pro2030; YSI, Inc., Yellow Springs, Ohio, USA; Table 1).

We repeated this experiment five times from 2014 to 2017 (A3–A7) to capture longer term variation in salinity and temperature (Table 1). In these additional rounds, we increased the experimental duration to three months to allow development of Dermo disease effects on oyster survival and introduced an additional factor: variation in initial oyster density. We manipulated oyster density because higher densities may increase predator attack rate (Sih 1984) and transmission of disease (Packer et al. 2003). Initial oyster density consisted of three levels (one, three, and five oysters), which were based on the minimum, average, and maximum densities of adult oysters (per unit area) on the primary commercial reefs in 2012 (FWCC 2013). The three levels of initial density were randomly assigned among the nine rebar frames within each zone of Apalachicola Bay so that each frame had the same initial density on all experimental units. Next, we randomly assigned the three experimental units of each rebar frame among the treatments: (1) control, (2) cage, and (3) cage-control.

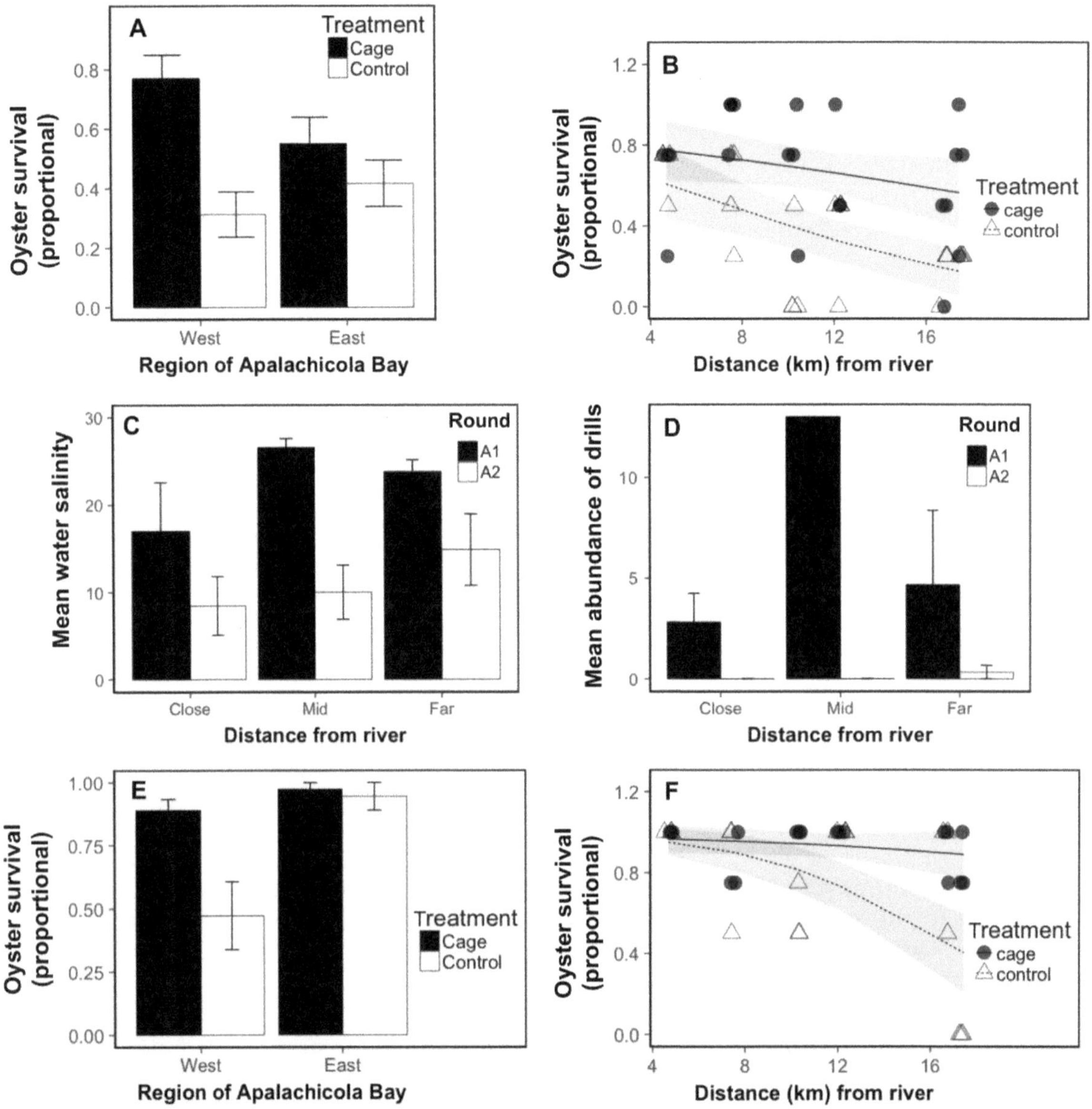

Fig. 2. Results of repeated field experiments and monitoring in Apalachicola Bay, Florida. Oyster survival in experiment A1 on reefs (A) toward the east and west of the Apalachicola River and (B) as a function of distance (km) from Apalachicola River as well as cage (closed circles) and control (open triangles) treatments. (C) Mean (±SE) of water salinity in Apalachicola Bay during experiment A1 (closed bars) and during a prolonged precipitation event prior to experiment A2 (open bars). (D) Mean (±SE) of drill abundance per rebar frame (1.2 m × 0.9 m × 0.6 m) in Apalachicola Bay during experiment A1 (closed bars) and experiment A2 (open bars). Oyster survival in experiment A2 on reefs (E) toward the east and west of the Apalachicola River and (F) as a function of distance (km) from Apalachicola River as well as cage (closed circles) and control (open triangles) treatments.

In experiments A3–A7, predator abundance within each rebar frame was monitored monthly and oyster survival was measured at the end. Water properties were quantified by conducting monthly conductivity–temperature–depth (CTD) profiles at each zone with a Sea-Bird Electronics (SBE 19plus V2; Table 1). For each CTD cast, we calculated the median for temperature and salinity across depths. Next, we calculated monthly median values of salinity and temperature at each zone. For each experiment, these monthly medians were averaged to create zone-specific values of salinity and temperature. The mean (± standard deviation, SD) salinity and temperature, as well as mean (±SD) number of drills per rebar frame, are presented in Appendix S2.

Data analysis for experiments A1–A7

Oyster survival.—We used a generalized linear model (GLM) with binomial error and logit link (i.e., logistic regression) to test whether final oyster survival in each experiment depended on treatment (cage vs. control), region (east or west), or distance (km) from the river mouth. Before proceeding with this analysis, we tested whether the control and cage-control treatments differed within any of the six zones (unique region × distance combinations). This analysis was performed by excluding the cage treatments and using a model with the factors "treatment" (levels = control or cage-control), "zone" (categorical, six levels), and their interaction. A procedural artifact of the caging material was identified if either the term for treatment or the treatment × zone interaction was statistically significant. We did not detect any significant procedural artifacts in experiments A1–A7. Thus, we excluded the cage-control treatments from each round and analyzed only the data from cage and control treatments.

Because experiments A3–A7 included variation in initial density but A1–A2 did not (Table 1), we next tested for an effect of initial density on survival. Given that the full, four-factor GLM failed to converge to a maximum-likelihood solution (and would have also been difficult to interpret had it converged), we constructed three separate GLMs. Each GLM included the initial density factor, one other factor (region, distance, or experimental treatment), and their interaction so that all possible two-factor models with density were evaluated. Despite increasing the type I error rate by taking this approach, we never detected a significant effect of initial density ($P > 0.05$) either by itself or in an interaction term in any model. Therefore, we excluded the density factor, and the analysis of A3–A7 was thus the same as that for A1–A2.

After testing for procedural artifacts and density effects, we used GLM to test for the effects of region, distance (km), and treatment on oyster survival. If the treatment effect was significant either by itself or in interaction with region, then we used Tukey's post hoc test to compare means among treatment × region pairs. If there was a significant treatment × distance interaction, we used the coefficients of the model to predict the separate logistic relationships between oyster survival and distance for cages and controls. In some cases, the full model contained treatments with either 100% survival or 100% mortality, leading to perfect separation of that factor in the logistic regression (Heinze and Ploner 2002). To circumvent the bias in regression parameter estimates associated with separation, for those cases (i.e., experiment A2), we split our analysis into two simpler models consisting of either treatment and region or treatment and distance from the river; neither simpler model had perfect separation.

Salinity effects on predation.—For each experiment, we calculated the average strength of predation at each zone and used separate linear regressions to evaluate whether predation strength (calculated from survival as [cage − control]/cage) was related to distance from the river (km) and average salinity (generated from point samples, Table 1). Additionally, we conducted two more rigorous tests of the effects of salinity on predation by combining the standardized predation effect sizes across multiple experimental rounds. First, average effect sizes of predation at each zone during experiments A1–A2 were combined to evaluate whether the striking difference between the results of the two experiments was associated with salinity in a linear regression. Next, average effect sizes of predation at each zone from all seven experiments were combined into a single data set to evaluate the relationship between predation intensity and salinity as well as temperature over a four-year time frame. Because the predation effect sizes were constrained between values of 0 and 1, we applied an arcsine square root transformation to the effect sizes.

For both analyses, we obtained higher resolution (daily) salinity and temperature values for each experimental zone from a hydrodynamic model of Apalachicola Bay (see Appendix S3 for model validation details). We used these interpolated values to better represent the environmental conditions, which vary on tidal, daily, and weekly scales. For experiments A1–A2, we used hydrodynamic model output for year 2013 directly in our model. Hydrodynamic model outputs were not available for years 2014–2017, when experiments A3–A7 occurred, so we obtained zone-specific salinity estimates by interpolating from synoptic salinity and temperature data recorded daily at monitoring stations elsewhere in the bay (see Appendix S3 for further details). We then used regression to test whether predation strength depended on the mean daily salinity during experiments A1–A2 and on the interaction between mean daily salinity and temperature during experiments A1–A7. Because multiple experimental rounds differed in duration (Table 1), we tested the duration of each experiment (days) as a covariate. That covariate was non-significant ($P = 0.45$) so we excluded it from the final analysis.

Hypothesis 2: The spatial extent of drill predation differs between two bays that share the same regional drought but experience different freshwater withdrawals

Between-bay differences in drought and freshwater input.—To test for between-bay differences in precipitation, we obtained daily precipitation records from 1970 to 2016 from the National Climatic Data Center for three weather stations: the Apalachicola Regional Airport (GHCND: USC00080211), Panacea Florida (GHCND: USC00086828; a town adjacent to Ochlockonee Bay), and Smith Creek Florida (GHCND: USC00088290; a town adjacent to Ochlockonee Bay). Smith Creek data were available from 1970 to 1983, and Panacea data were available from 2000 to 2016; we combined the two to obtain a single data set for Ochlockonee Bay. For both bays, we summed daily precipitation totals at a monthly scale and calculated mean totals for August, September, and October of each year to produce an annual time series of average monthly precipitation rates for the late low-flow season, when bay salinities and drill abundance are highest. We focused on 1970–2016 because it was a period of pronounced growth in human population and water extraction from the ACF basin (Marella and Fanning 2011).

To evaluate river flow, we obtained data for 1970–2016 at Chattahoochee, Florida (Apalachicola River, USGS station 02358000) and Bloxham, Florida (Ochlockonee River, USGS station 02330000). The Chattahoochee station is 173 km above the mouth of the Apalachicola River, and flow rates at this station are highly correlated with downstream river flow into Apalachicola Bay (Morey et al. 2009, Petes et al. 2012). The Bloxham USGS station is 67.1 km north of Ochlockonee Bay. For both data sets, we averaged daily flow rates to generate a mean daily flow rate for each month. We then averaged the monthly means for August, September, and October, producing an average daily flow rate for the late low-flow season of each year in each river. To compare temporal trends in flow between the two rivers, we standardized the seasonal average flow in each year by its long-term average ([seasonal average − long-term average]/long-term average), producing a data set of seasonal flow anomalies in each year for each river.

Experiments.—We replicated three rounds of our experiment in Ochlockonee Bay: O1 (fall 2014; "O" for Ochlockonee), O2 (summer 2015), and O3 (fall 2015), which corresponded to the timing of experiments A3, A5, and A6, respectively (Table 1). Before these experiments, we assigned subtidal reefs to a zone based on their distance from the river mouth (close, mid, far; Fig. 1C). Because Ochlockonee Bay is smaller and essentially limited to one dimension (i.e., east–west), we did not partition its reefs further in either east–west or north–south dimensions. Experiments in Ochlockonee only used oysters collected in that bay, and all methods were identical to those of experiments A3–A7.

Data analysis for experiments O1–O3.—The analysis of oyster survival proceeded as in experiments A3–A7 and involved the use of GLM with binomial error and logit link. We detected a significant caging effect in experiment O2 (summer 2015), so those data were excluded from further analysis. Because we failed to detect an effect of initial oyster density in experiments O1 and O3, we excluded that factor from future consideration. Next, we combined the results of experiments A3, A6, O1, and O3, which were conducted simultaneously (Table 1), into a single data set and proceeded with the main analysis. This analysis used a GLM with a binomial error distribution to test whether final survival depended on the fixed factors of bay (Apalachicola or Ochlockonee), distance from the river, round, and treatment. Instead of using raw distance (km), we used proportional distance of each zone from the respective river because of the size difference between Apalachicola Bay and Ochlockonee Bay. For instance, the "close" zones in Apalachicola Bay and Ochlockonee Bay were 4.69 km and 5.01 km from the river, respectively. To calculate the proportional distance of the reefs in these zones, we divided their raw distances by the raw distances from the river of the farthest reefs in the "far" zones (e.g., close reefs in Apalachicola 4.69 km/14.49 km = 0.32). Finally, we combined the predation effect sizes for experiments O1 and O3 with those of experiments A3 and A6 and used linear regression to evaluate the relationship between predation strength (arcsine square root transformed) and water salinity. Because we lacked a hydrodynamic model for Ochlockonee Bay, we deployed Onset HOBO conductivity loggers (model no: U24-002) to record salinity and temperature at 15-min intervals. These instruments failed during experiment O3. As a result, the mean salinity for the Ochlockonee sites during experiment O3 consists of point samples that were collected at an average frequency of 8–10 d for each site (Table 1). All analyses were performed in R 3.4.1 (R Core Team 2016).

Results

Hypothesis I

In A1, oyster survival depended on an interaction between the east–west region of Apalachicola Bay and treatment ($P = 0.01$; Appendix S4; Fig. 2A), with a significant difference between cage and control treatments in the west but not in the east (Tukey's HSD, $P < 0.001$ and $P = 0.60$, respectively). In addition, survival in the cage and control treatments decreased significantly with distance (km) from the river ($P = 0.005$; Appendix S4; Fig. 2B). Predation strength was high across sites so there was no difference in predation with increasing distance from the river ($F_{1,4} = 1.53$, $P = 0.28$; Appendix S5). However, there was a positive relationship between predation strength and salinity ($F_{1,4} = 3976.9$, $P < 0.01$; $y = 0.05x - 0.54$; $R^2 = 0.99$; Appendix S5). In A1, dead oysters in the control treatments were not damaged by a crushing predator (e.g., stone crab). Therefore, predation was attributed to the abundant drills.

Prior to experiment A2, precipitation had caused 55–61% reductions in salinity across zones relative to salinities during experiment A1 (Appendix S2; Fig. 2C). We also observed decreased drill abundance, including 100% reductions in the zones close and mid-distance from the river, and a 93% reduction at zones far from the river (Appendix S2; Fig. 2D). In A2, oyster survival again depended on the interaction between east–west region and treatment ($P = 0.04$; Appendix S6); survival was significantly lower in controls than in cages in the west, but not in the east (Tukey's HSD, $P = 0.002$, $P = 0.94$, respectively; Fig. 2E), where survivorship was high in both controls and cages. In general, oyster survival in cages was higher (88–97%) in A2 than in A1 (55–75%). Oyster survival in A2 was also affected by an interaction between treatment and distance from the river ($P = 0.03$, Appendix S6; Fig. 2F). Close to the river, oyster survival in the cage treatments was no different than that in the control treatments. But with increased distance from the river, survival decreased more in the control treatments (Fig. 2F). This difference between the cages and controls represented predation strength, which was lower at nearly all sites relative to experiment A1 and strengthened with increasing water salinity ($F_{1,4} = 21.1$, $P = 0.01$; $R^2 = 0.84$; $y = 0.06x - 0.26$; Appendix S5), with all predation attributed to oyster drills. Over the course of both experiments, spatial variation in predation strength was correlated with increasing salinity ($F_{1,10} = 14.5$, $P = 0.003$; $R^2 = 0.59$; $y = 0.04x - 0.19$; Fig. 3A).

In experiments A3–A7, we detected predation effects that increased with distance from the river and correlated positively with increasing water salinity, except for experiment A6 (Appendix S7). In addition, 99%, 99%, 100%, 92%, and 100% of the predation in experiments A3–A7 (respectively) were attributed to oyster drills. Because experiments A3–A7 were conducted during a wider range of water temperatures, we also detected a seasonal temperature effect, with less predation in cooler water. Across all seven experiments, the relationship between predation strength and salinity depended on temperature ($F_{3,38} = 4.78$, $P = 0.0001$, $R^2 = 0.42$; salinity × temp coefficient, $P = 0.03$; Fig. 3B). During these experiments, the mean ± length of drills was 44.78 ± 12.75.

Hypothesis 2

There was no linear trend over time in precipitation in Apalachicola Bay ($F_{1,45} = 0.203$, $P = 0.65$, $R^2 = 0.004$) or Ochlockonee Bay ($F_{1,26} = 0.072$,

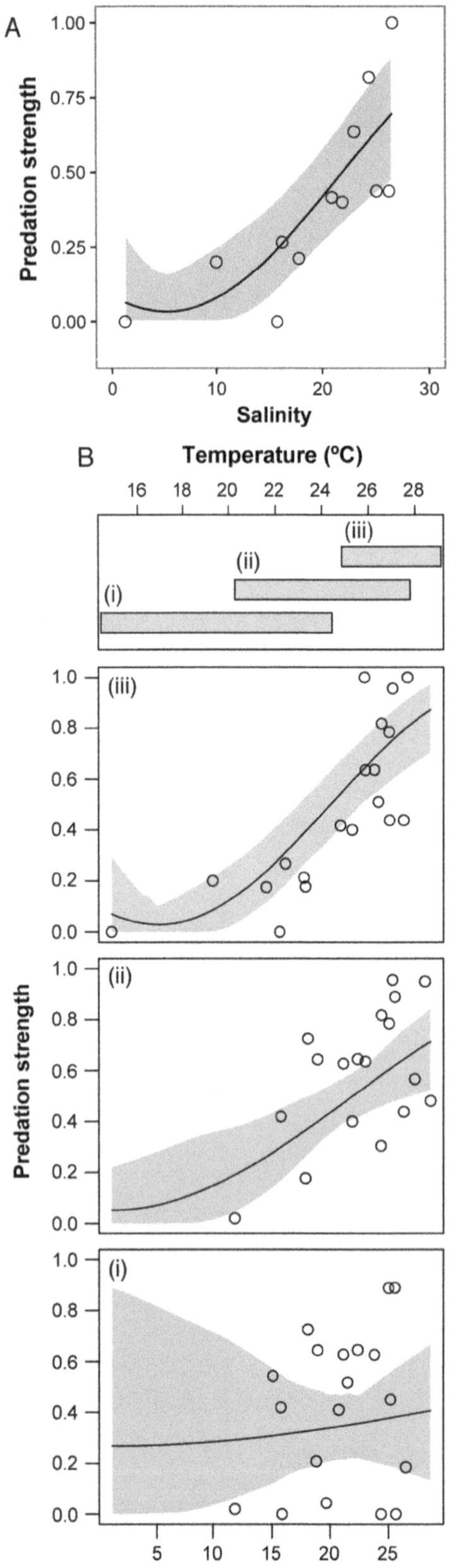

Fig. 3. (A) The relationship between the strength of predation (back-transformed) and salinity during experiments A1–A2. (B) Coplot of the relationship between predation strength (back-transformed) and salinity at three different ranges of temperature during experiments A1–A7.

$P = 0.79$, $R^2 = 0.003$). However, precipitation at the two bays in 1970–2015 was correlated (Pearson's correlation; $R = 0.65$, df = 26, $P = 0.0002$). The seasonal flow anomaly (log transformed) declined over time in Apalachicola Bay ($F_{1,45} = 9.83$, $P = 0.003$, $R^2 = 0.18$), but not in Ochlockonee Bay ($F_{1,45} = 2.18$, $P = 0.15$, $R^2 = 0.04$).

In experiments O1 and O3, 81% of oyster predation was attributed to the drill. Because our analysis of both data sets demonstrated no significant interaction between treatment and round (Appendix S8), we pooled the results of experiments A3, A6, O1, and O3 and detected a significant three-way interaction among treatment, estuary, and proportional distance from the river (Appendix S8). Oyster survival was lower in controls than in cages, and this difference (which reflected predation) increased with proportional distance from the river (Fig. 4A). In Apalachicola Bay, predation was statistically detectable throughout the entire estuary, while in Ochlockonee Bay, predation was not detectable in the upper half of the estuary. Across both experiments, predation increased significantly with increasing salinity (with outlier, $F_{1,16} = 14.93$, $P = 0.001$, $R^2 = 0.48$, $y = 0.05x - 0.58$; without outlier $F_{1,15} = 21.51$, $P = 0.0003$, $R^2 = 0.59$, $y = 0.06x - 0.66$; Fig. 4B). (One site in the high-salinity zone lacked drills during one experimental round, possibly in response to a temporary lack of ambient oysters.)

Discussion

Our results are consistent with the hypothesis that the oyster fishery collapse in Apalachicola Bay, Florida, was caused by an interaction between local- and regional-scale processes that facilitated a salinization-induced outbreak of predatory drills. In May 2013, we demonstrated that water salinity, drill abundance, and drill predation on oysters were high throughout the entire bay and increased with distance from the primary

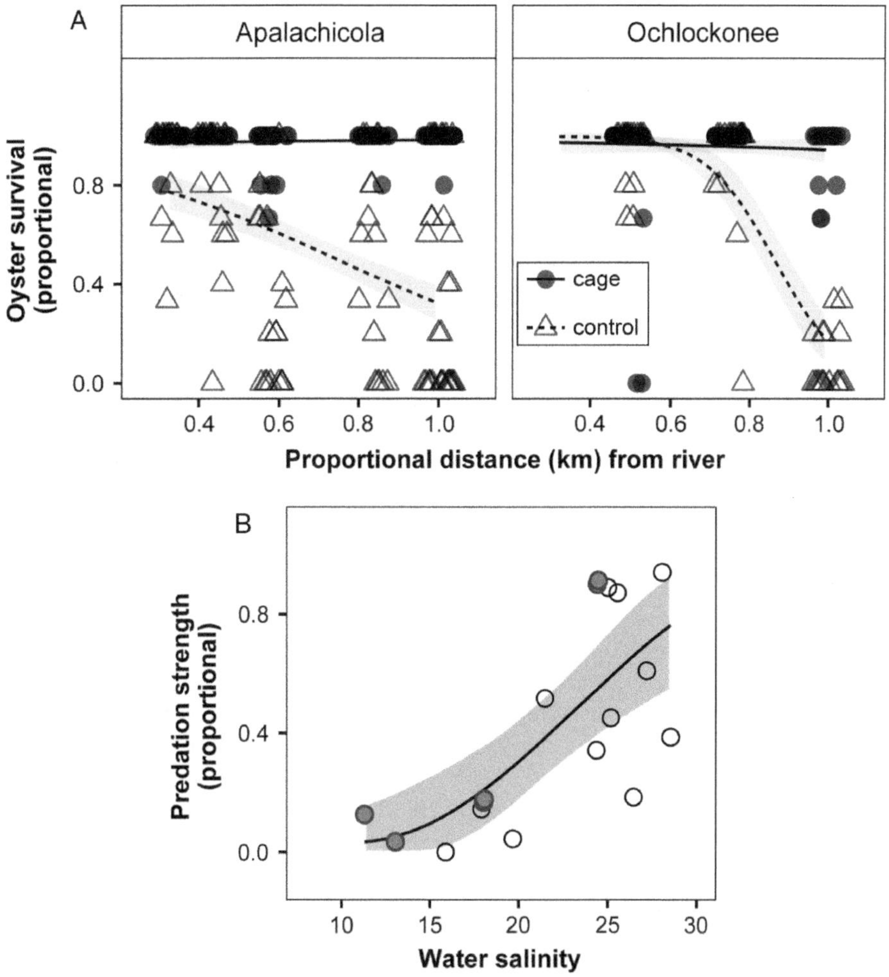

Fig. 4. (A) Results of experiments simultaneously conducted in Apalachicola Bay (left panel) and Ochlockonee Bay (right panel). In both panels, data represent oyster survival as a function of proportional distance from the river as well as cage (closed circles) and control (open triangles) treatments. (B) The relationship between the strength of predation (back-transformed) and salinity during the experiments in Apalachicola Bay (open circles) and Ochlockonee Bay (gray circles). Removal of outlier increased R^2 from 0.48 to 0.59.

source of freshwater, the Apalachicola River. By August 2013, intense precipitation had lowered salinity, and we detected baywide reductions in drill abundance and predation on oysters. These experiments suggested causal links among water salinity, predator abundance, foraging activity, and oyster mortality. By repeating this experiment over four years, we confirmed that predation on oysters consistently intensified with increasing salinity during non-winter months. Given that a regional-scale drought promoted unusually high salinity in Apalachicola Bay leading up to the 2013 oyster fishery collapse (Fig. 1D), it is reasonable to surmise that the fishery collapse could have been precipitated by a localized outbreak of predatory drills. Furthermore, results from a set of parallel observations and experiments in nearby Ochlockonee Bay revealed a lower intensity and extent of salinity-dependent predation. Therefore, a predator outbreak in Apalachicola Bay was driven by factors local to that bay, such as declining freshwater inflow due to upstream water withdrawal, in addition to regional factors affecting both bays, such as rainfall.

There was a similar positive relationship between salinity and drill predation in Ochlockonee

Bay and Apalachicola Bay. Further, predation on oysters intensified with increasing distance from the river (corresponding with increasing salinity) in both bays. These patterns support the general paradigm that stenohaline predators restrict oysters to lower-salinity areas of estuaries (Galtsoff 1964, Pollard 1973; Breithaupt and Dugas 1979, Wilber 1992, Kirby 2000). However, our experimental results also showed a key difference between the two bays: While drill predation occurred throughout all of Apalachicola Bay during warm summer months, 66% of the oyster reefs in Ochlockonee Bay (those nearest the Ochlockonee River) did not experience significant predation. Thus, river input and low salinity near the river mouth protected oysters from predation in Ochlockonee Bay, but not in Apalachicola Bay.

The presence of a predator refuge for oysters in Ochlockonee Bay—but not Apalachicola Bay—cannot be explained by differences in precipitation between the bays, because they experienced similar rainfall patterns over the past 45 yr. In contrast, there were differences in river flow into the two bays over the same period. The seasonal average of river flow during the driest portion of each year has declined since 1970 for the Apalachicola River, but not for the Ochlockonee River. Given similar precipitation conditions for the two bays, this difference in river flow stands out as a factor unique to Apalachicola Bay that would have increased salinity, particularly near the river mouth and in turn facilitated higher drill predation in the upper bay than that observed in Ochlockonee Bay. Because upstream water withdrawals in the ACF watershed have increased dramatically since 1970 (Marella and Fanning 2011), it is possible that water withdrawals interacted with the regional-scale drought to promote a salinization-induced outbreak of predatory drills in Apalachicola Bay and in turn the oyster fishery collapse.

While this conclusion is supported by our circumstantial evidence from monitoring and experiments, the conclusion would be more strongly supported if our results were based on a BACI experimental approach (Underwood 1993). Such an approach would have required research in both bays to have begun before the observed increase in upstream water withdrawal, maintained into the present, thus providing a consistent time series of when predator outbreaks developed relative to temporal changes in upstream water withdrawals, water salinity, and the oyster fishery. Then, if results demonstrated that a salinization-induced outbreak of predatory drills intensified more in Apalachicola Bay than in Ochlockonee Bay in 2012, we could more directly attribute the fishery collapse to upstream water withdrawals. Because our research began during the collapse, we adopted a space-for-time substitution approach, which assumes that the causes of spatial differences in predation between the two bays also reflect the causes of changes in predation within Apalachicola Bay over time. This can be a powerful approach, particularly when "before" data are unavailable. For example, in a recent test, a space-for-time approach was 70% as accurate as a before–after approach in predicting how the taxonomic composition of plant communities will be altered by climate change (Blois et al. 2013). A similar approach could be used to investigate multi-scale causes of sudden ecological shifts (e.g., predator outbreak) in other systems that also lack the requisite data to support a BACI approach.

Even if upstream withdrawals caused a predator outbreak by intensifying salinization, it remains unclear whether predation was the primary cause of the oyster fishery collapse. For instance, less freshwater input and higher salinity could also have increased the prevalence of the Dermo pathogen (Hoffmann et al. 1995, Powell et al. 2003). In fact, low oyster survival in protective cages during our first experiment suggests that disease may have also been a major contributor to mortality during the fishery collapse. However, all subsequent experiments failed to show a strong influence of disease. Less freshwater may have also decreased the nutrient supply needed by phytoplankton, which are ultimately consumed by oysters, thereby essentially starving oysters. Commercial fishing is another potential cause of oyster decline. Despite the management of the Apalachicola Bay oyster fishery being commended for facilitating one of the most productive and sustainable oyster fisheries worldwide (Zu Ermgassen et al. 2012), fishing has historically been a driver of oyster losses in north and mid-Atlantic estuaries (Kirby 2004). Thus, while pairing observations with field experiments—as we have done—is a powerful approach, there are limitations to the numbers and types of factors that can be manipulated. Furthermore, if predation, recruitment, growth, and harvesting fluctuate over time, then short-term experiments may not accurately reveal factors that shaped population-scale patterns of oysters. As Hastings and Boettiger (2013) argued, mechanistic models are required to reveal how causal factors interact and to predict the conditions under which large-scale systems or populations are likely to cross a dynamic threshold into a state of degradation. Consequently, our next research step will involve the integration of our empirical data (monitoring and experiments) with a mathematical model to produce the most reliable inference for the cause of the oyster fishery collapse.

If the combination of empirical data and modeling determines that salinity-induced predation was the primary cause of the oyster fishery collapse, then it behooves us to understand the mechanism by which salinity affects drills. Salinity could control drill abundance by inhibiting snail reproduction, if like *Melongena corona*, larvae of *Stramonita haemastoma* fail to metamorphose at reduced salinities (Garland and Kimbro 2015). Salinity could also exert population-scale control if sudden salinity reductions—such as the intense precipitation event of July 2013—overwhelm the osmoregulatory capacity of the drill and cause death or non-lethal effects that reduce their local populations (e.g., egress from oyster reefs). The ability of oysters to tolerate such conditions more than drills would align with the environmental stress model (Menge and Sutherland 1976) as well as the long-held (but previously untested) suggestion that oysters use inner portions of estuaries as refuge from stenohaline predators (Galtsoff 1964, Pollard 1973, Breithaupt and Dugas 1979, Wilber 1992, Kirby 2000). Less severe reductions in salinity could also operate at the individual level by causing non-lethal stress that impairs drill foraging. A combination of laboratory and field experiments is also needed to evaluate how realistic variation in water salinity influences drill populations and drill behavior.

It is clear that natural systems are hierarchically organized by forces operating at local and regional scales (Gouhier et al. 2010, Menge et al. 2015). Indeed, our study demonstrated that the development of a predator outbreak associated with the decline of an oyster fishery required both regional-scale and localized stressors in Apalachicola Bay. Furthermore, our study highlights how experiments and observations can be combined to disentangle the relative influence of regional- and local-scale forces even in the absence of a traditional BACI approach. Although predator outbreaks are a natural phenomenon observed in a diversity of systems (Silliman et al. 2013), their frequency and duration are predicted to increase with global climate change and the intensification of environmental stress. Accordingly, predicting and mitigating outbreaks fundamentally depends on the development of theoretical and empirical research approaches that account for the effects of multiple processes operating at local and regional scales.

Acknowledgments

We thank the staff of Apalachicola National Estuarine Research Reserve, T. Hanley, and C. Commander for field assistance and logistical support. We also thank T. Ward for access to oyster reefs on his private lease in Apalachicola and S. Hartsfield for facilitating initial research efforts and site selection. The staff and facilities at the Florida State University Coastal and Marine Laboratory were instrumental in our research. We thank L. Edmiston, G. Lewis, P. Montagne, and C.H. Peterson for advice during the design of the project. This project was funded by the Florida Department of Environmental Protection (award no. 1338372 to DLK) and Florida Sea Grant (award no. 0000023910 to DLK). This is contribution no. 357 from the Northeastern University Marine Science Center.

Literature Cited

Beck, M. W., et al. 2011. Oyster reefs at risk and recommendations for conservation, restoration, and management. BioScience 61:107–116.

Blois, J. L., J. W. Williams, M. C. Fitzpatrick, S. T. Jackson, and S. Ferrier. 2013. Space can substitute for time in predicting climate-change effects on biodiversity. Proceedings of the National Academy of Sciences 110:9374–9379.

Breithaupt, R. L., and R. J. Dugas. 1979. A study of the southern oyster drill (*Thais haemastoma*) distribution and density on the oyster seed grounds. Technical Bulletin 30:1–20.

Cook, B. I., R. L. Miller, and R. Seager. 2009. Amplification of the North American "Dust Bowl" drought through human-induced land degradation. Proceedings of the National Academy of Sciences USA 106:4997–5001.

Fabricius, K. E., K. Okaji, and G. De'ath. 2010. Three lines of evidence to link outbreaks of the crown-of-thorns seastar *Acanthaster planci* to the release of larval food limitation. Coral Reefs 29:593–605.

Florida Department of Agriculture and Consumer Services. 2012*a*. Oyster Assessment Report. http://www.floridaaquaculture.com/publications/Oyster_Res_Assessment_Report_August_2012.pdf

Florida Department of Agriculture and Consumer Services (FDACS). 2012*b*. Shellfish Harvesting Area Maps. http://www.freshfromflorida.com/Business-Services/Aquaculture/Shellfish-Harvesting-Area-Classification/Shellfish-Harvesting-Area-Maps

Florida Fish and Wildlife Conservation Commission (FFWCC). 2013. 2012–2013 Florida Gulf coast oyster disaster report. http://www.floridajobs.org/docs/default-source/2015-community-development/2015-cmty-plan-acsc/20122013floridagulfcoastoysterdisasterreport.pdf?sfvrsn=2

Florida Department of Environmental Protection's (FDEP) Office of Coastal and Aquatic Managed Areas (CAMA). 2013. Apalachicola national estuarine research reserve management plan. https://www.doi.gov/sites/doi.gov/files/migrated/deepwaterhorizon/adminrecord/upload/Apalachicola-National-Estuarine-Research-Reserve-Management-Plan-undated.pdf

Florida v. Georgia. 2014. Complaint – Florida v. Georgia, No. 142, Original. http://www.pierceatwood.com/floridavgeorgia142original

Galtsoff, P. S. 1964. The American oyster *Crassostrea virginica* Gmelin. Bulletin 64. U.S. Department of the Interior, Fish and Wildlife Service, Bureau of Commercial Fisheries, United States Government Printing Office, Washington, D.C., USA.

Garland, H. G., and D. L. Kimbro. 2015. Drought increases consumer pressure on oyster reefs in Florida, USA. PLoS ONE 10:e0125095. https://doi.org/10.1371/journal.pone.0125095

Gouhier, T. C., F. Guichard, and A. Gonzalez. 2010. Synchrony and stability of food webs in metacommunities. American Naturalist 175:E16–E34.

Grabowski, J. H., R. D. Brumbaugh, R. F. Conrad, A. G. Keeler, J. J. Opaluch, C. H. Peterson, M. F. Piehler, S. P. Powers, and A. R. Smyth. 2012. Economic valuation of ecosystem services provided by oyster reefs. BioScience 62:900–909.

Hastings, A., and C. Boettiger. 2013. From patterns to predictions. Nature 493:157–158.

Heinze, G., and M. Ploner. 2002. SAS and SPLUS programs to perform Cox regression without convergence problems. Computer Methods and Programs in Biomedicine 67:217–223.

Hoffman, E. E., E. N. Powell, J. M. Klinck, and G. Saunders. 1995. Modeling *Perkinsus marinus* infections in oysters. Journal of Shellfish Research 14:121–151.

Kaul, L. W., and P. N. Froelich Jr. 1984. Modeling estuarine nutrient geochemistry in a simple system. Geochimica et Cosmochimica Acta 48:1417–1433.

Kirby, M. X. 2000. Paleoecological differences between Tertiary and Quaternary *Crassostrea* oysters, as revealed by stable isotope sclerochronology. Palaios 15: 132–141.

Kirby, M. X. 2004. Fishing down the coast: historical expansion and collapse of oyster fisheries along continental margins. Proceedings of the National Academy of Sciences of the USA 101:13096–13099.

Kurz, W. A., C. C. Dymond, G. Stinson, G. J. Rampley, E. T. Neilson, A. L. Carroll, T. Ebata, and L. Safranyik. 2008. Mountain pine beetle and forest carbon feedback to climate change. Nature 452:987–990.

Livingston, R. J., et al. 2000. Modelling oyster population response to variation in freshwater input. Estuarine, Coastal and Shelf Science 50:655–672.

Marella, R. L., and J. L. Fanning. 2011. Water withdrawals, wastewater discharge, and water consumption in the Apalachicola-Chattahoochee-Flint River Basin, 2005, and water-use trends, 1970–2005. U.S. Geological Survey Scientific Investigations Report 2011–5130:1–34.

Menge, B. A., T. C. Gouhier, S. D. Hacker, F. Chan, and K. J. Nielsen. 2015. Are meta-ecosystems organized hierarchically? A model and test in rocky intertidal habitats. Ecological Monographs 85: 213–233.

Menge, B. A., and J. P. Sutherland. 1976. Species diversity gradients: synthesis of the roles of predation, competition, and temporal heterogeneity. American Naturalist 110:351–369.

Menge, B. A., et al. 2003. Coastal oceanography sets the pace of rocky intertidal community dynamics. Proceedings of the National Academy of Sciences 100:12229–12234.

Menzel, R. W., N. C. Hulings, and R. R. Hathaway. 1966. Oyster abundance in Apalachicola Bay, Florida in relation to biotic associations influenced by salinity and other factors. Gulf and Caribbean Research 2:73–96.

Morey, S. L., D. S. Dukhovskoy, and M. A. Bourassa. 2009. Connectivity of the Apalachicola River flow variability and the physical and bio-optical oceanic properties of the northern West Florida Shelf. Continental Shelf Research 29:1264–1275.

Mortazavi, B., R. L. Iverson, W. R. Huang, F. G. Lewis, and J. M. Caffrey. 2000. Nitrogen budget of Apalachicola Bay, a bar-built estuary in the northeastern Gulf of Mexico. Marine Ecology Progress Series 195:1–14.

Packer, C., R. D. Holt, P. J. Hudson, K. D. Lafferty, and A. P. Dobson. 2003. Keeping the herds healthy and alert: implications of predator control for infectious disease. Ecology Letters 6:797–802.

Paine, R. T. 1966. Food web complexity and species diversity. American Naturalist 100:65–75.

Petes, L. E., A. J. Brown, and C. R. Knight. 2012. Impacts of upstream drought and water withdrawals on the health and survival of downstream estuarine oyster populations. Ecology and Evolution 2:1712–1724.

Pollard, J. F. 1973. Experiments to reestablish historical oyster seed grounds and to control the southern oyster drill. Technical Bulletin 6. Louisiana Wildlife and Fisheries Commission, Baton Rouge, Louisiana, USA.

Powell, E. N., J. M. Klinck, E. E. Hofmann, and M. A. McManus. 2003. Influence of water allocation and freshwater inflow on oyster production: a hydrodynamic–oyster population model for Galveston Bay, Texas, USA. Environmental Management 31:100–121.

Putland, J. N., B. Mortazavi, R. L. Iverson, and S. W. Wise. 2013. Phytoplankton biomass and composition in a river-dominated estuary during two summers of contrasting river discharge. Estuaries and Coasts 37:664–679.

R Core Team. 2016. R: a language and environment for statistical computing. R Foundation for Statistical Computing, Vienna, Austria.

Scheffer, M., and S. R. Carpenter. 2003. Catastrophic regime shifts in ecosystems: linking theory to observation. Trends in Ecology and Evolution 18: 648–656.

Schmitz, O. J. 2008. Effects of predator hunting mode on grassland ecosystem function. Science 319:952–954.

Sheng, Y. P., B. Tutak, J. R. Davis, and V. Paramygin. 2008. Circulation and flushing in the lagoonal system of the Guana Tolomato Matanzas National Estuarine Research Reserve (GTMNERR), Florida. Journal of Coastal Research SI55:9–25.

Sih, A. 1984. Optimal behavior and density-dependent predation. American Naturalist 123:314–326.

Silliman, B. R., M. W. McCoy, C. Angelini, R. D. Holt, J. N. Griffin, and J. van de Koppel. 2013. Consumer fronts, global change, and runaway collapse in ecosystems. Annual Review of Ecology, Evolution, and Systematics 44:503–538.

Trussell, G. C., P. J. Ewanchuk, and C. Matassa. 2006. Habitat effects on the relative importance of trait- and density-mediated indirect interactions. Ecology Letters 9:1245–1252.

Twichell, D. C., B. D. Andrews, H. L. Edmiston, and W. R. Stevenson. 2007. Geophysical mapping of oyster habitats in a shallow estuary, Apalachicola Bay, Florida. Estuarine Coastal and Shelf Science 88:385–394.

Underwood, A. J. 1993. The mechanics of spatially replicated sampling programs to detect environmental impacts in a variable world. Australian Journal of Ecology 18:99–116.

Wilber, D. H. 1992. Associations between freshwater inflows and oyster productivity in Apalachicola Bay, Florida. Estuarine, Coastal and Shelf Science 35:179–190.

Zu Ermgassen, P. S. E., et al. 2012. Historical ecology with real numbers: past and present extent and biomass of an imperilled estuarine habitat. Proceedings of the Royal Society B 279:3393–3400.

5

The effect of lizards on spiders and wasps: variation with island size and marine subsidy

Jonah Piovia-Scott,[1,†] Louie H. Yang,[2] Amber N. Wright,[3] David A. Spiller,[4] and Thomas W. Schoener[4]

[1]*School of Biological Sciences, Washington State University, 14204 NE Salmon Creek Avenue, Vancouver, Washington 98686 USA*
[2]*Department of Entomology and Nematology, University of California, One Shields Avenue, Davis, California 95616 USA*
[3]*Department of Biology, University of Hawaii, Manoa, 2500 Campus Road, Honolulu, Hawaii 96822 USA*
[4]*Department of Evolution and Ecology, University of California, One Shields Avenue, Davis, California 95616 USA*

Abstract. Introduced predators can have dramatic effects on island ecosystems, the magnitude of which are likely to vary with island characteristics. We investigated the influence of two important properties of islands—size and amount of resource subsidy—on the effects of an introduced predatory lizard (*Anolis sagrei*) on three groups of arthropod prey. Lizards were experimentally introduced to 16 islands that spanned gradients in vegetated area and seaweed deposition (a marine resource subsidy); 16 similar islands served as lizard-free controls. The abundance of web spiders, salticid spiders, and wasps was estimated prior to lizard introduction and again four months after lizard introduction. Lizard introduction reduced the average abundance of all three groups of arthropods. The effect of lizards on salticid spiders—which was very large (94% reduction in salticid abundance)—decreased with island size. In contrast, the effect of lizards on wasps—which was also very large (88% reduction in wasp abundance)—tended to increase with island size, but with only marginal significance. There was no evidence for variation in the effect of lizards on web spiders with island size. This variation between prey taxa may be related to the relative importance of environmental stress (such as wind and wave exposure, which tend to be more pronounced on smaller islands) in determining abundance. Salticids seem to tolerate the stressful environmental conditions that characterize smaller islands, allowing for larger lizard effects; wasps seem to be limited by these conditions (either directly or indirectly via reduced prey availability), minimizing lizard effects on smaller islands. There was a marginally significant tendency for the effect of lizards on salticid spiders to be weaker on islands with more seaweed deposition, suggesting that subsidies may play a role in reducing predator effects on islands. Our results highlight the importance of ecological context in determining the top-down effects of introduced predators and underscore the need to extend existing theories relating island area and community characteristics toward an explicit consideration of species interactions.

Key words: context dependence; food web; predator–prey interactions; resource subsidy; top-down control.

† **E-mail:** jonah.piovia-scott@wsu.edu

Introduction

Introduced predators can have dramatic effects on ecological communities (Salo et al. 2007, Simberloff 2013). While there is increasing recognition that the strength of species interactions depends on environmental context (Bronstein 1994, Agrawal et al. 2007, Afkhami et al. 2014, Hoeksema and Bruna 2015), our understanding of how habitat characteristics influence the impact of introduced predators is still being developed (Baxter et al. 2004, Snyder and Evans 2006). The top-down effects of introduced predators are often particularly pronounced on islands (Schoener and

Spiller 1996, Wiles et al. 2003, Croll et al. 2005, Fukami et al. 2006, Rogers et al. 2012). For island food webs, two of the most important axes of environmental variation are island size (Schoener 1989, Holt et al. 1999, Takimoto et al. 2008, Holt 2010) and resource subsidies, that is, resources that move into the island ecosystem from the surrounding aquatic habitat (Polis and Hurd 1996*b*, Croll et al. 2005, Fukami et al. 2006, Young et al. 2010). Here, we use experimental introductions of predatory lizards on small islands to investigate the influence of island size and resource subsidies on predator–prey interactions.

Relatively few studies have investigated how the strength of predator–prey interactions changes across gradients of island size. Elton (1958) argued that invasive species have greater impacts on island food webs than those of continental ecosystems, and Ebenhard (1988) found that invasive mammals and birds tend to have larger impacts on prey on oceanic islands than on continental shelf islands and continents. More recently, mathematical modeling studies have shown that the top-down effects of consumers may be more pronounced in smaller habitat patches, primarily because the capacity for spatial heterogeneity to stabilize consumer–resource interactions is lost (McCann et al. 2005, Tunney et al. 2012). While there is some empirical support for larger predator effects on smaller islands (Lomolino 1984, Schoener and Spiller 1999*b*, 2010), Schoener et al. (2016) recently proposed a hump-shaped relationship between island area and the strength of predator effects. In their conceptual model, interaction strength on large islands is diminished primarily by biological complexity (e.g., compensating predators, which can limit the effect of any single type of predator by maintaining lower prey abundance) and spatial heterogeneity (which can increase the availability of prey refugia), while interaction strength on small islands is diminished by stressful environmental conditions, such as increased exposure to wind, waves, and storm surges (which can limit both predator and prey abundance), demographic stochasticity (which can decouple abundance from ecological interactions), and resource subsidies from surrounding habitats (which can provide alternative resources for predators).

There has been increasing recognition that resource subsidies influence food-web processes. For example, subsidies can have major impacts on the top-down effects of predators in the recipient community (Polis et al. 2004, Piovia-Scott et al. 2011). Subsidies can both increase and decrease the strength of these predator effects (Takimoto et al. 2009). The impact of subsidies on top-down effects of predators effects can depend on which trophic level(s) consume the subsidy (Leroux and Loreau 2008), as well as the type of predator response to subsidy—increases in predator effects are usually driven by increases in predator density (Polis and Hurd 1996*a*, Henschel et al. 2001, Murakami and Nakano 2002), while decreases are usually driven by changes in predator diet and behavior (Nakano et al. 1999, Sabo and Power 2002, Baxter et al. 2005, Spiller et al. 2010). In short-term studies where predator behavior is likely to be more important than predator numerical responses, subsidies are expected to decrease the effect of predators on in situ prey by causing the predators to focus on the alternate resource (Takimoto et al. 2009, Piovia-Scott et al. 2011).

To investigate the influence of island area and resource subsidies on the strength of top-down effects, we performed a predator-introduction experiment across gradients of island area and seaweed deposition in a Bahamian island system. We focus on the common carnivorous lizard *Anolis sagrei*, and the term "lizard" will hereafter refer to this particular species. We evaluated the effects of lizards on three groups of arthropods: web spiders, salticid spiders, and wasps. The effects of lizards on web spiders have been extensively documented in this system (Spiller and Schoener 1988, 1994) and vary with island size (Schoener and Spiller 1999*b*, 2010, Schoener et al. 2016), showing a positive relationship with island area over the range of island sizes used in the current study (Schoener et al. 2016). Reduced lizard effects on web spiders on the smallest islands are thought to be due to marine subsidies, stressful abiotic conditions, and demographic stochasticity, but the relative importance of these factors is unknown. By decoupling island size from seaweed deposition, the current study will shed light on the role of subsidies in driving the relationship between island size and lizard effects. Salticid spiders have not been the focus of previous studies in this system but are also likely to be strongly affected by lizard predation because they (1) occupy the same habitats as lizards (e.g., stems, branches, ground), (2) are a suitable size for consumption by lizards, and (3) are active during the day. We expect the effect of lizards on salticids to be constrained by the same factors influencing lizard effects on web spiders. The effect of lizards on parasitoid wasps is more equivocal, with positive or neutral relationships between lizards and wasps observed in some cases (Schoener et al. 1995, Schoener and Spiller 1999*a*, Spiller and Schoener 2007). Parasitoid wasp abundance increases with island area (Schoener et al. 1995), perhaps due to more pronounced effects of environmental disturbance on smaller islands. Thus, we expected the effects of lizards on wasps to increase in strength with island size.

Our study was abruptly terminated when Hurricane Sandy, a Category 1 storm, hit our field site approximately five months after lizards were introduced (and one month after our first post-introduction survey). Hurricane Sandy extirpated all of the introduced lizard populations and most of the web spiders (L. H. Yang et al., *unpublished data*). Hence, we report lizard effects over a ~4-month time period. The relatively short duration of this study is important, as certain lizard effects (e.g., on web spiders, Schoener and Spiller 1996, 1999*a*) can take longer to develop. In addition, changes in lizard behavior and diet in response to seaweed deposition occur rapidly (on the scale of days or months), while reproductive responses generally take a year or longer (Spiller et al. 2010, Wright et al. 2013; A. N. Wright et al., *unpublished data*), suggesting that seaweed is likely to reduce the top-down effects of lizards in the current study (as in Piovia-Scott et al. 2011).

Methods

Study system and experimental design

Our study was conducted on 32 small islands (vegetated area: 4–157 m^2; Appendix S1: Table S1, Data S1) fringing the much larger island of Great Abaco, Bahamas, between May and September 2012. Importantly, seaweed deposition and island size were not strongly correlated on the set of islands used in our study ($\rho = -0.16$), allowing us to disentangle the relative impacts of these environmental drivers on lizard effects. At the beginning of our experiment, none of the study islands had lizards on them, although lizards are known to occur sometimes on similar islands in this area (Schoener et al. 2001). Half of the 32 islands were randomly selected for lizard introductions, while the other half served as lizard-free controls. In order to ensure that lizard treatments were evenly distributed across the gradient of island sizes, we established four blocks of eight islands, each with similarly sized islands, and randomly assigned four of the eight islands in each block to the lizard-introduction treatment. We introduced lizards to each island in May 2012. In order to maintain roughly equivalent lizard densities on all lizard-introduction islands, the number of lizards introduced to each island scaled with vegetated area (roughly 0.2 lizards/m^2), with a female-to-male target ratio of 2:1 (mean across all lizard-introduction islands: 2.001:1). For islands with very small vegetated area, we introduced a minimal population of colonists consisting of two females and one male lizard. Because we did not introduce fewer than three lizards to any island, the smallest block of islands had higher lizard densities than the other three blocks (Appendix S1). Because of this, we repeated the analyses described below with a dataset from which we had removed the smallest block of islands. The results of this set of analyses were generally similar to those with all islands (Appendix S1: Table S2); we indicate in the results where the significance of a particular test (for $\alpha = 0.05$) differed between the two sets of analyses.

We used vegetated area as our measure of island size. We calculated vegetated area as the product of total area and the proportion of the total area that was vegetated for each island. We estimated total area by measuring the length and width of each island and assuming islands were elliptical in shape (total area = $0.25 \times \text{length} \times \text{width} \times \pi$), and we estimated proportion vegetation by taking the mean of at least two independent (different people) visual estimates. We used visual estimates to measure seaweed biomass (Spiller et al. 2010) on each island in May and September 2012; these estimates were divided by vegetated area to calculate seaweed density. We used the mean of the May and September seaweed density estimates to assign a seaweed deposition category: low seaweed (0–0.15 kg/m^2) or moderate seaweed (0.15–0.80 kg/m^2); we call this second category moderate because much higher levels of seaweed deposition (>5 kg/m^2) have been observed in this system (Spiller et al. 2010). We chose the threshold of 0.15 kg/m^2 dividing low and moderate seaweed deposition because it effectively distinguished the islands with notably higher seaweed density and because it created a balanced experimental design—while we did not take seaweed deposition into account when we assigned lizard treatments, we did end up with an even split between lizard and no-lizard islands in both the moderate seaweed category (four lizard, four no-lizard) and the low seaweed category (12 lizard, 12 no-lizard; Appendix S1: Table S1).

Data collection

We conducted web-spider censuses on each island by counting the number of intact webs present and recording species identity. We estimated salticid spider abundance using bowl traps. Yellow plastic bowls filled with 500 mL of water and a trace amount of detergent were placed on the ground in vegetated areas of each island. After 24 h, the traps were collected, and the number of salticid spiders was counted. In order to attain representative samples from each island, we set more traps on larger islands (approximately one trap per 15 m^2 of vegetated area). We estimated wasp abundance using sticky traps—22 × 14 cm sheets of clear plastic coated with Tangle-Trap adhesive (Contech, Victoria, British Columbia, Canada). Sticky traps were tied onto the branches of shrubs about 0.5 m above ground. After 24 h, the traps were collected, and the number of wasps was counted. The number of sticky traps set on each island was equal to the number of bowl traps.

Statistical analyses

We evaluated how the effect of lizard introduction on the abundance of salticid spiders, wasps, and web spiders was influenced by vegetated area and seaweed density using generalized linear models. The models featured log(vegetated area), seaweed density category (moderate vs. low), and lizard treatment (introduced vs. absent) as predictors. In order to test for effects of island area and seaweed deposition on the strength of lizard predation, we evaluated lizard × log(vegetated area) and lizard × seaweed interactions. Because arthropod abundances were represented by counts, we used generalized linear models with a log link function, a Poisson error distribution, and an offset term for sampling effort (number of traps for salticid spiders and wasps, island area for web spiders). This approach allows for analyses of abundance per unit sampling effort (i.e., density) without the biases introduced by using a ratio as a response variable and features an error distribution that is appropriate for count data. For hypothesis testing, we used chi-squared likelihood ratio tests. All analyses were performed using R version 3.3.2 (R Development Core Team 2016); raw data and code are available in the supporting information (Data S1 and Data S2). We used Cameron and Trivedi's (1990) method to test for overdispersion. We did not find evidence for overdispersion in the analyses of salticid spiders and wasps ($P > 0.26$ in both cases), but there was evidence for overdispersion in the analysis of web spiders ($P = 0.02$). Because of this, we added an island-level random effect to the model of web spiders (Harrison 2014); the model with an island-level random effect had a better Akaike's information criterion score than a zero-inflated model.

In order to determine whether or not there were pre-existing differences in salticid spider, wasp, and web-spider abundance between islands selected for lizard introductions and control islands, we used the statistical approach described above to model abundances in May 2012, prior to lizard introductions. While the May 2012 model included lizard treatment, vegetated area, and seaweed category as predictors, it did not include interactions between lizard treatment and the other predictor variables, as the lizard treatment had not yet taken place. This analysis found no evidence for pre-existing differences in arthropod abundance between islands selected for lizard introduction and controls ($P > 0.34$ in all cases).

Results

Lizard introduction led to a 94% reduction in salticid spider abundance (Fig. 1; lizard effect: $\chi^2 = 13.1$, df = 1, $P = 0.003$; log response ratio = −2.9). The effect of lizards was greater on smaller islands than it was on larger islands (Fig. 2; lizard × area interaction: $\chi^2 = 5.2$, df = 1, $P = 0.02$). This change in lizard effect with island area was associated with reduced salticid abundance on larger islands in the absence of lizards (area

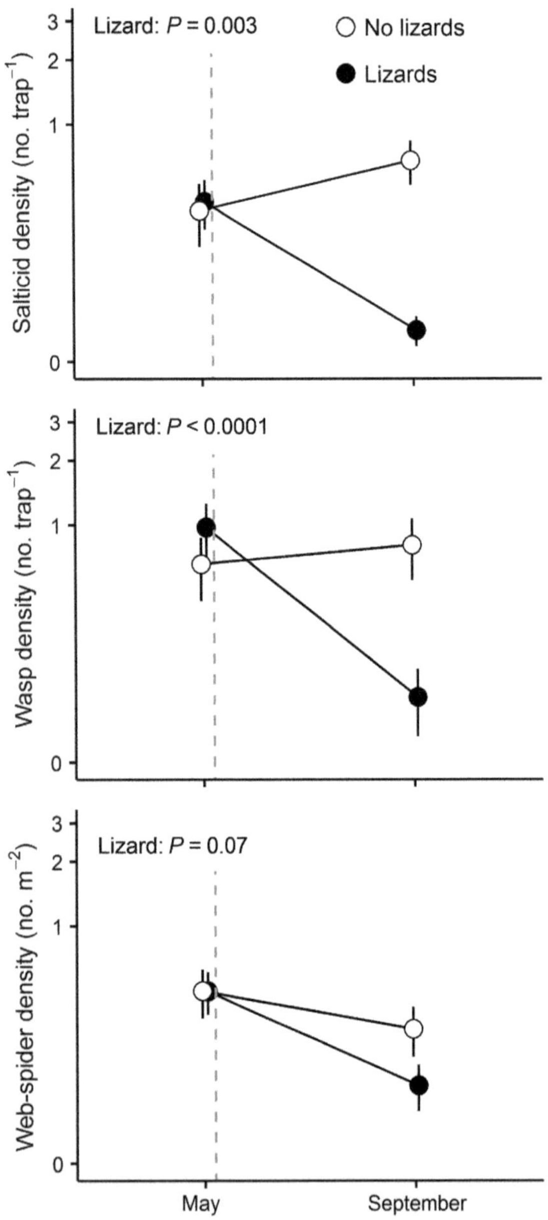

Fig. 1. Effects of lizards on the mean abundance of salticid spiders, wasps, and web spiders. Means and standard errors are shown for 16 islands to which lizards were introduced (solid circles) and 16 no-lizard control islands (open circles) in May 2012 (just prior to lizard introduction) and September 2012 (four months after lizard introduction). The dashed vertical line denotes the date of lizard introduction. *P*-values for lizard effects are from models described in the text. Note log scale on *y*-axis.

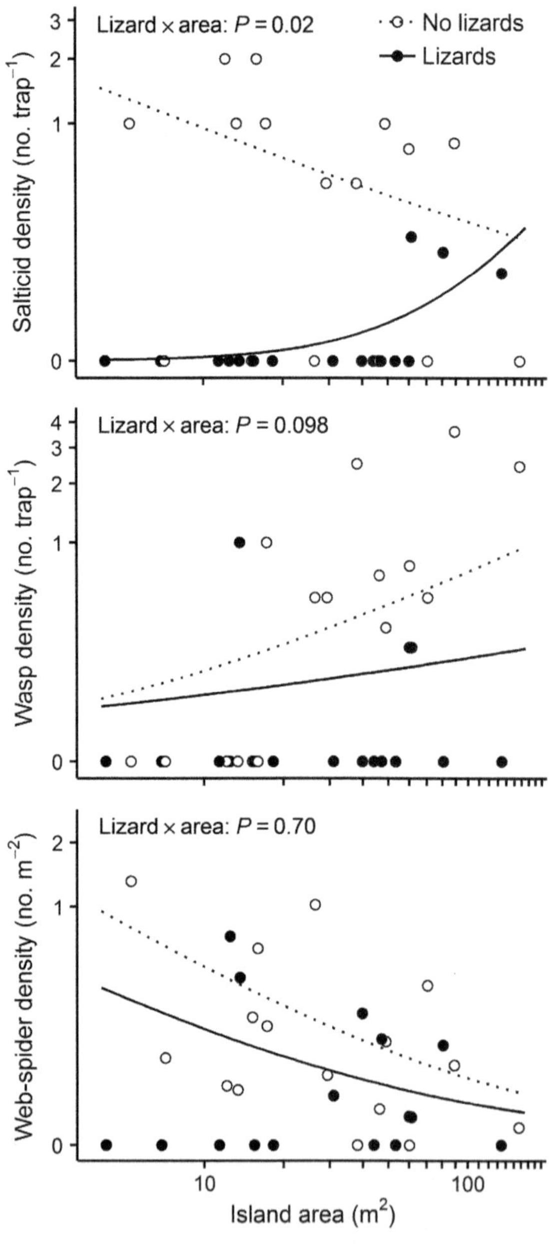

Fig. 2. Effects of lizards on the abundance of salticid spiders, wasps, and web spiders across gradients of island size. The symbols represent individual islands (solid circles for lizard-introduction islands and open circles for no-lizard islands); *P*-values for the lizard × area interaction and prediction lines are derived from generalized linear models with a log link function and a Poisson error distribution as described in the text. Note log scale on both axes.

effect in the absence of lizards: $\chi^2 = 5.1$, df = 1, $P = 0.02$) and a tendency for higher salticid abundance on larger islands in the presence of lizards, although this latter result was only marginally significant (area effect in the presence of lizards: $\chi^2 = 3.1$, df = 1, $P = 0.08$). There was also a marginally significant trend for the effect of lizards on salticids to be greater on islands with lower seaweed deposition (Fig. 3; lizard × seaweed interaction: $\chi^2 = 3.3$, df = 1, $P = 0.07$; lizard effect on low seaweed islands: $\chi^2 = 14.4$, df = 1, $P = 0.0001$; lizard effect on moderate seaweed islands: $\chi^2 = 0.15$, df = 1, $P = 0.70$).

Lizard introduction led to an 88% reduction in wasp abundance (Fig. 1; lizard effect: $\chi^2 = 41.9$, df = 1, $P < 0.0001$; log response ratio = −2.1). There was a marginally significant tendency for greater lizard effects on large islands (Fig. 2; lizard × area interaction: $\chi^2 = 2.7$, df = 1, $P = 0.098$); this effect was significant ($P = 0.039$) when the smallest block of islands was removed from the analysis (Appendix S1: Table S2). Wasp density was higher on larger islands (Fig. 2; area effect: $\chi^2 = 16.7$, df = 1, $P < 0.0001$), and there was a marginally significant tendency for higher wasp density on islands with low seaweed deposition (Fig. 3; seaweed effect: $\chi^2 = 3.4$, df = 1, $P = 0.06$). The lizard × seaweed interaction was not significant for wasps ($\chi^2 = 0.9$, df = 1, $P = 0.35$).

Lizard introduction led to a 58% reduction in web-spider abundance (Fig. 1; lizard effect: $Z = 1.8$, $P = 0.07$; log response ratio = −0.88); this effect was significant ($P = 0.04$) when the smallest block of islands was removed from the study (Appendix S1: Table S2). Web-spider density was higher on islands with more seaweed deposition (Fig. 3: seaweed effect: $\chi^2 = 4.5$, df = 1, $P = 0.03$). We did not detect an effect of area on web-spider abundance (Fig. 2; area effect: $\chi^2 = 0.65$, df = 1, $P = 0.42$) nor interactions between lizard treatment and island area ($\chi^2 = 0.15$, df = 1, $P = 0.70$) nor seaweed deposition ($\chi^2 = 0.26$, df = 1, $P = 0.61$). A single species, *Eustala cazieri*, accounted for the majority (68%) of spiders and webs seen in our study.

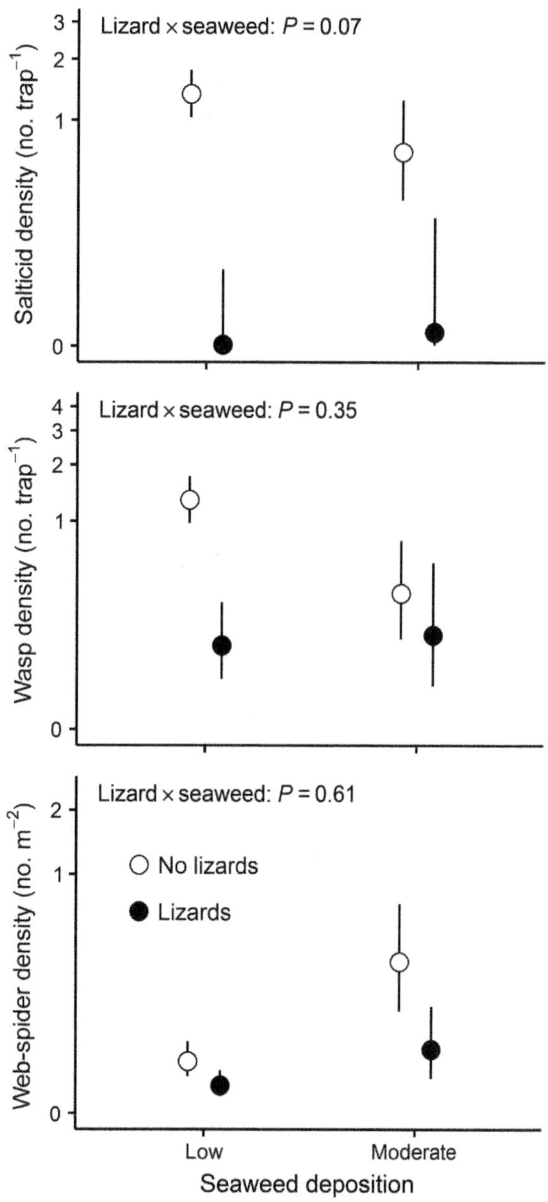

Fig. 3. Effects of lizards on the abundance of salticid spiders, wasps, and web spiders on islands with different levels of seaweed deposition. Estimates (solid circles for lizard-introduction islands and open circles for no-lizard islands), standard errors, and P-values for the lizard × seaweed interaction are derived from the generalized linear models described in the text. Note log scale on y-axis.

Discussion

Lizard introduction reduced the abundance of all three taxa of arthropod predators and parasitoids. Island area had important consequences for the strength of this top-down effect. For salticid spiders, the effect of lizards was reduced on larger islands. In contrast, the effect of lizards on wasps tended to be greater on larger islands, and there was no indication that the lizard effect varied with island area for web spiders. There was only limited support for the hypothesis that moderate levels of seaweed deposition influenced lizard effects. Specifically, there was a marginally significant trend for a reduced effect of lizards on salticid spider abundance on islands with more seaweed.

The influence of island area on the top-down effects of lizards was not consistent across arthropod taxa. For salticids, lizard effects decreased with island area. This pattern is consistent with certain models (McCann et al. 2005, Tunney et al. 2012), but did not match our expectations. We expected the effects of lizards on salticids to be weaker on smaller islands due to stressful environmental conditions and demographic stochasticity. In contrast, we found that lizard effects on salticids were stronger on small islands, in part because salticids achieve high densities on small islands in the absence of lizards and in part because lizards are not as effective at eliminating salticids on larger islands. We hypothesize that high dispersal rates and low abundance of alternate predators (such as wasps; Fig. 2; Schoener et al. 2016) help explain high salticid abundance on small, lizard-free islands and thus are partially responsible for the stronger effects of lizards on small islands. In addition, increased spatial and resource heterogeneity, including increased refuge availability, on larger islands (McCann et al. 2005, Schoener and Spiller 2010, Tunney et al. 2012, Schoener et al. 2016) likely contributed to the persistence of some salticids in the presence of lizards and hence weaker lizard effects. In contrast to salticids, the effect of lizards on wasps tended to increase with island size. This pattern was associated with the fact that wasp abundance was low on smaller islands, giving lizards little opportunity to reduce their abundance further. It seems likely that low wasp abundance on smaller islands is due to an increased effect of environmental disturbance, in the form of tropical storms (Spiller and Schoener 2007), and perhaps also to a lack of suitable prey resources (Schoener et al. 1995). Thus, stressful environmental conditions may help explain weak effects of lizards on wasps on small islands (Appendix S2).

Why did we not observe a hump-shaped relationship in effect sizes, as hypothesized by Schoener et al. (2016)? One explanation is that the current study did not consider large islands (i.e., >1000 m^2; the range of island sizes in the current study was 4–157 m^2), where the decreasing relationship between island area and lizard effects was most apparent. For example, we detected a trend for increasing lizard effects on wasps on larger islands, but were not able to determine whether this trend was reversed on still larger islands. In addition, our short-term experiment may not have allowed enough time for stressful environmental conditions and demographic stochasticity, factors mentioned by Schoener et al. (2016), to limit the size of lizard effects on small islands. For example, the strong effects of lizards on salticids on the smallest islands were associated with the high abundance of salticids on lizard-free islands in this size class; over longer time scales, average salticid abundance on these islands may be reduced by stressful environmental conditions and demographic processes.

Our results suggest that marine subsidies may play a role in reducing the effects of lizard predators on arthropod prey. Because our study was conducted over a relatively short time period (which limits lizard reproductive responses) and because our study took place on small islands (which limits aggregative responses), we predicted that lizard effects would be reduced on islands with more seaweed due to changes in diet (Spiller et al. 2010, Piovia-Scott et al. 2013) and foraging behavior (A. N. Wright et al., *unpublished data*). Consistent with this hypothesis, there was a marginally significant trend for weaker lizard effects on salticids on islands with more seaweed. There was a similar trend for wasps, but the lizard × seaweed interaction term was not significant, and there was no support for seaweed-based differences in top-down effects for web spiders. We suggest that the absence of stronger relationships between seaweed deposition and lizard effects in the current study is likely due to the relatively low rate of seaweed deposition on our study islands. Previous studies applied seaweed experimentally at a rate of 2.5 kg/m^2 to mimic the amounts of seaweed deposited during "pulse" events (often associated with large storms). In contrast, the maximum rate of seaweed deposition in the current study (which did not include any pulse events) was only 0.8 kg/m^2, and it may take more seaweed to cause more substantial reductions in the effects of lizards on arthropod prey.

In conclusion, we have demonstrated that the short-term effects of introduced lizards on certain arthropod predators and parasitoids depend on island size. However, the nature of this relationship was not consistent across taxa—lizard effects on salticid spiders are greater on smaller islands, whereas effects on wasps tended to be greater on larger islands. These differences were associated with pronounced relationships between island size and arthropod abundance—salticids can achieve high abundance on small islands in the absence of lizards, facilitating stronger lizard effects, while wasps have very low abundance on small islands, weakening lizard effects—although there was also some evidence that lizards are not as effective at limiting salticid spider abundance on larger islands. We also found some evidence that seaweed deposition can reduce the short-term top-down effects of lizards, suggesting that marine subsidies may play a role in weakening predator effects on small islands. Thus, while introduced predators can have devastating impacts on island communities, the strength of those effects can vary dramatically based on both island characteristics and the biology of the specific prey species. More generally, our study underscores the need to extend existing theories relating island area and community characteristics such as diversity (Connor and McCoy 1979, Schoener 2009) and food-chain length (Schoener 1989, Post 2002) toward an explicit consideration of species interactions (Holt 2010). This endeavor is likely to provide valuable insight into ecosystem function that cannot be gleaned from descriptive community metrics alone.

Acknowledgments

This research was supported by a grant from the National Science Foundation (DEB-11968). Marshall McMunn, Shahla Farzan, and Buddy Pinder provided invaluable support in the field, and James Whitney, Wanda Bonneville, Ryan Schemrich, and Laura Morgan helped process samples in the laboratory. We thank the Bahamas Environment, Science, and Technology Commission for permission to conduct the research. This research was approved by the UC Davis Institutional Animal Care and Use Committee.

Literature Cited

Afkhami, M. E., J. A. Rudgers, and J. J. Stachowicz. 2014. Multiple mutualist effects: conflict and synergy in multispecies mutualisms. Ecology 95: 833–844.

Agrawal, A. A., et al. 2007. Filling key gaps in population and community ecology. Frontiers in Ecology and the Environment 5:145–152.

Baxter, C. V., K. D. Fausch, and W. Carl Saunders. 2005. Tangled webs: Reciprocal flows of invertebrate prey link streams and riparian zones. Freshwater Biology 50:201–220.

Baxter, C. V., K. D. Fausch, M. Murakami, and P. L. Chapman. 2004. Fish invasion restructures stream and forest food webs by interrupting reciprocal prey subsidies. Ecology 85:2656–2663.

Bronstein, J. L. 1994. Conditional outcomes in mutualistic interactions. Trends in Ecology & Evolution 9:214–217.

Cameron, A. C., and P. K. Trivedi. 1990. Regression-based tests for overdispersion in the poisson model. Journal of Econometrics 46:347–364.

Connor, E. F., and E. D. McCoy. 1979. Statistics and biology of the species-area relationship. American Naturalist 113:791–833.

Croll, D. A., J. L. Maron, J. A. Estes, E. M. Danner, and G. V. Byrd. 2005. Introduced predators transform subarctic islands from grassland to tundra. Science 307:1959–1961.

Ebenhard, T. 1988. Introduced birds and mammals and their ecological effects. Swedish Wildlife Research 13:1–107.

Elton, C. S. 1958. The ecology of invasions by animals and plants. Wiley and Sons, New York, New York, USA.

Fukami, T., D. A. Wardle, P. J. Bellingham, C. P. H. Mulder, D. R. Towns, G. W. Yeates, K. I. Bonner, M. S. Durrett, M. N. Grant-Hoffman, and W. M. Williamson. 2006. Above- and below-ground impacts of introduced predators in seabird-dominated island ecosystems. Ecology Letters 9:1299–1307.

Harrison, X. A. 2014. Using observation-level random effects to model overdispersion in count data in ecology and evolution. PeerJ 2:e616.

Henschel, J. R., D. Mahsberg, and H. Stumpf. 2001. Allochthonous aquatic insects increase predation and decrease herbivory in river shore food webs. Oikos 93:429–438.

Hoeksema, J. D., and E. M. Bruna. 2015. Context-dependent outcomes of mutualistic interactions. Pages 181–202 *in* J. L. Bronstein, editor. Mutualism. Oxford University Press, Oxford, UK.

Holt, R. D. 2010. Toward a trophic island biogeography: reflections on the interface of island biogeography and food web ecology. Pages 143–185 *in* J. B. Losos and R. E. Ricklefs, editors. The theory of island biogeography revisited. Princeton University Press, Princeton, New Jersey, USA.

Holt, R. D., J. H. Lawton, G. A. Polis, and N. D. Martinez. 1999. Trophic rank and the species-area relationship. Ecology 80:1495–1504.

Leroux, S. J., and M. Loreau. 2008. Subsidy hypothesis and strength of trophic cascades across ecosystems. Ecology Letters 11:1147–1156.

Lomolino, M. V. 1984. Immigrant selection, predation, and the distributions of *Microtus pennsylvanicus* and *Blarina brevicauda* on islands. American Naturalist 123:468–483.

McCann, K. S., J. B. Rasmussen, and J. Umbanhowar. 2005. The dynamics of spatially coupled food webs. Ecology Letters 8:513–523.

Murakami, M., and S. Nakano. 2002. Indirect effect of aquatic insect emergence on a terrestrial insect population through predation by birds. Ecology Letters 5:333–337.

Nakano, S., H. Miyasaka, and N. Kuhara. 1999. Terrestrial-aquatic linkages: Riparian arthropod inputs alter trophic cascades in a stream food web. Ecology 80:2435–2441.

Piovia-Scott, J., D. A. Spiller, and T. W. Schoener. 2011. Effects of experimental seaweed deposition on lizard and ant predation in an island food web. Science 331:461–463.

Piovia-Scott, J., D. A. Spiller, G. Takimoto, L. H. Yang, A. N. Wright, and T. W. Schoener. 2013. The effect of chronic seaweed subsidies on herbivory: Plant-mediated fertilization pathway overshadows lizard-mediated predator pathways. Oecologia 172:1129–1135.

Polis, G. A., and S. D. Hurd. 1996*a*. Allochthonous input across habitats, subsidized consumers, and apparent trophic cascades: examples from the ocean-land interface. Pages 275–285 *in* G. A. Polis and K. O. Winemiller, editors. Food webs: integration of pattern and dynamics. Chapman and Hall, New York, New York, USA.

Polis, G. A., and S. D. Hurd. 1996*b*. Linking marine and terrestrial food webs: Allochthonous input from the ocean supports high secondary productivity on small islands and coastal land communities. American Naturalist 147:396–423.

Polis, G. A., F. Sánchez-Pinero, P. T. Stapp, W. B. Anderson, and M. D. Rose. 2004. Trophic flows from water to land: Marine input affects food webs of islands and coastal ecosystems worldwide. Pages 200–216 *in* G. A. Polis, M. E. Power, and G. R. Huxel, editors. Food webs at the landscape level. University of Chicago Press, Chicago, Illinois, USA.

Post, D. M. 2002. The long and short of food-chain length. Trends in Ecology & Evolution 17:269–277.

R Development Core Team. 2016. R: a language and environment for statistical computing. R Foundation for Statistical Computing, Vienna, Austria.

Rogers, H., J. H. R. Lambers, R. Miller, and J. J. Tewksbury. 2012. 'Natural experiment' demonstrates top-down control of spiders by birds on a landscape Level. PLoS ONE 7:e43446.

Sabo, J. L., and M. E. Power. 2002. River-watershed exchange: effects of riverine subsidies on riparian lizards and their terrestrial prey. Ecology 83:1860–1869.

Salo, P., E. Korpimaki, P. B. Banks, M. Nordstrom, and C. R. Dickman. 2007. Alien predators are more dangerous than native predators to prey populations. Proceedings of the Royal Society of London B: Biological Sciences 274:1237–1243.

Schoener, T. W. 1989. Food webs from the small to the large. Ecology 70:1559–1589.

Schoener, T. W. 2009. The MacArthur-Wilson equilibrium model: what it said and how it was tested. Pages 52–87 *in* J. B. Losos and R. E. Ricklefs, editors. The theory of island biogeography revisited. Princeton University Press, Princeton, New Jersey, USA.

Schoener, T. W., and D. A. Spiller. 1996. Devastation of prey diversity by experimentally introduced predators in the field. Nature 381:691–694.

Schoener, T. W., and D. A. Spiller. 1999*a*. Indirect effects in an experimentally staged invasion by a major predator. American Naturalist 153:347–358.

Schoener, T. W., and D. A. Spiller. 1999*b*. Variation in the magnitude of a predator's effect from small to large islands. Pages 35–66 *in* Ecologia de les Illes. Societat d'Història Natural de les Balears, Palma, Spain.

Schoener, T. W., and D. A. Spiller. 2010. Trophic cascades on islands. Pages 179–202 *in* J. Terborgh and J. Estes, editors. Trophic cascades: predators, prey, and the changing dynamics of nature. Island Press, Washington, D.C., USA.

Schoener, T. W., D. A. Spiller, and J. B. Losos. 2001. Natural restoration of the species-area relation for a lizard after a hurricane. Science 294:1525–1528.

Schoener, T. W., D. A. Spiller, and L. W. Morrison. 1995. Variation in the hymenopteran parasitoid fraction on Bahamian islands. Acta Oecologica 16:103–121.

Schoener, T. W., D. A. Spiller, and J. Piovia-Scott. 2016. Variation in ecological interaction strength with island area: theory and data from the Bahamian archipelago. Global Ecology and Biogeography 25:891–899.

Simberloff, D. 2013. Invasive species: what everyone needs to know. Oxford University Press, Oxford, UK.

Snyder, W. E., and E. W. Evans. 2006. Ecological effects of invasive arthropod generalist predators. Annual Review of Ecology Evolution and Systematics 37:95–122.

Spiller, D. A., J. Piovia-Scott, A. N. Wright, L. H. Yang, G. Takimoto, T. W. Schoener, and T. Iwata. 2010. Marine subsidies have multiple effects on coastal food webs. Ecology 91:1424–1434.

Spiller, D. A., and T. W. Schoener. 1988. An experimental study of the effect of lizards on web-spider communities. Ecological Monographs 58:57–77.

Spiller, D. A., and T. W. Schoener. 1994. Effects of top and intermediate predators in a terrestrial food web. Ecology 75:182–196.

Spiller, D. A., and T. W. Schoener. 2007. Alteration of island food-web dynamics following major disturbance by hurricanes. Ecology 88:37–41.

Takimoto, G., T. Iwata, and M. Murakami. 2009. Timescale hierarchy determines the indirect effects of fluctuating subsidy inputs on *in situ* resources. American Naturalist 173:200–211.

Takimoto, G., D. A. Spiller, and D. M. Post. 2008. Ecosystem size, but not disturbance, determines food-chain length on islands of the Bahamas. Ecology 89:3001–3007.

Tunney, T. D., K. S. McCann, N. P. Lester, and B. J. Shuter. 2012. Food web expansion and contraction in response to changing environmental conditions. Nature Communications 3:1105.

Wiles, G. J., J. Bart, R. E. Beck, and C. F. Aguon. 2003. Impacts of the brown tree snake: patterns of decline and species persistence in Guam's avifauna. Conservation Biology 17:1350–1360.

Wright, A. N., J. Piovia-Scott, D. A. Spiller, G. Takimoto, L. H. Yang, and T. W. Schoener. 2013. Pulses of marine subsidies amplify reproductive potential of lizards by increasing individual growth rate. Oikos 122:1496–1504.

Young, H. S., D. J. McCauley, R. Dunbar, and R. Dirzo. 2010. Plants cause ecosystem nutrient depletion via the interruption of bird derived spatial subsidies. Proceedings of the National Academy of Sciences USA 107:2072–2077.

6

Seasonal increases in fish trophic niche plasticity within a flood-pulse river ecosystem (Tonle Sap Lake, Cambodia)

THOMAS POOL,[1,†] GORDON HOLTGRIEVE,[1] VITTORIA ELLIOTT,[2] KEVIN MCCANN,[3] BAILEY MCMEANS,[4] NEIL ROONEY,[5] ADRIANNE SMITS,[6] THACH PHANARA,[2] MICHAEL COOPERMAN,[7] SYDNEY CLARK,[8] CHHENG PHEN,[2] AND SAMOL CHHUOY[2]

[1]*School of Aquatic and Fishery Sciences (SAFS), University of Washington, Seattle, Washington 98105 USA*
[2]*Inland Fisheries Research and Development Institute (IFReDI), Fisheries Administration, Phnom Penh, Cambodia*
[3]*Department of Integrative Biology, University of Guelph, Guelph, Ontario, Canada*
[4]*Department of Biology, University of Toronto, Mississauga, Ontario, Canada*
[5]*School of Environmental Sciences, University of Guelph, Guelph, Ontario, Canada*
[6]*Department of Biology, University of Nevada, Reno, Nevada 89557 USA*
[7]*Gordon and Betty Moore Center for Science, Conservation International, Arlington, Virginia 22202 USA*
[8]*Earth, Environmental, and Planetary Sciences, Brown University, Providence, Rhode Island 02912 USA*

Abstract. Species' responses to seasonal environmental variation can influence trophic interactions and food web structure within an ecosystem. However, our ability to predict how species' interactions will vary spatially and temporally in response to seasonal variation unfortunately remains inadequate within most ecosystems. Fish assemblages in the Tonle Sap Lake (TSL) of Cambodia—a dynamic flood-pulse ecosystem—were studied for five years (2010–2014) using stable isotope and Bayesian statistical approaches to explore both within- and among-species isotopic niche variation associated with seasonal flooding. Roughly 600 individual fish specimens were collected during 19 sampling events within the lake. We found that fishes within the same species tended to have a broader isotopic niche during the wet season, likely reflecting assimilation of resources from either a wider range of isotopically distinct prey items or a variety of habitats, or both. Furthermore, among-species isotopic niches tended to overlap and range more broadly during the wet season, suggesting that floodplain inundation promotes exploitation of more diverse and similar resources by different species in the fish community. Our study highlights that the flood-pulse dynamic that is typical of tropical aquatic ecosystems may be an essential element supporting freshwater fish community structure and the fish diversity that underpins the TSL food web. This flow regime is currently threatened by regional dam development, which may in turn impact the natural function and structure of the fishery food web.

Key words: flood-pulse; food webs; Mekong; omnivory; stable isotopes; trophic dynamics.

† **E-mail:** tpool@uw.edu

INTRODUCTION

An organism's trophic niche is the evolutionary result of its morphological, physiological, and behavioral adaptations to its environment (Winemiller and Jepsen 1998). At a broader community scale, identifying disparities in species' trophic roles can delineate the unique scaffolding that forms an ecosystem in terms of energy transfer and resource utilization. Studying spatial and temporal variation in species' trophic niches within a community is also of conservation interest because that variation is often related to the stability of ecosystem functions and resulting ecosystem services (Hooper et al. 2005). For example, high redundancy in species' trophic niches may reduce the likelihood that natural or human-caused disturbances may alter ecosystem processes. In environments that have a complex mosaic of habitat types and/or substantial seasonal variability, an improved understanding of how species' trophic roles shift within food webs may be useful for devising management strategies to preserve biodiversity and ecosystem services (Tallis et al. 2010).

Current approaches for characterizing species' trophic niche dimensions have advanced from static dietary profiles (Grimm 1995, Tilman et al. 1997) to dynamic and highly parameterized food web models that incorporate multiple interactions among species across complex habitats. Quantifying trophic niche overlap between species can provide an estimate of resource segregation within an environment, as well as provide a descriptive measure of assemblage organization (Corrêa et al. 2011). For example, some isotope tracer studies have broadly shown that omnivore species commonly share dietary items with other species within a community, in contrast to feeding specialists whose diets tend to overlap less with other species (Tillberg and Breed 2004). Interestingly, additional studies using isotope tracers have shown that seasonal variation in the overlap of species' niches can occur within some systems (Jaeger et al. 2010, Pokharel et al. 2015). As a result, there have been renewed efforts to explore species' trophic plasticity and the influence natural and/or human factors have on the adaptability of food web structures (Corrêa et al. 2011).

Omnivory is a common feeding strategy for many fish species in freshwater ecosystems, as many consumers utilize a range of prey items (Prejs and Prejs 1987, Mérona and Rankin-de-mérona 2004, Pouilly et al. 2006). Fishes, in particular, tend to have high dietary adaptability, which makes them useful model organisms for exploring how seasonal changes in aquatic environments may impact food web structure. Furthermore, fish are widely documented to influence overall ecosystem structure, including a food web's trophic interactions (e.g., Carpenter and Kitchell 1993), nutrient cycling (e.g., Vanni 2002), and primary productivity (e.g., Schindler et al. 1997). The structure of fish communities in tropical freshwater ecosystems is of particular interest because they seasonally experience strong and predictable high flow events that inundate floodplain areas (i.e., the flood pulse; Zaret and Rand 1971, Junk et al. 1989, Ortega et al. 2015), increasing habitat complexity and potentially altering resource availability. As a result, increases in habitat availability and type during the wet season might provide important low trophic-level foraging opportunities for fishes (i.e., plants and insects), in contrast to the dry season when piscivory may be the more dominate strategy (Wantzen et al. 2002). Therefore, seasonal shifts in the physical, chemical, and biological environment have the potential to "reshuffle" fishes' trophic architecture within tropical river and lake ecosystems.

In the field of trophic ecology, scientists strive to understand resource partitioning as a mediator of species coexistence within ecosystems. The partitioning of resources within a food web can manifest as consumers feeding on distinctive prey items within a single habitat or consumers feeding on similar prey items across multiple habitats. Quantifying species' isotopic niches (a representative proxy for their trophic niche; Layman et al. 2007, Boecklen et al. 2011) has proven to be an effective approach to study resource partitioning, reflecting both a consumer's habitat and resource use demands within an ecosystem. Isotopic tracer analysis therefore provides insight associated with basal production, predation, and competition within an ecosystem by enumerating species' trophic roles (Bearhop et al. 2004). Particularly within freshwater ecosystems, isotopic niche approximations have assisted with defining species' roles within complex food web communities, nested in dynamic biogeochemical aquatic landscapes (Jepsen and Winemiller 2002, Willson et al. 2010).

The Lower Mekong River Basin in Southeast Asia is an archetypal seasonal flood-pulse ecosystem, one of the few large tropical river basins that is currently undammed for a large proportion of its length. Within the heart of this system, the Tonle Sap Lake (TSL) is a seasonal wetland in central Cambodia that undergoes substantial annual flooding from the Mekong River mainstem (Appendix S1). As observed in many tropical freshwater systems, the TSL has a diverse fish fauna that supports a highly productive fishery that provides food security for the region, but substantial gaps remain in our understanding of the TSL's basic ecology (but see Lim et al. 1999). Furthermore, concomitant with increasing agricultural and urban development, numerous water infrastructure projects (i.e., dams and canals) are proposed throughout the Lower Mekong Basin for hydropower, irrigation, flood control, and municipal use (MRC 2009; Lauri et al. 2012). Unfortunately, lessons learned from other flood-pulse systems underscore the importance of natural river flow dynamics in supporting ecosystem function (Junk et al. 1989, Thomaz et al. 2007, Wantzen et al. 2008). With this understanding, there is strong indication that dam development in the TSL region may alter the timing and extent of seasonal flooding with negative consequences on the lake's fishery (Arias et al. 2013, 2014). Therefore, quantitatively establishing the influence of seasonal flood cycles on the food web structural plasticity is needed for the current and future management of the TSL ecosystem (Zarfl et al. 2014, Benchimol and Peres 2015).

In this study, we used carbon (C), nitrogen (N), and sulfur (S) stable isotope tracers and a Bayesian niche breadth model (Swanson et al. 2015) to explore seasonal within- and among-species isotopic niche variation in the TSL. There are four scenarios that depict how species' isotopic niche spaces might change from the dry (Fig. 1a) to wet season (Fig. 1b–e): (1) increased niche breadth with increased niche overlap (Fig. 1b), (2) increased niche breadth with decreased niche overlap (Fig. 1c), (3) decreased niche breadth with increased niche overlap (Fig. 1d), and (4) decreased niche breadth and decreased niche overlap (Fig. 1e). We hypothesize that floodplain inundation during the wet season provides additional food resources for fishes across a more diverse array of habitats, increasing both the total available niche range for individual species and the isotopic range for the entire fish community (Fig. 1b or c). Although previous freshwater fish research has provided some evidence that increased resource partitioning may seasonally

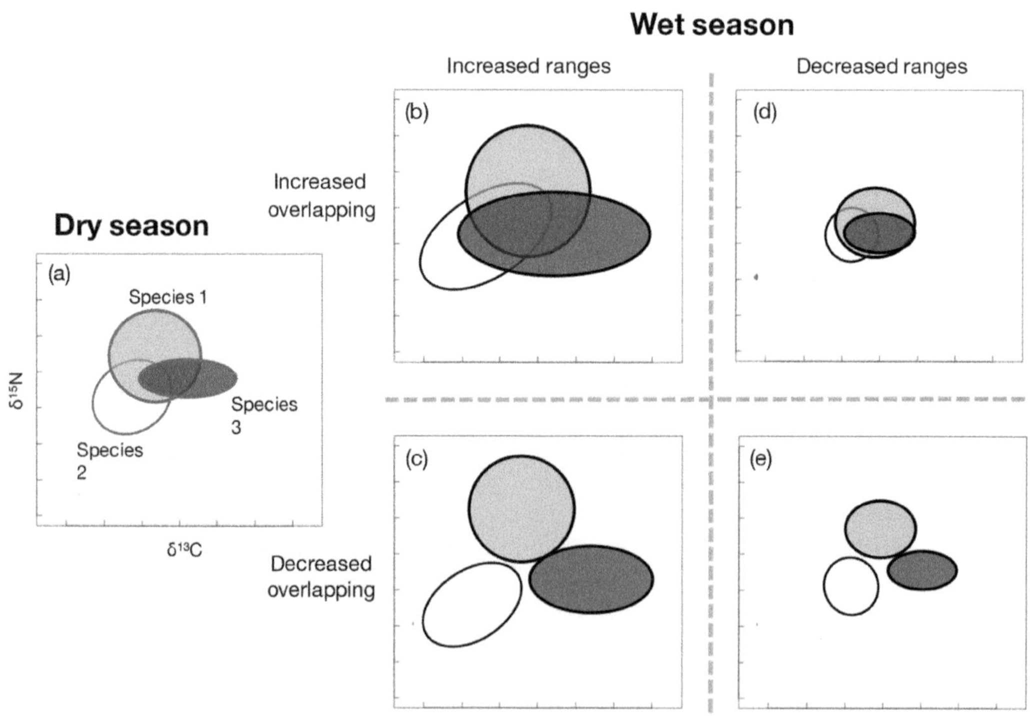

Fig. 1. Conceptual model of four possible scenarios associated with changes in species' trophic niche breadths from the dry season (part a) to the wet season (parts b–e). As flooding occurs during the wet season, species may display an increase (part b or c) or a decrease (part d or e) in trophic niche ranges. Regardless of the change in niche ranges, species can also display an increase (part b or d) or a decrease (part c or e) in niche overlap with other species.

occur in some systems (i.e., niche differentiation; Zaret and Rand 1971), we further hypothesize that increased niche overlap is common for TSL fishes during the wet season when resources are more abundant (Fig. 1b), consistent with observed decreases in resource partitioning by fish in other tropical freshwater environments (Goulding 1980, Meschiatti 1995). Our analyses provide an enhanced understanding of the resource structure of this poorly studied fish community and offers insight into how the fish community may respond to future anthropogenic disturbances within the basin.

Methods

Study system

The TSL is the largest lake and wetland system in the Mekong River Basin (UNESCO 2010). As an integrated aquatic and terrestrial landscape influenced by flooding from the mainstem Mekong River, the lake's surface area increases from approximately 2000 km^2 in the dry season (typically May–July), to over 10,000 km^2 during the wet season (typically September–February; MRC 2005). The average lake depth increases from less than one meter in the dry season, to more than 8 m at high water (MRC 2005). There is a diverse set of natural and agricultural habitats within the TSL floodplain, including forests, shrublands, grasslands, and cleared fields. This landscape is exceedingly important for Cambodian people economically, supporting a substantial proportion of Cambodia's rice production and ~60% of its total fishery production (Bonheur and Lane 2002). Evidence also suggests the lake's dynamic flood-pulse cycle is the primary driver supporting numerous plant, reptile, and bird species, many of which are of conservation interest for the region (Davidson 2006).

Field sampling

The TSL fish community is highly diverse and varies seasonally. There are at least 120 species targeted by the fishery, as well as numerous non-fishery species. This creates considerable logistical challenges for obtaining a representative sample using independent survey methods. To illustrate, local fishermen utilize over 200 gear types—including gill nets, basket traps, trawl and seine nets in numerous habitats throughout the lake and floodplain (Deap et al. 2003). To address this difficulty, we employed a multi-tiered sampling approach to acquire fish specimens from a diverse array of habitats. First, we collected specimens from five regional markets and/or associated fish landing areas located around the perimeter of the TSL (Appendix S1). Second, we approached fishermen and fish traders directly on the fishing grounds and within floating and stilted villages during daily peak trading periods to acquire additional samples that were not collected in the regional markets. Third, we acquired incidental specimens from multi-panel gill netting at six locations around the basin (Appendix S1). All these specimens were then pooled to characterize the fish community within the TSL. This combination of approaches ensured that substantial fishing effort was consistently incorporated spatially around the lake, as well as across seasons. In total, specimen collection involved 19 sampling events over 89 d during both dry (May–August) and wet seasons (September–March), from 2010 to 2014. During field collections, we also attempted to acquire specimens across the observed range of lengths for each species. Sampling efforts for our different collection methods were not balanced across space and time (i.e., extensive multi-panel netting took place exclusively during 2014 to increase the diversity of species included in our study). However, we are confident our approach provides a representative sampling of the most abundant fishery species in the lake during the dry and wet seasons (see Appendix S2 for sample size and collection details).

For this study, we included 18 common species utilizing at least five specimens per season, resulting in a minimum of 10 specimens per species (mean across all species = 21 specimens), with a range of life-history characteristics (i.e., max size, trophic position, and diet breadth), representing 14 families of fishes (Table 1). After collection, the fish were immediately stored on ice until processing at the Inland Fisheries Research and Development Institute of Cambodia. Specimens were measured to the nearest 1 mm for standard and total lengths and then weighed to the nearest 0.1 g total weight. Roughly 1 g dorsal muscle tissue was removed and frozen until the sample could be processed for stable isotope analysis. Samples were oven-dried (−60°C for 24 h) or

Table 1. The mean overlap of each species (n = 17) standard Bayesian ellipsoid (SEA_b) with all species ellipsoids using carbon ($\delta^{13}C$), nitrogen ($\delta^{15}N$), and sulfur ($\delta^{34}S$) data in three-dimensional (C–N–S) and two-dimensional space during the dry and wet seasons.

	C–N–S Overlap (%)		C–N Overlap (%)		C–S Overlap (%)		N–S Overlap (%)	
Species	Dry	Wet	Dry	Wet	Dry	Wet	Dry	Wet
Anabas testudineus	21.2	14.2	34.5	8.6	54.3	9.5	66.9	30.9
Barbonymus gonionotus	**3.2**	**27.8**	**8.2**	**43.7**	**31.7**	**70.1**	**33.5**	**51.3**
Boesemania microlepis	**3.4**	**11.4**	**7.8**	**17.2**	80.4	63.8	28.0	22.3
Channa micropeltes	**3.1**	**17.4**	**15.2**	**23.1**	**5.0**	**21.2**	**25.6**	**71.1**
Channa striata	**13.1**	**49.3**	**41.2**	**58.0**	**54.7**	**81.1**	60.9	57.2
Cirrhinus siamensis	**7.8**	**9.6**	23.7	12.1	44.1	39.8	25.7	21.5
Clarias macrocephalus	8.0	1.9	**11.6**	**32.6**	31.5	10.6	29.7	4.9
Cyclocheilichthys enoplos	12.3	11.1	21.8	21.4	36.0	35.4	46.4	30.4
Hemibagrus spilopterus	**1.3**	**33.7**	**16.8**	**52.2**	**25.1**	**54.1**	**1.5**	**44.3**
Kryptopterus apogon	**3.1**	**6.7**	8.8	8.7	**47.5**	**56.9**	**23.7**	**38.1**
Macrognathus siamensis	**17.2**	**77.9**	**69.6**	**83.3**	**54.9**	**89.8**	**60.9**	**83.1**
Notopterus notopterus	**18.1**	**48.6**	**22.5**	**56.9**	**65.9**	**87.4**	**64.2**	**82.3**
Pangasius larnaudii	**6.0**	**10.2**	**9.1**	**31.3**	49.7	24.1	61.5	26.1
Parambassis wolffii	**3.3**	**11.3**	**12.7**	**19.8**	**16.3**	**41.3**	**7.3**	**18.6**
Pristolepis fasciata	**4.5**	**52.8**	**6.8**	**62.1**	**33.7**	**79.8**	**27.6**	**68.5**
Puntioplites proctozystron	**15.0**	**19.2**	48.5	31.9	**32.9**	**47.0**	**54.0**	**54.7**
Thynnichthys thynnoides	**1.3**	**3.2**	**2.8**	**6.3**	8.6	8.5	**15.6**	**59.9**

Notes: Bold cells identify increases in species' isotopic overlap from the dry to wet season. We removed *Mystus albolineatus* since there were not enough data for this type of analysis.

freeze-dried (−80°C for 24 h) in Cambodia prior to shipping to the University of Washington (UW) for further processing and analysis.

Stable isotope analysis

We used the distribution of carbon ($^{13}C/^{12}C$), nitrogen ($^{15}N/^{14}N$), and sulfur ($^{34}S/^{32}S$) stable isotopes to explore food web structure within the TSL. Carbon stable isotopes indicate the origin of C sources consumed by an organism, providing insight into primary and secondary production. Nitrogen stable isotopes can reveal the trophic positioning of species, with elevated $^{15}N/^{14}N$ being suggestive of species higher in the trophic pyramid. Sulfur stable isotopes assist with identification of the habitats that species are utilizing because they differentiate benthic and pelagic producers, as well as marsh and phytoplankton producers. We included $^{34}S/^{32}S$ to allow us to discriminate between detrital and living algae resources, because it is more sensitive than $^{13}C/^{12}C$ (France 1996) and there is very little enrichment of ^{34}S over ^{32}S between trophic levels (McCutchan et al. 2015). Taken together, this suite of tracers provides information about the type of food items being consumed and spatial variability in where those resources are located across an aquatic landscape.

Upon arrival to UW, samples were again freeze-dried to remove any residual water and homogenized to a fine powder. Subsequently, between 0.4 and 6 mg of material was packed into tin or silver capsules, depending on the specific analysis. We then performed stable isotope analysis in the UW Δ × IsoLab by standard combustion and subsequent reduction (N), using a Costech ECS 4010 Elemental Analyzer coupled to Thermo Finnigan MAT 253 (Costech Analytical Technologies Inc., Valencia, California, USA) in continuous flow mode. Lipid extraction or correction of $^{13}C/^{12}C$ was not needed because the sample C:N ratios (muscle tissue = 3.47) were below the suggested cutoff of 3.5 by Post et al. (2007). Stable isotope compositions throughout our study are expressed in terms of delta values (δ) in per mil (‰) relative to the international standards of Vienna Pee Dee Belemnite, atmospheric air, and Vienna Canyon Diablo Troilite for C, N, and S, respectively.

Quantifying variation in species' trophic niche space across time and space using isotopes can be challenging because of variation in "baseline" values (i.e., specimens at the base of the food chain that provide a comparable starting point for higher trophic-level assessments of feeding position for species such as fish). If "trophic expansion" is observed for a particular species, using stable isotope values across seasons, that variation could be explained due to natural seasonal variation in the $\delta^{13}C$, $\delta^{15}N$, and $\delta^{34}S$ incorporated into baseline specimens in addition to genuine changes in prey items consumed by higher trophic-level species. In our study, we pooled together fish collections spanning multiple years and locations but baseline information was not available to incorporate into our analyses. Despite our inability to account for the possibility of resource baseline expansion across seasons, we believe our numerous and diverse collection of specimens from this region is extensive enough to identify general patterns of resource use for fishes within the TSL.

Statistical analyses

In the past decade, considerable progress has been made in quantifying species' trophic niches utilizing stable isotope data (Layman et al. 2007, Jackson et al. 2011). Researchers often use convex hulls to define the maximum isotopic area in two-dimensional space, but concern associated with the sensitivity of this approach to outliers and differences in sample size has led to the increased use of standard ellipse approaches (Jackson et al. 2011). We calculated standard ellipses with a standard sample size correction (SEA_c; Jackson et al. 2011), using pairwise combinations of $\delta^{13}C$, $\delta^{15}N$, and $\delta^{34}S$ data to estimate species' isotopic area in two-dimensional space (e.g., $\delta^{13}C$ with $\delta^{15}N$, and $\delta^{13}C$ with $\delta^{34}S$). First, we calculated within-species ellipse centroids for the dry and wet seasons to assess the mean shift in occupied isotopic space for each species. A two-tailed *t* test was used to determine whether the seasonal change in species' mean isotopic values (i.e., $\delta^{13}C$, $\delta^{15}N$, and $\delta^{34}S$) was significant for the community. Second, we used within-species ellipses to calculate niche range expansion and the significance of seasonal change for each species using all three isotopic tracers (two-tailed *t* tests). Third, we calculated a within-season beta-diversity value (i.e., the amount of dissimilarity) for species' ellipses, using the Jaccard's index (Villéger et al. 2008) to determine the extent of seasonal niche overlap within the community. Within the context of our study, the beta-diversity metric calculated for each species determined its overlap with all other species' isotopic ellipses, with lower beta-diversity values indicating reduced niche overlap and higher values indicating increased niche overlap. For each species, we assessed whether their mean niche overlap with all other species significantly increased from the dry to wet season.

Lastly, complementing the two-dimensional standard ellipse approach, we incorporated isotope data into a Bayesian framework that calculates species' ellipses in three-dimensional space (Swanson et al. 2015) for the dry and wet seasons. Using Bayesian ellipsoids (SEA_b) to explore the overlap between species allowed us to simultaneously incorporate the data from all three isotopes. This meant we could maximize our ability to infer how fish use habitat and resources seasonally. For each species, we calculated a mean overlap value with all other species to identify whether they were occupying unique or shared isotopic niche space within the community. Following the protocol established by Swanson et al. (2015), we used a default prior to construct a trophic niche region and utilized 10,000 samples from the normal distribution to obtain a single Monte Carlo estimate. Given that this approach expresses uncertainty about unknown parameters probabilistically, significance tests were not appropriate for these data. We removed *Mystus albolineatus* since there were not enough data for this type of analysis. We completed all analyses in R (R Development Core Team 2017) and used the following packages: ape, base, BiodiversityR, Geometry, Mvtnorm, nicheROVER, Pastecs, plyr, siar, and vegan.

Results

Seasonal shifts in species' mean $\delta^{13}C$, $\delta^{15}N$, and $\delta^{34}S$ values were not significant for the fish community as a whole ($\delta^{13}C$: $t_{17} = -1.866$, $P = 0.300$; $\delta^{15}N$: $t_{17} = -1.545$, $P = 0.141$; $\delta^{34}S$: $t_{17} = -0.709$, $P = 0.487$; see Fig. 2 for paired display of species shifts). However, within-species isotopic variation expanded significantly ($\delta^{13}C$: $t_{17} = 1.847$, $P = 0.041$; $\delta^{15}N$: $t_{17} = 2.930$, $P = 0.003$; $\delta^{34}S$: $t_{17} = 6.501$, $P < 0.001$) from the dry to wet season for most species (see Fig. 3 for all species; see Fig. 4 for selected species-specific $\delta^{13}C$ with $\delta^{15}N$ and $\delta^{34}S$ with $\delta^{15}N$ examples of

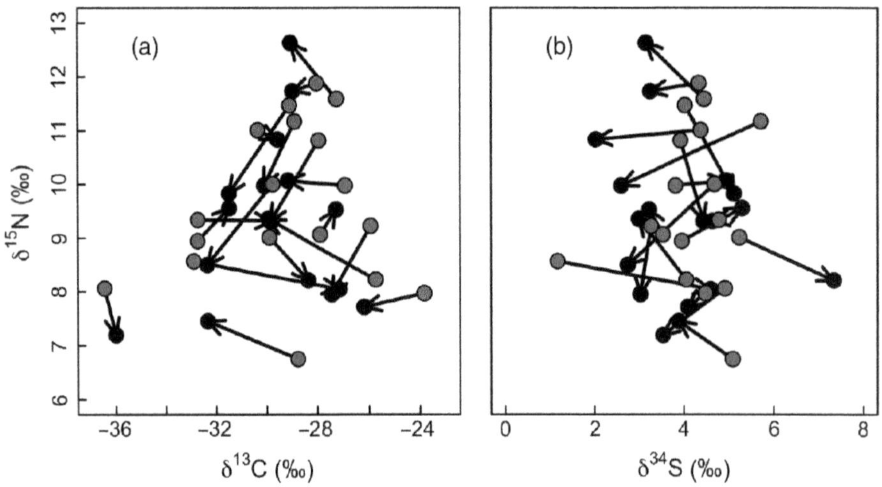

Fig. 2. Fish species' (n = 18) isotopic values for carbon, nitrogen, and sulfur. Points are mean values for each species and are linked via arrows displaying their temporal shift in trophic space from the dry season (gray) to the wet season (black).

seasonal niche expansion). Among all species and individuals, $\delta^{13}C$ values ranged by >20‰, from −39.4‰ to −18.4‰ during the dry season and increased from −41.0‰ to −18.3‰ during the wet season. $\delta^{15}N$ showed a smaller range among samples, with values from 5.5‰ to 13.0‰ during the dry season and increasing to a range of 4.8‰ to 13.5‰ during the wet season. $\delta^{34}S$ values spanned from 0.1‰ to 7.9‰ during the dry season, and subsequently increased to a range of 4.8‰ to 13.5‰ during the wet season. For many species, their niche range expansion encompassed more depleted isotopic tracer values (Fig. 3), although there were also examples of wet season niche range expansion along other axes (enriched $\delta^{15}N$ range, Fig 4a; enriched $\delta^{13}C$ range, Fig. 4d; or $\delta^{13}C$ and $\delta^{15}N$ range enrichment and depletion, Fig. 4b, c). When comparing the beta-diversity of species between seasons (i.e., the similarity of all species' isotopic niches), we found a greater overlap of species' standard ellipses (SEA_c) in the wet season relative to the dry (see Fig. 5a–d for $\delta^{13}C$ with $\delta^{15}N$ and $\delta^{15}N$ with $\delta^{34}S$ pairings; see Fig. 5e displaying 12 of 18 species as significantly increasing ($P > 0.05$) in $\delta^{13}C$ with $\delta^{15}N$ SEA_c overlap and Fig. 5f displaying 10 of 18 species significantly increasing ($P > 0.05$) in $\delta^{13}C$ with $\delta^{15}N$ SEA_c overlap).

Dry season, Bayesian isotopic niche ellipsoids (SEA_b) incorporating $\delta^{13}C$, $\delta^{15}N$, and $\delta^{34}S$ data displayed a mean C–N–S overlap of 8.1% in contrast to 25.7% during the wet season among all species. Individual species' overlaps—the proportion of a species' isotopic niche overlapping with all other species' isotopic niches—ranged from 1.3% (*Hemibagrus spilopterus* and *Thynnichthys thynnoides*) to 21.2% (*Anabas testudineus*) during the dry season and 1.9% (*Clarias macrocephalus*) to 77.9% (*Macrognathus siamensis*) during the wet season (Table 1). For some species, particular isotopic dimensions contributed more substantially than others to the three-dimensional overlap of species' niche ellipsoids (Table 1), but all three stable isotopes contributed significantly to the C–N–S overlap values (i.e., C: $R^2 = 0.515$, $P < 0.001$; N: $R^2 = 0.598$, $P < 0.001$; S: $R^2 = 0.205$, $P < 0.001$; see Appendix S3).

Discussion

Our study explored seasonal shifts in the within- and among-species isotopic niche variation of TSL fishes in wet vs. dry periods to elucidate the complex species habitat and resource associations that characterize flood-pulse river systems. Changes in a consumer's seasonal isotopic niche reflect an isotopically expanded or contracted prey base, which can arise from either utilizing more diverse prey items or accessing prey from different habitats with distinct isotopic

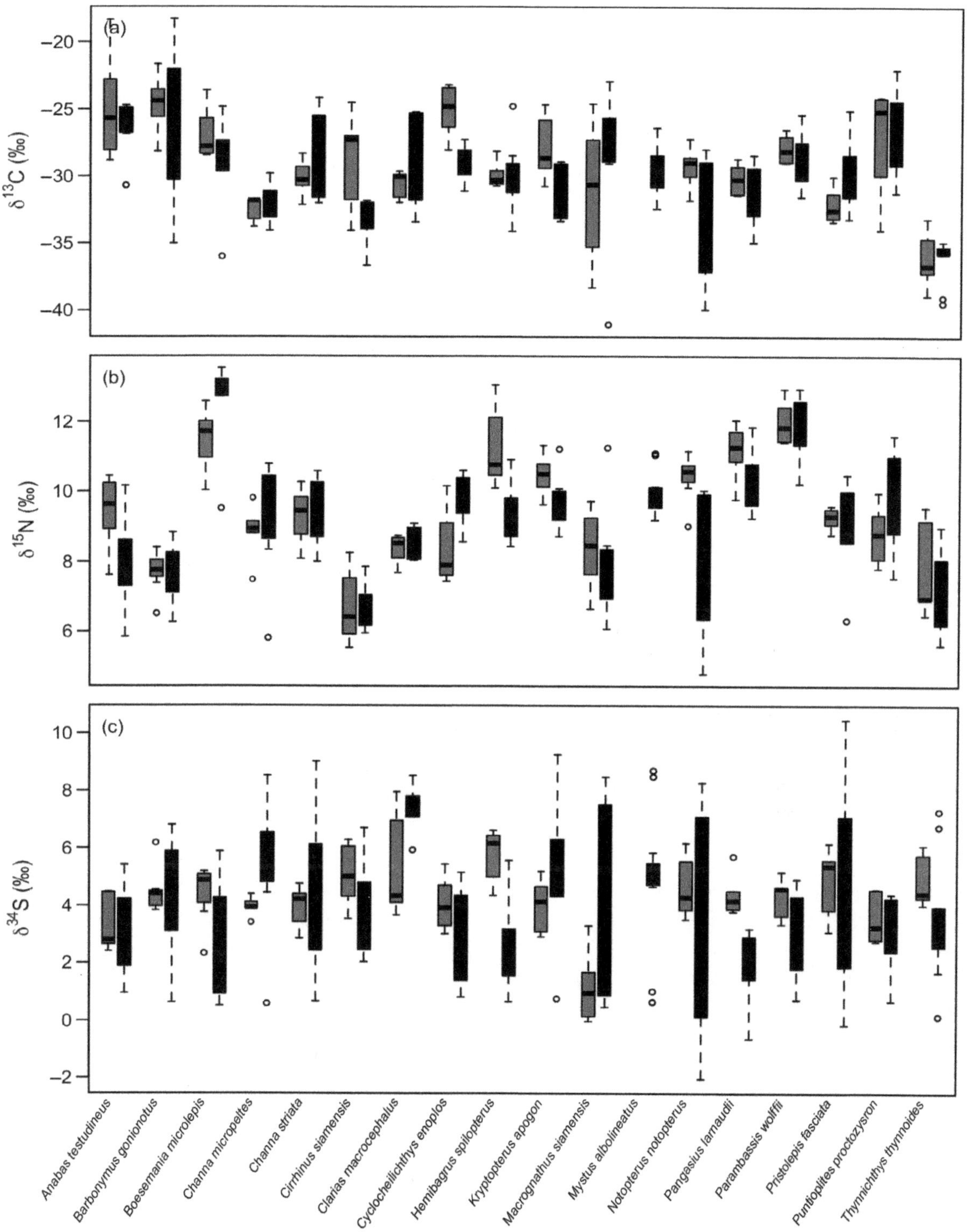

Fig. 3. Each species $\delta^{13}C$, $\delta^{15}N$, and $\delta^{34}S$ range of values during the dry (gray) and wet (black) seasons. For all 18 species, the maximum and minimum isotope values along with the first and third quartiles for nitrogen and carbon are displayed (outliers are identified with circles).

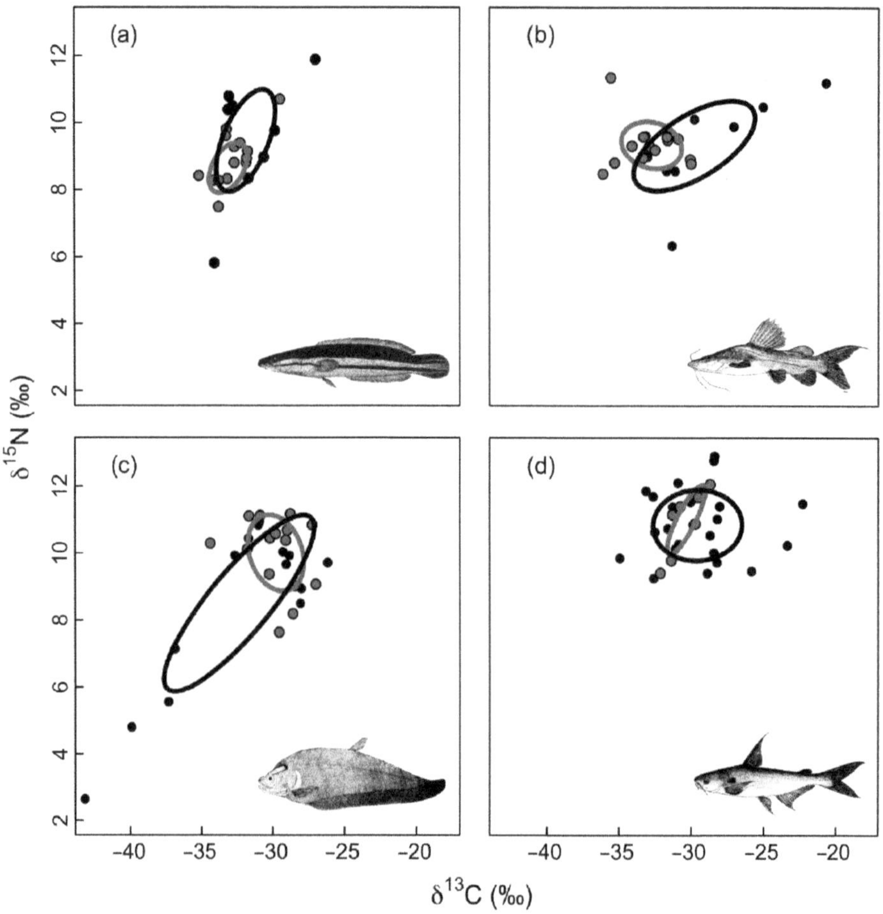

Fig. 4. Four species in the Tonle Sap Lake with points representing individual specimen $\delta^{13}C$ and $\delta^{15}N$ isotopic values for each species. Standard ellipses (SEA_c) identify the trophic niche space for (a) *Channa micropeltes*, (b) *Hemibagrus spilopterus*, (c) *Notopterus notopterus*, and (d) *Pangasius larnaudii* during the dry (gray) and wet (black) seasons. Sulfur data are not displayed here but display a similar pattern when plotted with the nitrogen isotope values.

signatures, or both (Newsome et al. 2007). In either case, a wider or narrower isotopic niche space reflects changes in available resources in response to environmental drivers. Our finding of generally higher isotopic niche widths during the wet season suggests that while omnivory appears to be widespread throughout the year, individuals within the same species appear to expand their resource base through consuming a broader range of prey items and/or utilizing a wider variety of habitats during the wet season in contrast to the dry season. Furthermore, among-species isotopic niches tend to overlap more during the wet season, suggesting that floodplain inundation promotes an increase in species utilization of similar resources and/or habitats within the fish community. Taken together, species' isotopic niche range and overlap increase during seasonal flooding (Fig. 1b). While most species' mean trophic positions did not substantially change between seasons, as indicated by their mean $\delta^{13}C$, $\delta^{15}N$, and $\delta^{34}S$ values, the isotopic variation we observed (i.e., species' isotopic ellipses) indicates a dynamic seasonal relationship between fishes' trophic structure and their freshwater environments.

The flood-pulse concept posits that fish will have access to a greater diversity of aquatic habitats and novel food sources during floodplain inundation within tropical freshwaters because of

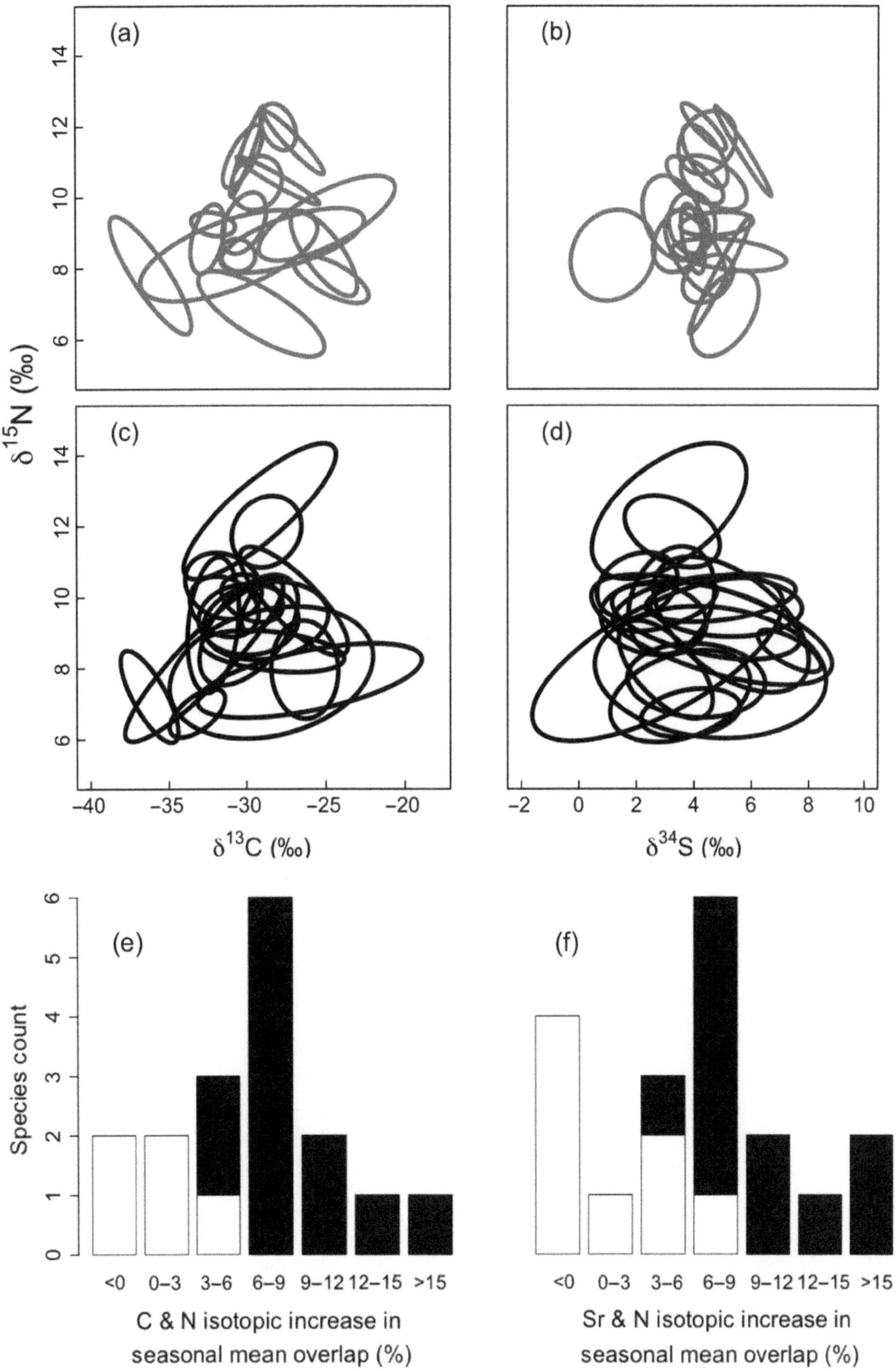

Fig. 5. Species (n = 18) isotope niches are represented by standard ellipses (SEA_c) for the dry (gray ellipses)

(Fig. 5. *Continued*)
and wet (black ellipses) seasons. The bivariate relationship between nitrogen and carbon isotope values (a, c), as well as nitrogen and sulfur isotopes values (b, d) is represented. The percentage increase in overlap of each species with all other species that occurred seasonally (dry to wet season) is displayed for each isotopic pairing (e, f) along with the significance of overlap change for each species (white bars are non-significant and black bars are significant; paired *t* test $P < 0.05$).

the inclusion of terrestrial food sources from the floodplain (Junk et al. 1989, Amoros and Bornette 2002, Galacatos et al. 2004). Within the TSL, an increase in resource availability during the wet season is highly likely, given that a substantial amount of terrestrial organic plant matter is annually submerged around the lake (Arias et al. 2013). Relatively nominal estimates of primary production within the TSL (Holtgrieve et al. 2013) paired with its high fishery output (MRC 2005) also support the theory that additional resources are seasonally derived from floodplain habitats and that these allochthonous resources are critical in maintaining fish/fishery productivity. In step with those observations, the total range in isotopic C values among all our samples (−39.4 to −18.4 δ^{13}C) was greater than the typical range for fishes in other ecosystems (−29 to −20 δ^{13}C in Hesslein et al. 1991, Vander Zanden and Casselman 1999, Fuller et al. 2012). Such an increase in the TSL within- and among-species isotopic ranges could be associated with reduced competitive interactions between species, due to greater resource availability in the floodplain during the wet season. Conversely during the dry season, specialized foraging strategies such as exclusive piscivory may be more common because of reduced basal resource availability (i.e., zooplankton and cyanobacteria may be the primary energy sources during low water periods; Samal 2009). While there are certainly advantages associated with dietary specialization at any time of the year, the diverse and abundant resources accessible to fish during the wet season may counterbalance the benefits of specialization for some fish.

In addition to the more typical C and N tracers incorporated into our study, S was also a significant contributor to the overlap of fishes' trophic niches during the wet season. Variation in δ^{34}S is driven primarily by redox conditions among habitats (Peterson and Howarth 1987), which suggests that many of the fishes are capitalizing not only on increased prey availability within floodplain habitat patches, but are also foraging across a variety of habitat locations or types. Although local assemblages of fishes and macroinvertebrates can characterize particular habitat types (Arrington and Winemiller 2006), the range in our δ^{13}C and δ^{34}S results suggests that fish species consume a diversity of prey items and/or forage in disparate habitat patches. Building on our understanding that the fish community utilizes floodplain habitat heterogeneity to support a diverse array of life-history strategies (such as reproduction and seasonal dormancy; Rainboth 1996), we infer that a mosaic of habitats may be required to support TSL fishes. However, while the S isotope results indicate that the fish community is exploiting resources from a range of habitats, they do not indicate the relative importance of specific habitats as dietary sources for each species. For example, the broad range in S isotope values found in *Notopterus notopterus* could be a result of individuals feeding on prey items in two different habitat types or feeding in the same habitat type in two different locations. Further work mapping the relationship between δ^{34}S of sulfur and lower trophic levels is necessary before we can use this tracer to identify specific habitat types within the TSL. Similarly, additional research exploring species seasonal dietary dependency on particular habitat types, such as flooded gallery forest and shrublands, is needed because those habitats may be disproportionately impacted by future anthropogenic disturbances associated with flow regime change (Arias et al. 2012).

Ellipsoid overlap (i.e., all three isotopic tracers combined) and the relative contribution of each tracer varied between species. However, integrating the ellipsoid results with existing knowledge about specific fishes permitted some speculation about the relative influence of foraging behavior vs. habitat use on explaining their niche breadth and overlap with other species. Using all the tracer data simultaneously also allowed us to infer additional details about species within a community context. For example, *Thynnichthys thynnoides* had a low C–N–S overlap value with other species across both seasons, suggesting that it occupies a unique trophic niche within the community, consistent with its classification as a periphyton grazer. However, this species had a relatively high N–S overlap value with the other fishes, suggesting it may be utilizing resources from similar habitats as many other species in the fish community. In contrast, *N. notopterus* overlapped substantially across all tracers, suggesting this species has a broad trophic niche utilizing common resources and habitats with numerous fishes in the community, consistent with its classification as a broad omnivore that displays movements between many floodplain habitat types. This improved understanding of TSL species, informed by our trophic niche overlap findings, provides a solid foundation for further studies of fish species' interactions within this region.

Understanding species' interactions within the TSL is further complicated because the biogeochemistry of the region appears to be spatially and temporally dynamic, potentially influencing our stable isotope values. Modeling, theory, and field data indicate changing redox conditions both horizontally and vertically through the floodplain, while the permanent lake maintains a generally mixed oxic conditions (Lamberts and Koponen 2008, Holtgrieve et al. 2013; B. Miller and G. Holtgrieve, *unpublished data*). Conditions of extreme hypoxia/anoxia can have pronounced effects on nutrient biogeochemistry and resource pathways, therefore confounding the standard interpretation of these isotopes in a food web context. For example, some individuals of *N. notopterus* show pronounced depletion of both $\delta^{13}C$ and $\delta^{15}N$ during the wet season. The low C values suggest a methanogenic resource pathway, and the associated depletion of N indicates a source other than nitrate for this resource pathway (such as ammonium). Nevertheless, these complexities do not change the general result that most species display broader trophic niche variation during the wet season, suggesting that the seasonal inundation of the complex network of floodplain habitats is sustaining the current food web structure of TSL fish community.

Throughout the tropics, freshwater flood-pulse systems are experiencing increased levels of human alteration (i.e., dam construction, channelization, water extraction, and wetland development) threatening the seasonal connectivity between terrestrial and aquatic environments and consequently the flora and fauna that they support. Reductions in the amount or length of floodplain inundation in the TSL are particularly concerning because our findings suggest the high fish diversity within the system is at least partially supported by increased resource availability during the wet season (see Taphorn and Lilyestrom 1985). Altering the composition, location, or quantity of seasonally available resources to fish may negatively impact the coexistence of resident species that appear to support this systems' food web structure (Chesson and Warner 1981). A reduction in the lake's floodplain connectivity may also influence the distribution and abundances of migratory fishes entering the system during the wet season that are believed to be subsidizing resident predatory fishes (Winemiller and Jepsen 1998). Curiously, previous tropical river studies exploring whether human alteration of the seasonal flow regime impacts fishes' ecological niches have produced equivocal results (Zaret and Rand 1971, Goulding 1980, Meschiatti 1995, Corrêa et al. 2011, Sá-Oliveira et al. 2014). Each river basin's unique environmental and human disturbance landscape undoubtedly regulates the impact that flow regime change has on the structure of aquatic food webs which, at least in part, accounts for the range of literature conclusions on this topic. As such, it remains unknown what the cumulative impact of the over 200 dams in development or scheduled for construction in the Mekong River Basin will have on the fishes of the TSL (MRC 2009). Ultimately, our research highlights that an assortment of floodplain habitats, potentially providing pathways for terrestrially derived C sources, may be critical for the preservation of species diversity and productivity in the TSL. Additional insights associated with the seasonal connectivity of habitat types are needed to successfully target species and habitats for protection within the ecosystem.

Acknowledgments

We thank the MacArthur Foundation and Belmont Forum for substantial funding support for this research. Furthermore, we thank the international team of researchers that have collaborated on this project for the past five years. In particular, we thank all the IFReDi scientists and technicians for their assistance working throughout the Lower Mekong River Basin ultimately making this work possible.

Literature Cited

Amoros, C., and G. Bornette. 2002. Connectivity and biocomplexity in waterbodies of riverine floodplains. Freshwater Biology 47:761–776.

Arias, M. E., T. A. Cochrane, M. Kummub, H. Lauric, G. Holtgrieve, J. Koponenc, and T. Piman. 2014. Impacts of hydropower and climate change on drivers of ecological productivity of Southeast Asia's most important wetland. Ecological Modelling 272:252–263.

Arias, M. E., T. A. Cochrane, D. Norton, T. J. Killeen, and P. Khon. 2013. The flood pulse as the underlying driver of vegetation in the largest wetland and fishery of the Mekong Basin. Ambio 42:864–876.

Arias, M. E., T. A. Cochrane, T. Piman, M. Kummu, B. S. Caruso, and T. J. Killeen. 2012. Quantifying changes in flooding and habitats in the Tonle Sap Lake (Cambodia) caused by water infrastructure development and climate change in the Mekong Basin. Journal of Environmental Management 112: 53–66.

Arrington, D. A., and K. O. Winemiller. 2006. Habitat affinity, the seasonal flood pulse, and community assembly in the littoral zone of a Neotropical floodplain river. Journal of the North American Benthological Society 25:126–141.

Bearhop, S., C. E. Adams, S. Waldron, R. A. Fuller, and H. Macleod. 2004. Determining trophic niche width: a novel approach using stable isotope analysis. Journal of Animal Ecology 73:1007–1012.

Benchimol, M., and C. A. Peres. 2015. Widespread forest vertebrate extinctions induced by a mega hydroelectric dam in lowland Amazonia. PLoS ONE 10:1–15.

Boecklen, W. J., C. T. Yarnes, B. A. Cook, and A. C. James. 2011. On the use of stable isotopes in trophic ecology. Annual Review of Ecology Evolution and Systematics 42:411–440.

Bonheur, N., and B. D. Lane. 2002. Natural resources management for human security in Cambodia's Tonle Sap Biosphere Reserve. Environmental Science and Policy 5:33e41.

Carpenter, S. R., and J. F. Kitchell. 1993. The trophic cascade in lakes. Cambridge University Press, Cambridge, UK.

Chesson, P., and R. Warner. 1981. Environmental variability promotes coexistence in lottery competitive systems. American Naturalist 117:923–943.

Corrêa, C. E., M. P. Albrecht, and N. S. Hahn. 2011. Patterns of niche breadth and feeding overlap of the fish fauna in the seasonal Brazilian Pantanal, Cuiabá River basin. Neotropical Ichthyology 9:637–646.

Davidson, P. J. 2006. The Biodiversity of the Tonle Sap Biosphere Reserve: status review. UNDP/GEF-funded Tonle Sap Conservation Project, Phnom Penh, Cambodia.

Deap, L., P. Degen, and N. Zalinge. 2003. Fishing gears of the Cambodian Mekong. Inland Fisheries Research and Development Institute, Phnom Penh, Cambodia.

France, R. L. 1996. Scope for use of stable carbon isotopes in discerning the incorporation of forest detritus into aquatic foodwebs. Hydrobiologia 325: 219–222.

Fuller, B. T., M. Gundula, W. Van Neer, and M. P. Richards. 2012. Carbon and nitrogen stable isotope ratio analysis of freshwater, brackish and marine fish from Belgian archaeological sites. Journal of Analytical Atomic Spectrometry 27:807–820.

Galacatos, K., R. Barriga-Salazar, and D. J. Stewart. 2004. Seasonal and habitat influences on fish communities within the lower Yasuni River basin of the Ecuadorian Amazon. Environmental Biology of Fishes 71:33–51.

Goulding, M. 1980. The fishes and the forest. University of California Press, Berkeley, California, USA.

Grimm, N. B. 1995. Why link species and ecosystems? A perspective from ecosystem ecology. Pages 5–15 *in* C. G. Jones and J. H. Lawton, editors. Linking species and ecosystems. Chapman and Hall, Boston, Massachusetts, USA.

Hesslein, R. H., M. J. Capel, D. E. Fox, and K. A. Hallard. 1991. Stable isotopes of sulfur, carbon, and nitrogen as indicators of trophic level and fish migration in the lower Mackenzie River Basin, Canada. Canadian Journal of Fisheries and Aquatic Sciences 48:2258–2265.

Holtgrieve, G. W., M. E. Arias, K. N. Irvine, D. Lamberts, E. J. Ward, M. Kummu, J. Koponen, J. Sarkkula, and J. E. Richey. 2013. Patterns of ecosystem metabolism in the Tonle Sap Lake, Cambodia with links to capture fisheries. PLoS ONE 8:e71395.

Hooper, D. U., F. S. Chapin, J. J. Ewel, A. Hector, P. Inchausti, and S. Lavorel. 2005. Effects of biodiversity on ecosystem functioning: a consensus of current knowledge. Ecological Monographs 75: 3–35.

Jackson, A. L., R. Inger, A. C. Parnell, and S. Bearhop. 2011. Comparing isotopic niche widths among and within communities: SIBER – Stable Isotope Bayesian Ellipses in R. Journal of Animal Ecology 80: 595–602.

Jaeger, A., M. Connan, and P. Richard. 2010. Use of stable isotopes to quantify seasonal changes of trophic niche and levels of population and individual specialisation in seabirds. Marine Ecology Progress Series 401:269–277.

Jepsen, D. B., and K. O. Winemiller. 2002. Structure of tropical river food webs revealed by stable isotope ratios. Oikos 96:46–55.

Junk, W. J., P. B. Bayley, and R. E. Sparks. 1989. The flood pulse concept in river-floodplain systems. Canadian Special Publication of Fisheries and Aquatic Sciences 106:110–127.

Lamberts, D., and J. Koponen. 2008. Flood-pulse alterations and productivity of the Tonle Sap ecosystem: a model for impact assessment. Ambio 37:178–184.

Lauri, H., H. De Moel, P. J. Ward, T. A. Räsänen, M. Keskinen, and M. Kummu. 2012. Future changes in Mekong River hydrology: impact of climate change and reservoir operation on discharge. Hydrology and Earth System Sciences 16:4603–4619.

Layman, C. A., D. A. Arrington, C. G. Montaña, and D. M. Post. 2007. Can stable isotope ratios provide for community-wide measures of trophic structure? Ecology 88:42–48.

Lim, P., S. Lek, S. Tana, S. Mao, and B. Chhouk. 1999. Diversity and spatial distribution of freshwater fish in Great Lake and Tonle Sap river (Cambodia, Southeast Asia). Aquatic Living Resources 12: 379–386.

McCutchan, J. H., W. M. Lewis, C. Kendall, and C. C. McGrath. 2015. Variation in trophic shift for stable isotope ratios of carbon, nitrogen, and sulfur. Oikos 102:378–390.

De Mérona, B., and J. Rankin-de-mérona. 2004. Food resource partitioning in a fish community of central Amazon floodplain. Neotropical Ichthyology 2:75–84.

Meschiatti, A. J. 1995. Alimentação da comunidade de peixes de uma lagoa marginal do Rio Mogi-Guaçu. Acta Limnologica Brasiliensia 7:115–137.

MRC. 2005. Overview of the hydrology of the Mekong Basin. MRC, Vientiane, Laos.

MRC. 2009. Mekong River Commission spatial database. MRC, Vientiane, Laos.

Newsome, S. D., C. M. del Rio, S. Bearhop, and D. L. Phillips. 2007. A niche for isotopic ecology. Frontiers in Ecology and the Environment 5:429–436.

Ortega, J. C. G., R. M. Dias, A. C. Petry, E. F. Oliveira, and A. A. Agostinho. 2015. Spatio-temporal organization patterns in the fish assemblages of a Neotropical floodplain. Hydrobiologia 745:31–41.

Peterson, B. J., and R. W. Howarth. 1987. Sulfur, carbon, and nitrogen isotopes used to trace organic matter flow in the salt-marsh estuaries of Sapelo Island, Georgia. Limnology and Oceanography 32:1195–1213.

Pokharel, K. P., E. Yohannes, and I. Salvarina. 2015. Isotopic evidence for dietary niche overlap between barking deer and four-horned antelope in Nepal. Journal of Biological Research-Thessaloniki 22:6.

Post, D. M., C. A. Layman, D. A. Arrington, G. Takimoto, J. Quattrochi, and C. G. Montana. 2007. Getting to the fat of the matter: models, methods and assumptions for dealing with lipids in stable isotope analyses. Oecologia 152:179–189.

Pouilly, M., S. Barrera, and C. Rosales. 2006. Changes of taxonomic and trophic structure of fish assemblages along an environmental gradient in the Upper Beni watershed (Bolivia). Journal of Fish Biology 68:137–156.

Prejs, A., and K. Prejs. 1987. Feeding of tropical freshwater fishes: seasonality in resource availability and resource use. Oecologia 71:397–404.

R Core Team. 2017. R: a language and environment for statistical computing. R Foundation for Statistical Computing, Vienna, Austria. https://www.R-project.org

Rainboth, W. J. 1996. Fishes of the Cambodian Mekong. FAO species identification field guide for fishery purposes. FAO, Rome, Italy.

Samal, S. 2009. Trophic linkage: the importance of micro-algae to the fisheries of Boeung Tonle Sap, Cambodia. Dissertation. Monash University, Clayton, Victoria, Australia.

Sá-Oliveira, J. C., R. Angelini, and V. J. Isaac-Nahum. 2014. Diet and niche breadth and overlap in fish communities within the area affected by an amazonian reservoir (Amapá, Brazil). Anais da Academia Brasileira de Ciencias 86:383–405.

Schindler, D. E., S. R. Carpenter, J. J. Cole, J. F. Kitchell, and M. L. Pace. 1997. Influence of food web structure on carbon exchange between lakes and the atmosphere. Science 277:248–251.

Swanson, H. K., M. Lysy, M. Power, A. D. Stasko, J. D. Johnson, and J. D. Reist. 2015. A new probabilistic method for quantifying n-dimensional ecological niches and niche overlap. Ecology 96:318–324.

Tallis, H., P. S. Levin, M. Ruckelshaus, S. E. Lester, K. L. McLeod, D. L. Fluharty, and B. S. Halpern. 2010. The many faces of ecosystem-based management: making the process work today in real places. Marine Policy 34:340–348.

Taphorn, D. C., and C. G. Lilyestrom. 1985. Los peces del modulo "Fernando Corrales" resultados ictiologicos del proyecto de investigacion del CONICIT-PIMA-18. Revue Scientifique et Technique 3:55–85.

Thomaz, S. M., L. M. Bini, and R. L. Bozelli. 2007. Floods increase similarity among aquatic habitats in river-floodplain systems. Hydrobiologia 579: 1–13.

Tillberg, C. V., and M. D. Breed. 2004. Placing an omnivore in a complex food web: dietary contributions to adult biomass of an ant. Biotropica 36:266–272.

Tilman, D., J. Knops, D. Wedin, P. Reich, M. Ritchie, and E. Siemann. 1997. The influence of functional diversity and composition on ecosystem processes. Science 277:1300–1302.

UNESCO. 2010. Biosphere Reserves eWorld Network. UNESCO-MAB (Man and the Biosphere). http://www.unesco.org/new/en/natural-sciences/environment/ecological-sciences/man-and-biosphere-programme/

Vander Zanden, J., and J. M. Casselman. 1999. Stable isotope evidence for the food web consequences of species invasions in lakes. Nature 401:1997–2000.

Vanni, M. J. 2002. Nutrient cycling by animals in freshwater ecosystems. Annual Review of Ecology and Systematics 33:341–370.

Villéger, S., N. W. H. Mason, and D. Mouillot. 2008. New multidimensional functional diversity indices for a multifaceted framework in functional ecology. Ecology 89:2290–2301.

Wantzen, K. M., W. J. Junk, and K. O. Rothhaupt. 2008. An extension of the floodpulse concept (FPC) for lakes. Hydrobiologia 613:151–170.

Wantzen, K. M., F. D. Machado, M. Voss, H. Boriss, and W. J. Junk. 2002. Seasonal isotopic shifts in fish of the Pantanal wetland, Brazil. Aquatic Sciences 64:239–251.

Willson, J. D., C. T. Winne, M. A. Pilgrim, C. S. Romanek, and J. W. Gibbons. 2010. Seasonal variation in terrestrial resource subsidies influences trophic niche width and overlap in two aquatic snake species: a stable isotope approach. Oikos 119: 1161–1171.

Winemiller, K. O., and D. B. Jepsen. 1998. Effects of seasonality and fish movement on tropical river food webs. Journal of Fish Biology 53:267–296.

Zaret, T. M., and A. S. Rand. 1971. Competition in tropical stream fishes: support for the competitive exclusion principle. Ecology 52:336–342.

Zarfl, C., A. E. Lumsdon, J. Berlekamp, L. Tydecks, and K. Tockner. 2014. A global boom in hydropower dam construction. Aquatic Sciences 77:161–170.

7

Predicting landscape effects of Mississippi River diversions on soil organic carbon sequestration

Hongqing Wang,[1,†] Gregory D. Steyer,[2] Brady R. Couvillion,[1] Holly J. Beck,[1] John M. Rybczyk,[3] Victor H. Rivera-Monroy,[4] Ken W. Krauss,[5] and Jenneke M. Visser[6]

[1]*Wetland and Aquatic Research Center, U.S. Geological Survey, Baton Rouge, Louisiana 70803 USA*
[2]*Southeastern Region, U.S. Geological Survey, Baton Rouge, Louisiana 70803 USA*
[3]*Department of Environmental Science, Western Washington University, Bellingham, Washington 98225 USA*
[4]*Department of Oceanography and Coastal Sciences, Louisiana State University, Baton Rouge, Louisiana 70803 USA*
[5]*Wetland and Aquatic Research Center, U.S. Geological Survey, Lafayette, Louisiana 70506 USA*
[6]*Institute for Coastal and Water Research, School of Geosciences, University of Louisiana at Lafayette, Lafayette, Louisiana 70506 USA*

Abstract. Large Mississippi River (MR) diversions (peak water flow >1416 m^3/s and sediment loads >165 kg/s) have been proposed as part of a suite of coastal restoration projects and are expected to rehabilitate and rebuild wetlands to alleviate the significant historic wetland loss in coastal Louisiana. These coastal wetlands are undergoing increasing eustatic sea-level rise, land subsidence, climate change, and anthropogenic disturbances. However, the effect of MR diversions on wetland soil organic carbon (SOC) sequestration in receiving basins remains unknown. The rate of SOC sequestration or carbon burial in wetlands is one of the variables used to assess the role of wetland soils in carbon cycling and also to construct wetland carbon budgets. In this study, we examined the effects of MR water and sediment diversions on landscape-scale SOC sequestration rates that were estimated from vertical accretion for the next 50 yr (2010–2060) under two environmental (moderate and less optimistic) scenarios. Our analyses were based on model simulations taken from the Wetland Morphology model developed for Louisiana's 2012 Coastal Master Plan. The master plan modeled a "future-without-action" scenario as well as eight individual MR diversion projects in two of the hydrologic basins (Barataria and Breton Sound). We examined the effects that discharge rates (peak flow) and locations of these individual diversion projects had on SOC sequestration rates. Modeling results indicate that large river diversions are capable of improving basin-wide SOC sequestration capacity (162–222 $g\ C{\cdot}m^{-2}{\cdot}yr^{-1}$) by up to 14% (30 $g\ C{\cdot}m^{-2}{\cdot}yr^{-1}$) in Louisiana deltaic wetlands compared to the future-without-action scenario, especially under the less optimistic scenario. When large river diversions are placed in the upper receiving basin, SOC sequestration rates are 3.7–10.5% higher (6–24 $g\ C{\cdot}m^{-2}{\cdot}yr^{-1}$) than when these structures are placed in the lower receiving basin. Modeling results also indicate that both diversion discharge and location have large effects on SOC sequestration in low-salinity (freshwater and intermediate marshes) as compared to high-salinity marshes (brackish and saline marshes).

Key words: Barataria Basin; Breton Sound Basin; Louisiana; Mississippi River; sea-level rise; sediment diversion; soil organic carbon sequestration; subsidence; vertical accretion.

† **E-mail:** wangh@usgs.gov

Introduction

Large Mississippi River (MR) diversions have been proposed for coastal restoration to rehabilitate and rebuild wetlands to alleviate significant historic wetland loss in coastal Louisiana (Day et al. 2009, Allison and Meselhe 2010, Allison et al. 2014). The Louisiana wetland loss crisis is mainly caused by the combination of eustatic sea-level rise (SLR), subsidence, and reduced sediment supply due to the construction of dams and levees along the MR (Blum and Roberts 2009). These dams and levees restrict the distribution of wetland-building mineral sediments and associated nutrients onto the deltaic plain. River diversions, akin to punching holes in existing levees, reintroduce freshwater and mineral sediments to degraded wetlands (Allison and Meselhe 2010, Allison et al. 2012, 2013, Kolker et al. 2012, Nittrouer et al. 2012) and, in the process, enhance marsh vertical accretion and surface elevation when sediment loads are sufficient (DeLaune et al. 2003, 2013, Lane et al. 2006, Day et al. 2011, Couvillion et al. 2013, Wang et al. 2014) which, in turn, could potentially mitigate the effects of climate change, specifically rising sea levels.

Other ecological benefits of freshwater and sediment diversions (hereafter, "diversions") include removing elevated nitrates in river water before discharge into the Gulf of Mexico (DeLaune et al. 2005, Wang et al. 2011, Rivera-Monroy et al. 2013, VanZomeren et al. 2013), increasing marsh primary productivity (both aboveground and belowground; Twilley and Nyman 2005, Day et al. 2012), and, potentially, increasing rates of soil organic carbon (SOC) sequestration (DeLaune and White 2012, DeLaune et al. 2013). Soil organic carbon sequestration is defined as "the process of transferring CO_2 from the atmosphere into the soil of land unit through unit plants, plant residue, and other organic solids, which are stored or retained in the unit as part of the soil organic matter (SOM)" (Olson et al. 2014). Generally, coastal wetland ecosystems are known to sequester carbon efficiently because of their high rates of primary productivity and low rates of SOM decomposition as well as accumulating soil and sediment carbon 40 times faster than the average terrestrial forest (Mcleod et al. 2011). This ecosystem service allows coastal wetlands to be entered into the voluntary carbon market, once a SOC sequestration rate is quantified (Chmura et al. 2003, Mitra et al. 2005, Crooks et al. 2011, Callaway et al. 2012, DeLaune and White 2012, Pendleton et al. 2012, Mitsch et al. 2013). As it stands now, due to net wetland loss in the MR Deltaic Plain, the estimated rates of SOC loss range from 1000 to 2050 g $C{\cdot}m^{-2}{\cdot}yr^{-1}$ for the period 1978–2000 (Markewich et al. 2007), while sequestration by vertical accretion occurs at a lower rate (100–380 g $C{\cdot}m^{-2}{\cdot}yr^{-1}$; DeLaune and White 2012). However, little is known about the influence of large river diversions on landscape-level sequestration of SOC. One major question is whether large MR diversions can maximize SOC sequestration and land-building potential simultaneously.

In this study, our objective was to examine the effect of MR diversions on the capacity of Louisiana's deltaic wetlands to sequester carbon in soils using a range of discharge rates and locations along the MR. Establishing discharge rates (including opening time and duration) and suitable locations for river diversion projects are the two major concerns in studying, planning, and implementing diversions to achieve maximum land building while simultaneously reducing flood risks to established wetlands (Day et al. 2012, Meselhe et al. 2012, 2013). Specifically, we evaluated SOC sequestration by vertical accretion and land area change under the influence of a series of individual proposed diversion projects from the lower MR in the Barataria and Breton Sound basins under future plausible environmental changes using simulation results of a Wetland Morphology model developed for the 2012 Coastal Master Plan (Steyer et al. 2012, Couvillion et al. 2013, Peyronnin et al. 2013, Wang et al. 2014). The Wetland Morphology model simulated multiple individual diversion projects for a 50-yr period (2010–2060) under future environmental conditions taking into account eustatic SLR, subsidence, MR discharge, and storm frequency and intensity (Peyronnin et al. 2013). The assessment of river diversion impacts on SOC sequestration in deltaic wetlands could provide critical information for the optimal design and implementation of future river diversion projects. Such information is crucial not only for coastal Louisiana, but also for other coastal regions in the world (e.g., the Yellow River Delta, China), where river diversions are used as a technique for coastal restoration (Cui et al. 2009, Wang et al. 2011).

Materials and Methods

Study area

The MR Deltaic Plain includes the Terrebonne, Barataria, Breton Sound, MR Delta, and Pontchartrain basins (Fig. 1). We focused our analyses on the coastal marsh within two major hydrological basins: Barataria and Breton Sound (Fig. 1). These locations were selected because the 2012 Louisiana Coastal Master Plan (http://coastal.la.gov/a-common-vision/2012-coastal-master-plan/) proposed a number of MR diversions in these two basins (Peyronnin et al. 2013). Vegetation across the region was classified into four types along a salinity gradient from low to high seaward in our model application: freshwater (0–0.5 practical salinity units [psu]), intermediate (0.5–5 psu), brackish (5–12 psu), and saline (12–20 psu) marshes (Stagg et al. 2016). The dominant species in these habitats are *Panicum hemitomon* and *Typha lancifolia* (freshwater marsh), *Sagittaria lancifolia* and *Schoenoplectus americanus* (intermediate marsh), *Spartina patens* and *S. americanus* (brackish marsh), and *Spartina alterniflora* and *Juncus roemerianus* (saline marsh; Visser et al. 2003, Sasser et al. 2008).

Calculation of landscape SOC sequestration rates

In this study, we used the simulation results of a spatially explicit Wetland Morphology model to assess and predict the landscape effects of MR diversions on SOC sequestration under future environmental conditions. The Wetland Morphology model was one of the seven integrated, coastwide predictive models that were developed in support of Louisiana's 2012 Coastal Master Plan (Peyronnin et al. 2013). The description of each predictive model and their integration appears in Appendix S1. Other models include the Eco-Hydrology, Barrier Shoreline Morphology, Vegetation, Ecosystem Services, Storm Surge and Wave, and Risk Assessment models (Appendix S1: Fig. S1). The Wetland Morphology model predicts coastwide land area and landscape configuration

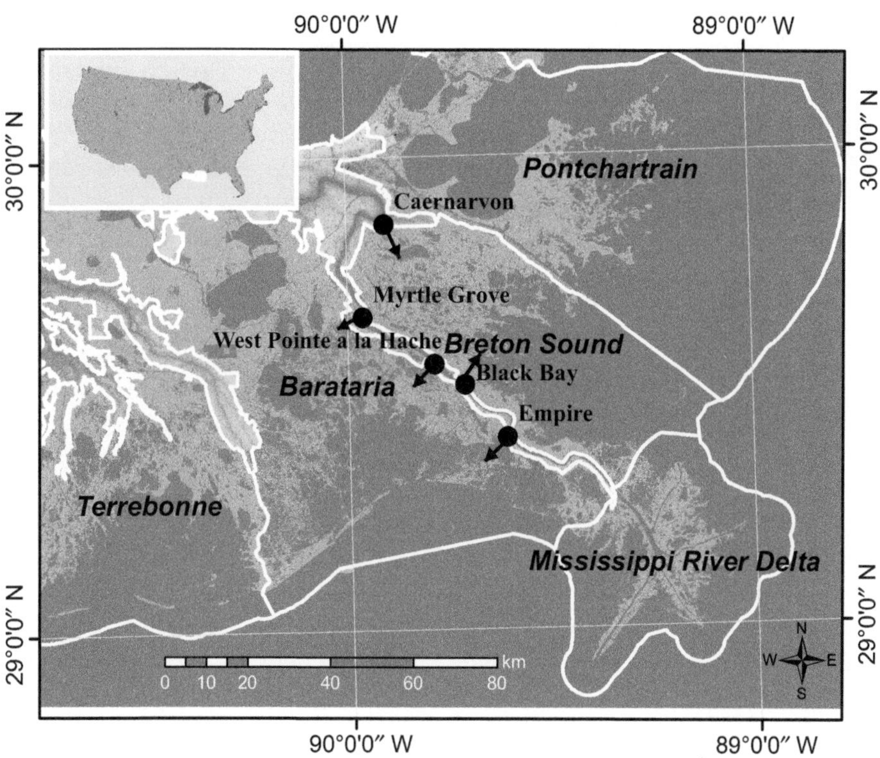

Fig. 1. Map of studied Louisiana deltaic wetlands and locations of selected diversions along the lower Mississippi River.

in response to surface elevation change, hurricanes, and implementation of protection and restoration projects. The model is driven by salinity, water level (stage), and sediment accumulation provided by the Eco-Hydrology model and plant community type and distribution from the Vegetation model. Other inputs for the Wetland Morphology model include historical land loss rates, wetland area, topography, bathymetry, land subsidence, and eustatic sea level. The outputs of the Wetland Morphology model include areas of land and water, vertical accretion, and surface elevation (Steyer et al. 2012). The Wetland Morphology model was calibrated and validated on long-term (decadal) vertical accretion derived from cesium-137 (^{137}Cs) dating of sediment cores from previous studies and the Coastwide Reference Monitoring System (CRMS; http://lacoast.gov/crms2/home.aspx). Appendix S1 also describes in detail the theory, assumptions, equations, inputs, outputs, calibration, and validation of the Wetland Morphology model.

The rate of SOC sequestration (ΔSOC) is estimated from model-simulated vertical accretion, depending on location, as:

$$\Delta\text{SOC} = \left(\frac{\text{BD} \times \text{OM} \times 0.01}{2.2}\right) \times H \times 10{,}000 \quad (1)$$

where ΔSOC is SOC sequestration rate (g $\text{C}\cdot\text{m}^{-2}\cdot\text{yr}^{-1}$), BD is soil bulk density (g/cm^3), OM is soil organic matter content (%), 2.2 is the organic carbon-to-organic matter conversion (OC-OM) factor, which was derived from CRMS data analysis (Steyer et al. 2012, Wang et al. 2016), and H is the rate of vertical accretion (cm/yr). In this study, SOC sequestration rates depend mainly upon the contribution of OM that is determined by the balance between plant growth (belowground) and OM decomposition in addition to inputs from other sources via river diversion. The OC-OM factor was derived from 60 soil cores collected during 2006–2007 to a depth of ~50 cm at 30 sites across costal Louisiana (Piazza et al. 2011). The linear regression between OC and OM in coastal Louisiana soils using this data set is described by the following equation (Steyer et al. 2012):

$$\begin{aligned} &\text{OC}\,(\%) = 0.4541 \times \text{OM}\,(\%) \\ &(n = 1142, R^2 = 0.93,\ P < 0.0001). \end{aligned} \quad (2)$$

The values of BD and OM were determined for each basin-vegetation type based on CRMS soil data (0–24 cm depth) and the set of 60 soil cores described above through a vertical accretion calibration process (Steyer et al. 2012). Bulk density and OM values vary correspondingly when vegetation types change in a particular basin as a result of fluctuations in salinity and flooding regimes (Visser et al. 2013).

The landscape processes of vegetation type change, vertical accretion, wetland/water area change, and associated SOC sequestration were represented at a grid cell resolution of 500 m. We first calculated SOC density ([BD × OM]/2.2) at each pixel (500-m resolution) for two time periods (2010–2035 and 2036–2060) based on the vegetation community distribution model output from the Vegetation model (Visser et al. 2013). We then estimated SOC sequestration rates for each diversion project and Future-Without-Action (FWOA) scenario under the two future environmental scenarios (moderate and less optimistic; see details in *Future environmental scenarios*) during the two periods by multiplying the derived SOC density by the vertical accretion rate from the Wetland Morphology model (Couvillion et al. 2013). Lastly, we used the areas of land and water distribution during the two periods simulated by the Wetland Morphology model (Couvillion et al. 2013) to define wetland areas across basins and vegetation types. These estimated areas served as the zones for the estimation of SOC sequestration rates in the selected basins (Barataria and Breton Sound) and vegetation types (freshwater, intermediate, brackish, and saline marshes) using the ESRI's ArcGIS zonal statistical analysis tool (ArcMap 10.2.2.; ESRI, Redlands, California, USA).

Individual sediment diversion projects

We selected eight individual river diversion projects (Table 1) from a series of proposed diversion projects evaluated for the 2012 Coastal Master Plan to assess their effects on basin-wide and vegetation-based SOC sequestration rates. The discharge rates (Q: peak flow through the structure and channel) of these diversion projects range from 142 to 7080 m^3/s. Three scales of MR diversions were selected using the peak flow rate: (1) large: diversion discharge ≥1416 up to 7080 m^3/s; (2) medium: >283–1415 m^3/s; and (3)

Table 1. Features of selected individual diversion projects in the 2012 Coastal Master Plan for the Barataria and Breton Sound basins.

Diversion	Description	Peak flow (m^3/s)
Barataria		
Myrtle Grove	Myrtle Grove Diversion, 142 m^3/s capacity (continuous operation at capacity for river flows above 5663 m^3/s; no operation below 5663 m^3/s)	142
Myrtle Grove	Myrtle Grove Diversion, 1416 m^3/s capacity (operation at capacity when Mississippi River (MR) flow exceeds 16,990 m^3/s; operation at 8% of river flow from 16,990 down to 5663 m^3/s; no operation below 5663 m^3/s)	1416
Myrtle Grove	Myrtle Grove Diversion, 7080 m^3/s capacity (operation at capacity when MR flow exceeds 25,485 m^3/s; operation at 1416 m^3/s for flows from 25,485 down to 16,990 m^3/s; operation at 8% of river flow for river flows from 16,990 down to 5663 m^3/s; no operation below 5663 m^3/s)	7080
West Point a la Hache	West Pointe a la Hache Diversion, 1416 m^3/s capacity (operation at capacity when MR flow exceeds 16,990 m^3/s; operation at 8% of river flow from 16,990 down to 5663 m^3/s; no operation below 5663 m^3/s)	1416
Empire	Empire Diversion, 1416 m^3/s capacity (operation at capacity when MR flow exceeds 16,990 m^3/s; operation at 8% of river flow from 16,990 down to 5663 m^3/s; no operation below 5663 m^3/s)	1416
Breton Sound		
Caernarvon	Caernarvon Diversion, 1416 m^3/s capacity (operation at capacity when MR flow exceeds 16,990 m^3/s; operation at 8% of river flow from 16,990 down to 5663 m^3/s; no operation below 5663 m^3/s)	1416
Caernarvon	Caernarvon Diversion, 7080 m^3/s capacity (70% Mississippi/30% Atchafalaya; operation at capacity when MR flow exceeds 25,485 m^3/s; operation at 1416 m^3/s for flows from 25,485 down to 16,990 m^3/s; operation at 8% of river flow for river flows from 16,990 down to 5663 m^3/s; no operation below 5663 m^3/s)	7080
Black Bay	Black Bay Diversion, 7080 m^3/s capacity (60% Mississippi/40% Atchafalaya; operation at capacity when MR flow exceeds 25,485 m^3/s; operation at 1416 m^3/s for flows from 25,485 down to 16,990 m^3/s; operation at 8% of river flow for river flows from 16,990 down to 5663 m^3/s; no operation below 5663 m^3/s)	7080

small: $\leq$283 m^3/s (Wang et al. 2014). According to the diversion protocols described in Table 1, there are power function relationships between peak flow rates and annual total water amounts and between peak rates and sediment load; thus, we used peak flow rates in our model evaluation. The higher the peak flow rate, the higher the delivered water amount and sediment load. For example, annual total flows through the Caernarvon Diversion at Breton Sound Basin with diversion flow regimes of 142, 1416, and 7080 m^3/s would be 4.5×10^9, 38.4×10^9, and 56.3×10^9 m^3, respectively. Under these three diversion regimes, the annual total sediment loads would be 0.1, 1.2, and 2.1 million metric tons, respectively.

Future environmental scenarios

A number of uncertainty parameters were identified in the 2012 Coastal Master Plan modeling study: eustatic SLR, subsidence, hurricane/storm intensity and frequency, MR discharge, rainfall, evapotranspiration, MR nutrient concentrations, and marsh collapse thresholds (Peyronnin et al. 2013). The ranges in uncertainty parameters in the next 50 yr (2010–2060) were determined from previous studies and best professional judgment (Table 2). Two future environmental change scenarios were selected: (1) *moderate* and (2) *less optimistic* (Couvillion et al. 2013, Meselhe et al. 2013, Visser et al. 2013). Full details on the plausible range of each uncertainty parameter under each scenario are described in Peyronnin et al. (2013) and Wang et al. (2014). In particular, eustatic SLR rates over the 50-yr simulation period were non-linear and resulted in 0.27 and 0.45 m of sea-level change, respectively. In the case of the moderate and less optimistic scenarios, subsidence rates varied spatially in the range of 0–1.9 cm/yr and 0–2.5 cm/yr. Consequently, the time-averaged rates of relative SLR (RSLR = eustatic SLR + subsidence) for Barataria Basin and Breton Sound Basin were 1.24 and 1.14 cm/yr under the moderate scenario and 2.00 and 1.80 cm/yr under the less optimistic scenario, respectively (Wang et al. 2014). As a

Table 2. Environmental uncertainty parameters and the two scenarios used in the 2012 Louisiana Master Plan model simulations.

Parameter	Plausible range	Moderate scenario	Less optimistic scenario
Sea-level rise	0.16–0.8 m over 50 yr	0.27 m over 50 yr	0.45 m over 50 yr
Subsidence	0–3.5 cm/yr; varies spatially	0–1.9 cm/yr (values vary spatially)	0–2.5 cm/yr (values vary spatially)
Storm intensity	Current storm intensities to +30% of current intensities	+10% of current intensities	+20% of current intensities
Storm frequency	−20% to +10% of current storm frequency	Current storm frequency (One Category 3 or greater storm every 19 yr)	+2.5% of current storm frequency (One Category 3 or greater storm every 18 yr)
Mississippi River (MR) discharge	−7% to +14% of annual mean discharge; adjusted for seasonality	Mean annual discharge (15,121 m^3/s)	−5% of mean annual discharge (14,413 m^3/s)
Rainfall	Historical monthly range; varies spatially	Variable percentage of historical monthly mean	Variable percentage of historical monthly mean
Evapotranspiration	Historical monthly range (±1 SD); varies spatially	Historical monthly mean (values vary spatially)	+0.4 SD from historical mean monthly (values vary spatially)
MR nutrient concentration	−45% to +20% of current nitrogen and phosphorus concentrations	−12% of current concentrations (mg/L) Phosphorus = 0.19 Nitrite + nitrate = 1.1 Ammonium = 0.038 Organic nitrogen = 0.67	Current concentrations (mg/L) Phosphorus = 0.22 Nitrite + nitrate = 1.3 Ammonium = 0.044 Organic nitrogen = 0.77
Marsh collapse threshold	Salinity (psu) Swamp: 4–7 Fresh marsh: 6–8 Inundation (water depth, cm) Intermediate marsh: 31–38 Brackish marsh: 20–26 Saline marsh: 16–23	Mid-range values of salinity and inundation result in collapse Salinity (psu) Swamp: 6 Fresh marsh: 7 Inundation (cm) Intermediate marsh: 34 Brackish marsh: 23 Saline marsh: 21	Lower 25th percentile values of salinity and/or inundation ranges result in collapse Salinity (psu) Swamp: 5 Fresh marsh: 7 Inundation (cm) Intermediate marsh: 33 Brackish marsh: 21 Saline marsh: 18

comparison, the average RSLR in coastal Louisiana based solely upon tide gauge trends at Grand Isle, Louisiana, during 1947–2006, was 0.92 cm/yr (Couvillion and Beck 2013). These rates would be higher if wetland submergence were included (Cahoon 2015).

Assessment of diversion effects on landscape SOC sequestration rates

The effects of sediment diversions on SOC sequestration were evaluated under the following criteria.

Discharge rate effect.—We selected Myrtle Grove in Barataria Basin and Caernarvon in Breton Sound Basin and used three master plan diversion peak flow rates in each basin: 142, 1416, and 7080 m^3/s as well as the FWOA condition to examine the basin-wide and vegetation type-based average SOC sequestration rates. Because historical flows from existing diversions at Davis Pond in Barataria and Caernarvon in Breton Sound basins were incorporated in both FWOA and master plan diversion conditions (Meselhe et al. 2013), the diversion discharge rates used in our analysis do represent FWOA water flows into the two basins.

Location effect.—We selected diversions in Barataria Basin with the same discharge peak flow rate of 1416 m^3/s (downriver order: Myrtle Grove, West Pointe a la Hache, and Empire) and in Breton Sound Basin with the same discharge rate of 7080 m^3/s (downriver order: Caernarvon and Black Bay) to examine basin-wide and vegetation type-based average SOC sequestration rates over the 50-yr simulation period (Fig. 1).

Statistical analysis

We conducted a three-way ANOVA with hydrologic basin, future environmental scenario, and diversion discharge and all possible interactions as explanatory variables for SOC sequestration rates in Barataria and Breton Sound basins. We also conducted a three-way ANOVA with vegetation type, future environmental scenario, and diversion discharge and their interactions on SOC sequestration rates. A three-way ANOVA was also used to examine the effects of diversion location, vegetation types, and future environmental scenario on SOC sequestration rates. Data were square-root-transformed to meet normality and homoscedasticity assumptions. The SAS 9.3 software package (SAS Institute, Cary, North Carolina, USA) was used for the statistical analyses. All the tests were two-tailed based on type III sums of squares and considered significant at $P < 0.05$.

Table 3. Statistical result of the effects of hydrologic basin, future environmental scenario, and diversion discharge on simulated SOC sequestration rate.

Factor	df	Type III SS	F	P
Basin	1	20,265.48	1438.15	***
Environmental scenario	1	66,914.92	4748.65	***
Discharge	3	20,681.75	489.23	***
Basin × environmental scenario	1	27,918.03	1981.22	***
Basin × discharge	2	20,452.09	725.70	***
Environmental scenario × discharge	3	31,780.32	751.77	***
Basin × environmental scenario × discharge	2	14,302.72	507.50	***

Note: SOC, soil organic carbon; SS, sums of squares; df, degrees of freedom.
*** $P < 0.001$.

Results

Discharge rate effect

Our results indicated that simulated SOC sequestration rates varied significantly ($P < 0.001$) with hydrologic basin, future environmental scenario, diversion discharge, and their interactions (Table 3). The highest average simulated SOC sequestration rates within Barataria Basin (mean ± standard error; moderate: 168 ± 1.0 g $C{\cdot}m^{-2}{\cdot}yr^{-1}$, $n = 10{,}325$; and less optimistic: 252 ± 1.4 g $C{\cdot}m^{-2}{\cdot}yr^{-1}$, $n = 9433$) were observed for the sediment diversion with a peak flow rate of 1416 m^3/s under both the moderate and less optimistic scenarios (Fig. 2a). Compared to the rates under the FWOA condition and the moderate scenario (162 ± 1.07 g $C{\cdot}m^{-2}{\cdot}yr^{-1}$, $n = 9699$), the

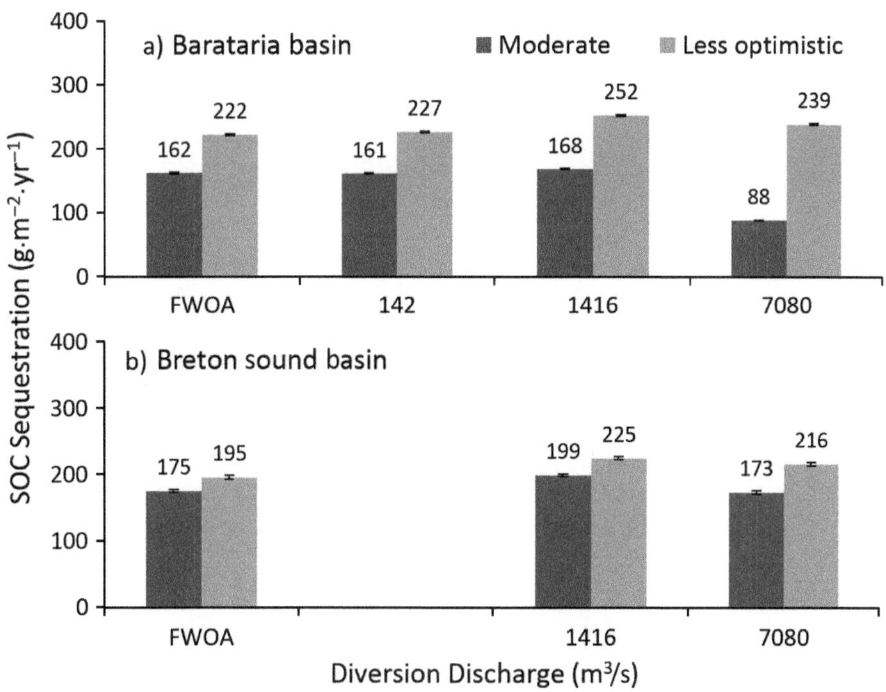

Fig. 2. Simulated soil organic carbon (SOC) sequestration rates with varying diversion discharge rates in the Barataria and Breton Sound basins under moderate and less optimistic scenarios. Means (±1 standard error) are significantly different ($P < 0.001$) among diversion discharge and future environmental scenario combinations.

simulated SOC sequestration rates increased by 3.8% at a 1416 m^3/s discharge, whereas simulated rates decreased by 0.6% and 45.5% when 142 and 7080 m^3/s discharge rates were used. In contrast, simulated SOC sequestration rates increased with discharge rates of 142 (2.1%), 1416 (13.7%), and 7080 m^3/s (7.5%), compared to rates under the FWOA and the less optimistic scenarios (222 ± 1.3 g $C{\cdot}m^{-2}{\cdot}yr^{-1}$, $n = 8914$).

Similarly, within the Breton Sound Basin, the highest averaged SOC sequestration rates were observed for the diversion with the peak flow rate of 1416 m^3/s under both scenarios (moderate: 199 ± 2.0 g $C{\cdot}m^{-2}{\cdot}yr^{-1}$; less optimistic: 225 ± 2.3 g $C{\cdot}m^{-2}{\cdot}yr^{-1}$; Fig. 2b). In the case of the FWOA condition under the moderate scenario (SOC sequestration rate: 175 ± 2.2 g $C{\cdot}m^{-2}{\cdot}yr^{-1}$), the simulated rate of SOC sequestration increased by 13.7% applying the 1416 m^3/s discharge, while the rate decreased by 1.1% using the 7080 m^3/s discharge rate. Further, SOC sequestration rates increased by 15.4% and 10.8% using the 1416 and 7080 m^3/s discharge rates as input when compared to simulated SOC sequestration rates under FWOA condition and the less optimistic scenario (i.e., 195 ± 2.8 g $C{\cdot}m^{-2}{\cdot}yr^{-1}$).

Simulated SOC sequestration rates varied significantly as a result of interactions ($P < 0.001$) among vegetation type, environmental scenario, and diversion discharge (Table 4). Plots of simulated SOC sequestration rates across vegetation types under various discharge rates and environmental scenarios in Barataria Basin and Breton Sound Basin are shown in Appendix S2. In Barataria Basin, brackish marsh soils had the highest SOC sequestration (374 ± 3.4 g $C{\cdot}m^{-2}{\cdot}yr^{-1}$) under diversion water flow of 1416 m^3/s in the moderate scenario followed by saline and intermediate marsh soils (280 ± 2.7 g $C{\cdot}m^{-2}{\cdot}yr^{-1}$ and 276 ± 3.6 g $C{\cdot}m^{-2}{\cdot}yr^{-1}$, respectively). Freshwater marsh soils showed the lowest SOC sequestration rates (166 ± 0.7 g $C{\cdot}m^{-2}{\cdot}yr^{-1}$) under the flow of 1416 m^3/s (Appendix S2: Fig. S1a).

Table 4. Statistical results of the effects of vegetation type, environmental scenario, and diversion discharge on simulated SOC sequestration rate.

Factor	df	Type III SS	F	P
Vegetation type	3	346,317.78	9928.71	***
Environmental scenario	1	29,669.47	2551.82	***
Discharge	3	29,230.39	838.02	***
Vegetation type × environmental scenario	3	7352.89	210.80	***
Vegetation type × discharge	9	27,529.09	263.08	***
Environmental scenario × discharge	3	19,783.86	567.19	***
Vegetation type × environmental scenario × discharge	9	17,711.46	169.26	***

Note: SOC, soil organic carbon; SS, sums of squares; df, degrees of freedom.
*** $P < 0.001$.

Under the less optimistic scenario, SOC sequestration capacity in Barataria Basin changed in response to the salinity gradient. Saline marsh soils showed the highest SOC sequestration rates (350 ± 3.1 g $C{\cdot}m^{-2}{\cdot}yr^{-1}$), followed by brackish and intermediate marsh soils (296 ± 4.5 and 282 ± 1.3 g $C{\cdot}m^{-2}{\cdot}yr^{-1}$); the lowest SOC sequestration capacity (177 ± 1.4 g $C{\cdot}m^{-2}{\cdot}yr^{-1}$) was registered in the freshwater marsh soils under diversion water flow of 1416 m^3/s (Appendix S2: Fig. S1a). Unlike the basin-wide increase in SOC sequestration capacity under the less optimistic scenario, the 1416 m^3/s diversion scenario resulted in decreasing SOC sequestration rates in brackish (−3.4%) and saline marshes (−23.5%) compared to the FWOA condition. Further, freshwater and intermediate marshes sequestered more carbon than saline and brackish marshes in Barataria Basin when the total area for each vegetation category was considered (Fig. 3a). Indeed, freshwater marshes (1373–1723 km^2) contributed 53–66% of the total annual sequestered soil carbon under the moderate scenario while intermediate marshes (941–991 km^2) contributed 54–62% of the total annual carbon sequestration under the less optimistic scenario (Fig. 3a).

In the case of Breton Sound Basin, a river diversion of 7080 m^3/s resulted in an increase in SOC sequestration in freshwater marsh when compared to the FWOA scenario (Appendix S2: Fig. S1b). The increase in simulated SOC sequestration rates per marsh type was 37.8% under the moderate scenario and 37.6% under the less optimistic scenario. Similar to the case in Barataria Basin, freshwater marshes in Breton Sound Basin (116–242 km^2) were also critical contributors (up to 55%) to total carbon sequestered, especially under medium- to large-scale diversions, although brackish marshes (20–249 km^2) contributed more carbon under FWOA (up to 70%) and with small freshwater flow (Fig. 3b). These

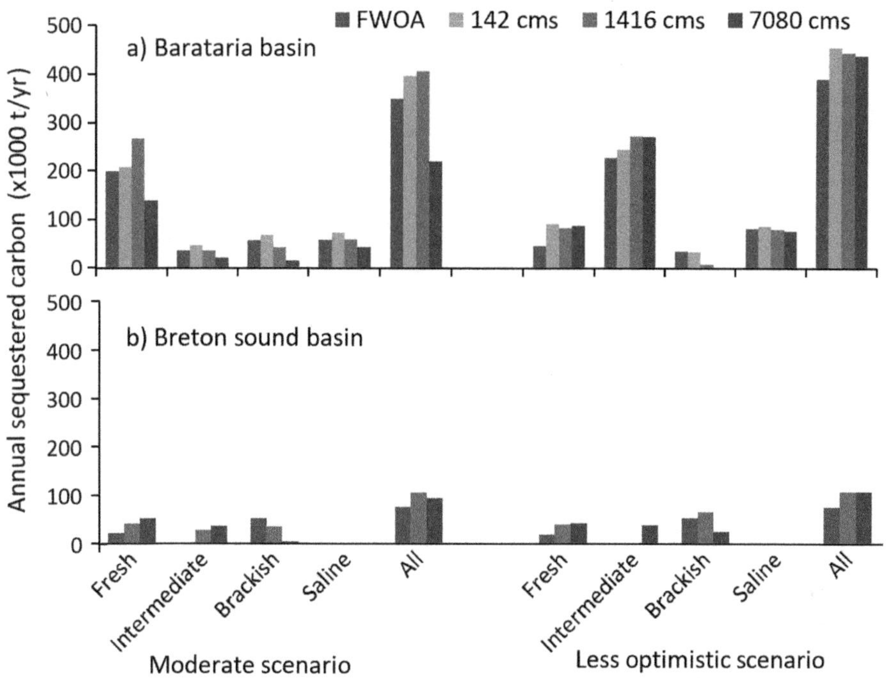

Fig. 3. Simulated annual total sequestered soil carbon (×1000 t/yr) within different marsh types with varying diversion discharge rates under moderate and less optimistic scenarios (cms = m^3/s).

simulation results for Breton Sound showed a larger effect of MR diversion on SOC sequestration in low-salinity (freshwater and intermediate marshes) than in high-salinity marshes (brackish and saline marshes).

Location effect

Simulated basin-wide SOC sequestration rates varied with the diversion structure location. The highest basin-wide SOC sequestration rates (168 ± 1.0 g $C{\cdot}m^{-2}{\cdot}yr^{-1}$ and 252 ± 1.4 g $C{\cdot}m^{-2}{\cdot}yr^{-1}$ under the moderate and less optimistic scenarios, respectively) were obtained when placing the diversion at or near Myrtle Grove in the upper receiving basin in Barataria Basin (Fig. 4a). Placing a diversion with a capacity flow of 1416 m^3/s near West Pointe a la Hache and Empire resulted in reduction in SOC sequestration ranging from 3.2% to 9.4% (159 ± 0.9 g $C{\cdot}m^{-2}{\cdot}yr^{-1}$ and 162 ± 1.1 g $C{\cdot}m^{-2}{\cdot}yr^{-1}$ under the moderate scenario and 231 ± 1.3 g $C{\cdot}m^{-2}{\cdot}yr^{-1}$ and 228 ± 1.3 g $C{\cdot}m^{-2}{\cdot}yr^{-1}$ under the less optimistic scenario) compared to the case when the diversion structure was placed next to Myrtle Grove (Fig. 4a). Placing a diversion of 7080 m^3/s near Black Bay in the lower receiving basin in the Breton Sound Basin resulted in an increase in SOC sequestration rates by 7.5% and 1.0% under the moderate (i.e., 186 ± 2.5 g $C{\cdot}m^{-2}{\cdot}yr^{-1}$) and less optimistic scenarios (i.e., 218 ± 2.5 g $C{\cdot}m^{-2}{\cdot}yr^{-1}$) when compared to building the diversion structure at or near Caernarvon (173 ± 2.5 g $C{\cdot}m^{-2}{\cdot}yr^{-1}$ and 216 ± 2.5 g $C{\cdot}m^{-2}{\cdot}yr^{-1}$ under the two scenarios; Fig. 4b).

Simulated SOC sequestration rates varied significantly ($P < 0.001$) with diversion structure location, vegetation type, environmental scenario, and their interactions in Barataria Basin (Table 5) and Breton Sound Basin (Table 6). The effect of diversion location on carbon sequestration rates varied with the type of vegetation present within each of the two hydrological basins. Plots of simulated SOC sequestration rates across vegetation types under different diversion structure locations and environmental scenarios in Barataria Basin and Breton Sound Basin are shown in Appendix S2: Fig. S2. Placing the 1416 m^3/s diversion at or near Empire in the lower receiving basin in Barataria Basin slightly increased SOC sequestration rates in saline marsh soils under the

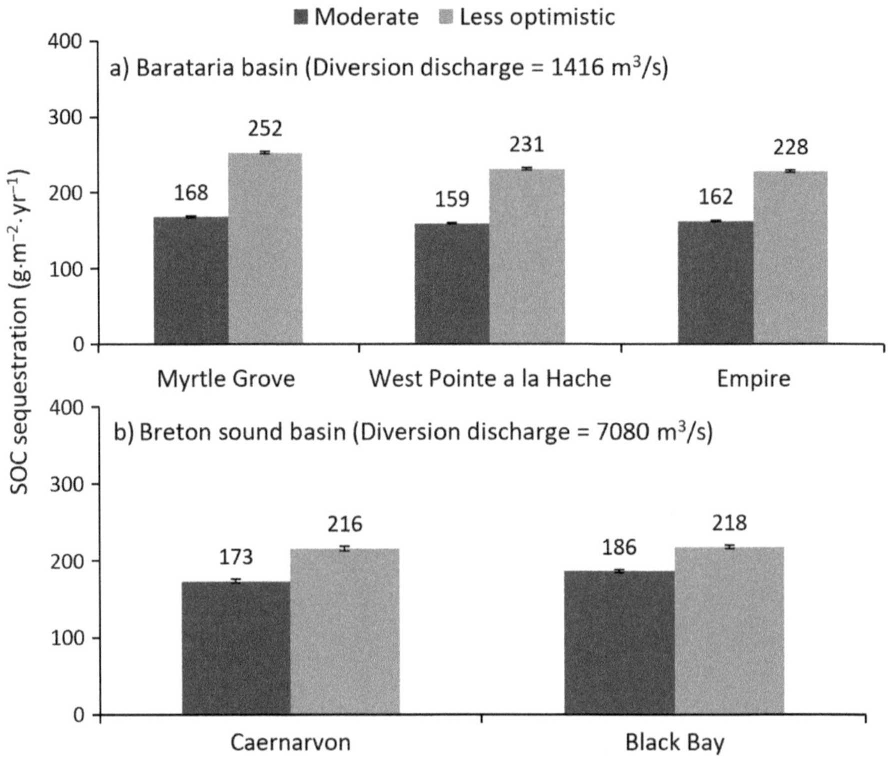

Fig. 4. Simulated soil organic carbon (SOC) sequestration rates under moderate and less optimistic scenarios when placing diversion structures at Barataria Basin (with diversion discharge of 1416 m^3/s) and Breton Sound Basin (with diversion discharge of 7080 m^3/s) at different locations. Means (±1 standard error) are significantly different ($P < 0.001$) among diversion location and future environmental scenario combinations.

Table 5. Statistical results of the effects of diversion location, vegetation type, and environmental scenario on simulated SOC sequestration rate in Barataria Basin.

Factor	df	Type III SS	*F*	*P*
Vegetation type	3	289,712.37	10,384.90	***
Environmental scenario	1	240.38	25.85	***
Location	2	5318.96	285.99	***
Vegetation type × environmental scenario	3	6223.75	223.09	***
Vegetation type × location	6	2880.69	51.63	***
Environmental scenario × location	2	57.87	3.11	*
Vegetation type × environmental scenario × location	6	2653.17	47.55	***

Note: SOC, soil organic carbon; SS, sums of squares; df, degrees of freedom.
$^*P < 0.05$, $^{***}P < 0.001$.

Table 6. Statistical results of the effects of diversion location, vegetation type, and environmental scenario on simulated SOC sequestration rate in Breton Sound Basin.

Factor	df	Type III SS	*F*	*P*
Vegetation type	3	8629.18	335.75	***
Environmental scenario	1	722.56	84.34	***
Location	1	445.08	51.95	***
Vegetation type × environmental scenario	3	761.16	29.62	***
Vegetation type × location	3	4955.04	192.79	***
Environmental scenario × location	1	45.48	5.31	*
Vegetation type × environmental scenario × location	3	646.37	25.15	***

Note: SOC, soil organic carbon; SS, sums of squares; df, degrees of freedom.
$^*P < 0.05$, $^{***}P < 0.001$.

moderate scenario (284 ± 2.3 g $C{\cdot}m^{-2}{\cdot}yr^{-1}$) and in brackish marsh soils under the less optimistic scenario (309 ± 3.2 g $C{\cdot}m^{-2}{\cdot}yr^{-1}$) compared to placing the diversion next to Myrtle Grove (280 ± 3.4 g $C{\cdot}m^{-2}{\cdot}yr^{-1}$ for saline marsh and 296 ± 4.5 g $C{\cdot}m^{-2}{\cdot}yr^{-1}$ for brackish marsh; Appendix S2: Fig. S2a). This is in contrast with the decreasing basin-wide SOC rates with locations at Myrtle Grove, West Point a la Hache and Empire (Fig. 4a). In the case of Breton Sound Basin, placing a 7080 m^3/s large diversion at Caernarvon resulted in higher SOC sequestration rates for freshwater (220 ± 2.2 g $C{\cdot}m^{-2}{\cdot}yr^{-1}$ and 229 ± 2.4 g $C{\cdot}m^{-2}{\cdot}yr^{-1}$) and intermediate (245 ± 3.6 g $C{\cdot}m^{-2}{\cdot}yr^{-1}$ and 329 ± 3.6 g $C{\cdot}m^{-2}{\cdot}yr^{-1}$) marsh soils when compared to placing the diversion at or near Black Bay under the moderate and less optimistic scenarios (Appendix S2: Fig. S2b). In contrast, installing the 7080 m^3/s river diversion at or near Black Bay resulted in increased SOC sequestration rates for brackish marshes (243 ± 2.2 g $C{\cdot}m^{-2}{\cdot}yr^{-1}$ and 277 ± 2.1 g $C{\cdot}m^{-2}{\cdot}yr^{-1}$) compared to placing the diversion at Caernarvon (215 ± 7.9 g $C{\cdot}m^{-2}{\cdot}yr^{-1}$ and 223 ± 2.4 g $C{\cdot}m^{-2}{\cdot}yr^{-1}$) under the two scenarios (Appendix S2: Fig. S2a). In addition, relatively stable SOC sequestration rates were estimated for saline marsh soils (~130 and 144 g $C{\cdot}m^{-2}{\cdot}yr^{-1}$ under the two future scenarios; Appendix S2: Fig. S2b). These simulation results underscore the significant impact of diversion location on basin-wide and marsh type-specific SOC sequestration rates in Louisiana deltaic wetland soils. Locations of large diversions affect to a larger degree SOC sequestration in freshwater, intermediate, and brackish marshes than in saline marshes.

Discussion

Our results show that increasing discharge from 142 to 1416 m^3/s could significantly enhance SOC sequestration rates in Barataria and Breton Sound basins. However, the benefit in soil carbon gain due to a large diversion would decline as discharge rate increases beyond a threshold. This threshold appears between 1416 and 7080 m^3/s (Figs. 2, 3) depending on a hydrologic basin's geomorphological features and change in environmental conditions (e.g., salinity, inundation, and temperature). The diminishing marginal benefit of large diversions on soil carbon sequestration is primarily attributed to changes in vegetation structure (e.g., species composition, spatial distribution) as a result of the direct changes in salinity and hydroperiod regimes associated with large river diversions (Visser et al. 2013). Visser et al. (2013) found that brackish marsh area would shrink and saline marsh area would remain relatively stable, while areas of freshwater and intermediate marshes would expand with restoration including river diversions and SLR over the next 50 yr. The rate of SOC sequestration in this study was calculated as the product of OC density and vertical accretion (Eq. 1). Thus, estimated SOC sequestration rates depend largely upon the OC density regulated by the type of wetland vegetation growing in situ especially if communities are dominated by wetlands with high peat production as is the case of intermediate (mean OC density = 0.041 g/cm^3), brackish (0.039 g/cm^3), and saline (0.036 g/cm^3) marshes (Morris et al. 2013, Wang et al. 2016). Indeed, the lowest OC density has been reported for freshwater marshes (0.028 g/cm^3) based on the Louisiana CRMS soil data (over the 0–24 cm depth) compiled during 2006–2009 (Wang et al. 2016). This difference in SOC densities was explicitly included in the Wetland Morphology model, which assigned SOC values to the four marsh categories included in our simulations. The interaction between SOC density and plant composition is clearly observed when large diversions (i.e., 1416 to 7080 m^3/s) promote the conversion of non-freshwater marshes to freshwater marshes as a result of low surface water and porewater salinity (DeLaune et al. 2003). Once freshwater wetlands are dominant, total carbon sequestration decreases on a per unit area basis. Large diversions (i.e., 7080 m^3/s) could lead to a significant decrease in the percentage of SOC density when marshes switch from brackish (48%) and intermediate (28%) marshes to freshwater marshes while vertical accretion tended to increase (<20%) under the 7080 m^3/s diversion (Wang et al. 2014).

Further, marsh plant productivity also tends to have a parabolic relationship with water level or elevation (Morris et al. 2002, Mudd et al. 2009, Kirwan et al. 2012) with maximum growth and production at the peak of the parabola. In coastal Louisiana, relatively high plant productivity indicated by high Normalized Difference Vegetation Index was found to be at an elevation relative to mean water level: 4.7–19.0 cm for intermediate marsh, 12.6–26.8 cm for brackish marsh, and 16.3–31.9 cm for saline marsh, respectively (Couvillion and Beck 2013). This relationship between productivity and elevation indicates optimal water depths for marsh plant productivity, and once these optimal depths are exceeded, plant growth and biomass significantly decrease, likely resulting in a decrease in SOM accumulation. For instance, both aboveground biomass and belowground biomass of *Spartina alterniflora* and *Spartina patens* were found to decrease significantly in response to increased flooding in Breton Sound Basin (Snedden et al. 2015).

The location of a diversion structure relative to the coastline and the river watershed also plays a critical role in SOC sequestration. This role is further influenced by the discharge rate regime and the in situ geomorphological properties of the receiving hydrologic basin (e.g., configuration of land and water areas, vegetation distribution, topography, and bathymetry). A representative example of this effect on SOC rates in our modeling exercise is the diversion placed near Myrtle Grove in Barataria Basin that used a peak water flow of 1416 m^3/s (a large diversion). High SOC sequestration rates (Fig. 4a) correlated with the highest vertical accretion and reduction in elevation loss when placing the 1416 m^3/s diversion at or near Myrtle Grove (Wang et al. 2014). In contrast, when placing the diversion structure and setting a flow rate of 7080 m^3/s (a mega-diversion) within Breton Sound Basin at or near Caernarvon, SOC sequestration was not the highest at the basin-wide level (Fig. 4b) although this large diversion at this location could lead to the highest vertical accretion (>0.7 cm/yr) and reduction in wetland surface elevation loss compared to FWOA (Wang et al. 2014). Previous studies show that a diversion at an upriver location would reduce the energy, water, and sediment capture at downriver distributaries (Allison and Meselhe 2010, Allison et al. 2013, Meselhe et al. 2013). Normally, along the upper → middle → lower estuary gradient within a hydrologic basin, the land-to-water ratio decreases while the percentage of vegetated marsh area decreases, and marsh elevation decreases while water depth increases. All of these factors contribute to the lower retention rates of delivered sediment and decreased vertical accretion (Couvillion et al. 2013, Wang et al. 2014). Even with relatively higher OC densities (Wang et al. 2016), SOC sequestration in brackish and saline marshes would decrease because of lower vertical accretion in the lower basin (Wang et al. 2014).

Overall, the discharge and location effects of MR diversion projects identified by this study underscore the need to evaluate management priorities in obtaining desired ecosystem services from carbon sequestration under brackish or freshwater dominated marshes characterized by significant mineral sediment input that promotes vertical accretion in the MR Delta Plain.

Conclusions

Landscape-level SOC sequestration rates (both basin-wide and vegetation type averaged) under three MR sediment diversion discharge rates (peak flow: 142, 1416, and 7080 m^3/s) and under different diversion structure locations along the lower MR were evaluated using simulation results of a Wetland Morphology model developed, calibrated, and validated for the Louisiana's 2012 Coastal Master Plan. A diversion with a water discharge rate of 1416 m^3/s and placed in the upper reach of the lower MR resulted in the highest SOC sequestration at the landscape level. Simulation results indicate that large MR diversions (peak flow ≥1416 m^3/s and sediment loads >165 kg/s) are capable of improving landscape-scale SOC sequestration capacity in Louisiana deltaic wetlands. Additionally, large diversions could help mitigate wetland salinity stress (salinization) due to saltwater intrusion from SLR and land subsidence (Herbert et al. 2015), thus improving SOC sequestration capacity in freshwater marshes.

Additional individual diversion projects with large discharge rates ranging from 2000 to 6000 m^3/s should be considered in future feasibility modeling studies to explore the specific discharge thresholds in specific basins, vegetation types, and future environmental conditions including a range of relative SLR rates. Since most existing river diversions are small scale (≤283 m^3/s) and sea level is rising at an accelerating rate induced by climate change (Williams 2013), large diversions would be required if soil carbon sequestration capacity in Louisiana deltaic wetlands is to be enhanced or maintained under future climate and environmental change.

Since SOC sequestration rates may decrease once a potential threshold of discharge is exceeded, it is paramount to evaluate the economic and ecological consequences due to the decline of diversion benefit in sequestering carbon via vertical accretion. Model results imply that the diversion discharge threshold, if present, would vary with basin, vegetation type, future environmental conditions, and possibly location of diversion structure as well.

Acknowledgments

This research was supported by funds from the State of Louisiana's Coastal Protection and Restoration Authority (LACPRA) in support of Louisiana's 2012 Coastal Master Plan to USGS Wetland and Aquatic Research Center (WARC). We would like to thank Ehab Meselhe, Alex McCorquodale, Guerry Holm Jr., and Sijan Sapkota for their assistance in data analysis. We would like to thank Michael J. Osland, Julie Zinnert, and two anonymous reviewers for their constructive review for improving this manuscript. Participation of VRHM was partially supported by the NSF-Coupled Natural and Human Systems program under grant no. DBCS 1212112. HW and KWK were also partially supported by USGS Climate and Land Use Research & Development Program and the USGS Ecosystem Mission Area. Any use of trade, firm, or product names is for descriptive purposes only and does not imply endorsement by the U.S. Government.

Literature Cited

Allison, M. A., C. R. Demas, B. A. Ebersole, B. A. Kleiss, C. D. Little, E. A. Meselhe, N. J. Powell, T. C. Pratt, and B. M. Vosburg. 2012. A water and sediment budget for the lower Mississippi-Atchafalaya River in flood years 2008-2010: implications for sediment discharge to the oceans and coastal restoration in Louisiana. Journal of Hydrology 432–433:84–97.

Allison, M. A., and E. A. Meselhe. 2010. The use of large water and sediment diversions in the lower Mississippi River (Louisiana) for coastal restoration. Journal of Hydrology 387:346–360.

Allison, M. A., M. T. Ramirez, and E. A. Meselhe. 2014. Diversion of Mississippi River water downstream of New Orleans, Louisiana, USA to maximize sediment capture and ameliorate coastal land loss. Water Resources Management 28:4113–4126.

Allison, M. A., B. M. Vosburg, M. T. Ramirez, and E. A. Meselhe. 2013. Mississippi River channel response to the Bonnet Carre Spillway opening in the 2011 flood and its implications for the design and operation of river diversions. Journal of Hydrology 477:104–118.

Blum, M. D., and H. H. Roberts. 2009. Drowning of the Mississippi Delta due to insufficient sediment supply and global sea level rise. Nature Geoscience 2:488–491.

Cahoon, D. R. 2015. Estimating relative sea-level rise and submergence potential at a coastal wetland. Estuaries and Coasts 38:1077–1084.

Callaway, J. C., E. L. Borgnis, R. E. Turner, and C. S. Milan. 2012. Carbon sequestration and sediment accretion in San Francisco Bay tidal wetlands. Estuaries and Coasts 35:1163–1181.

Chmura, G. L., S. C. Anisfeld, D. R. Cahoon, and J. C. Lynch. 2003. Global carbon sequestration in tidal, saline wetland soils. Global Biogeochemical Cycles 17:1111–1123.

Couvillion, B. R., and H. Beck. 2013. Marsh collapse thresholds for coastal Louisiana estimated using topography and vegetation index data. Journal of Coastal Research 63(Special Issue):58–67.

Couvillion, B. R., G. D. Steyer, H. Wang, H. J. Beck, and J. M. Rybczyk. 2013. Forecasting the effects of coastal protection and restoration projects on wetland morphology in coastal Louisiana under multiple environmental uncertainty scenarios. Journal of Coastal Research 67(Special Issue):29–50.

Crooks, S., D. Herr, J. Tamelander, D. Laffoley, and J. Vandever. 2011. Mitigating climate change through restoration and management of coastal wetlands and near-shore marine ecosystems: challenges and opportunities. Environment Department Paper 121. World Bank, Washington, D.C., USA.

Cui, B., Q. Yang, Z. Yang, and K. Zhang. 2009. Evaluating the ecological performance of wetland restoration in the Yellow River Delta, China. Ecological Engineering 35:1090–1103.

Day, J. W., G. P. Kemp, D. Reed, D. R. Cahoon, R. M. Boumans, J. M. Suhayda, and R. Gambrell. 2011. Vegetation death and rapid loss of surface elevation in two contrasting Mississippi delta salt marshes: the role of sedimentation, autocompaction and sea level rise. Ecological Engineering 37:229–240.

Day, J. W., et al. 2009. The impacts of pulsed reintroduction of river water on a Mississippi Delta coastal basin. Journal of Coastal Research 54:225–243.

Day, J. W., et al. 2012. Ecological response of forested wetlands with and without large Mississippi River input: implications for management. Ecological Engineering 46:57–67.

DeLaune, R. D., A. Jugsujinda, G. W. Peterson, and W. H. Patrick Jr. 2003. Impact of Mississippi River freshwater reintroduction on enhancing marsh accretionary processes in a Louisiana estuary. Estuarine, Coastal and Shelf Science 58:653–662.

DeLaune, R. D., A. Jugsujinda, J. L. West, C. B. Johnson, and M. Kongchum. 2005. A screening of the capacity of Louisiana freshwater wetlands to process nitrate in diverted Mississippi River water. Ecological Engineering 25:315–321.

DeLaune, R. D., M. Kongchum, J. R. White, and A. Jugsujinda. 2013. Freshwater diversions as an ecosystem management tool for maintaining soil organic matter accretion in coastal marshes. Catena 107:139–144.

DeLaune, R. D., and J. R. White. 2012. Will coastal wetlands continue to sequester carbon in response to an increase in global sea level? A case study of the rapidly subsiding Mississippi River Deltaic Plain. Climate Change 110:297–314.

Herbert, E. R., P. Boon, A. J. Burgin, S. C. Neubauer, R. B. Franklin, M. Ardón, K. N. Hopfensperger, L. P. M. Lamers, and P. Gell. 2015. A global perspective on wetland salinization: ecological consequences of a growing threat to freshwater wetlands. Ecosphere 6:206.

Kirwan, M. L., J. A. Langley, G. R. Guntenspergen, and J. P. Megonigal. 2012. The impact of sea-level rise on organic matter decay rates in Chesapeake Bay brackish tidal marshes. Biogeosciences 9:14689–14708.

Kolker, A. S., M. D. Miner, and H. D. Weathers. 2012. Depositional dynamics in a river diversion receiving basin: the case of the West Bay Mississippi River Diversion. Estuarine, Coastal and Shelf Science 106:1–12.

Lane, R. R., J. W. Day Jr., and J. N. Day. 2006. Wetland surface elevation, vertical accretion, and subsidence at three Louisiana estuaries receiving diverted Mississippi River water. Wetlands 26: 1130–1142.

Markewich, H. W., G. R. Buell, L. D. Britsch, J. P. McGeehin, J. A. Robbins, J. H. Wrenn, D. L. Dillon, T. L. Fries, and N. R. Morehead. 2007. Organic-carbon sequestration in soil/sediment of the Mississippi River deltaic plain—data; landscape distribution, storage, and inventory; accumulation rates; and recent loss, including a post-Katrina preliminary analysis. *In* H. W. Markewich, editor. Soil carbon storage and inventory for the continental United States. Professional Paper 1686-B, U.S. Geological Survey, Reston, Virginia, USA.

Mcleod, E., G. L. Chmura, S. Bouillon, R. Salm, M. Björk, C. M. Duarte, C. E. Lovelock, W. H. Schlesinger, and B. R. Silliman. 2011. A blueprint for blue carbon: toward an improved understanding of the role of vegetated coastal habitats in sequestering CO_2. Frontiers in Ecology and the Environment 9:552–560.

Meselhe, E. A., I. Georgiou, M. A. Allison, and J. A. McCorquodale. 2012. Numerical modelling of hydrodynamics and sediment transport in lower Mississippi at proposed delta building diversion. Journal of Hydrology 472–473:340–354.

Meselhe, E., J. A. McCorquodale, J. Shelden, M. Dortch, T. S. Brown, P. Elkan, M. D. Rodrigue, J. K. Schindler, and Z. Wang. 2013. The Eco-Hydrology Model for coastal master plan. Journal of Coastal Research 67(Special Issue):16–28.

Mitra, S., R. Wassmann, and P. L. G. Vlek. 2005. An appraisal of global wetland area and its organic carbon stock. Current Science 88:25–35.

Mitsch, W. J., B. Bernal, A. M. Nahlik, U. Mander, L. Zhang, C. J. Anderson, S. E. Jorgensen, and H. Brix. 2013. Wetlands, carbon, and climate change. Landscape Ecology 28:583–597.

Morris, J. T., G. P. Shaffer, and J. A. Nyman. 2013. Brinson Review: perspectives on the influence of nutrients on the sustainability of coastal wetlands. Wetlands 33:975–988.

Morris, J. T., P. V. Sundareshwar, C. T. Nietch, B. Kjerfve, and D. R. Cahoon. 2002. Responses of coastal wetlands to rising sea level. Ecology 83:2869–2877.

Mudd, S. M., S. M. Howell, and J. T. Morris. 2009. Impact of dynamic feedbacks between sedimentation, sea-level rise, and biomass production on near-surface marsh stratigraphy and carbon accumulation. Estuarine, Coastal and Shelf Science 82:377–389.

Nittrouer, J. A., J. L. Best, C. Brantley, R. W. Cash, M. Czapiga, P. Kumar, and G. Parker. 2012. Mitigating land loss in coastal Louisiana by controlled diversion of Mississippi River sand. Nature Geoscience 5:534–537.

Olson, K. R., M. M. Al-kaisi, R. Lal, and B. Lowery. 2014. Experimental consideration, treatments, and methods in determining soil organic carbon sequestration rates. Soil Science Society of America Journal 78:348–360.

Pendleton, L., et al. 2012. Estimating global "blue carbon" emissions from conversion and degradation of vegetated coastal ecosystems. PLoS ONE 7: e43542.

Peyronnin, N. S., M. Green, C. P. Richards, A. Owens, D. Reed, D. Groves, J. Chamberlain, K. Rhinehart, and K. Belhadjali. 2013. Louisiana's 2012 Coastal Master Plan: overview of a science-based and publicly-informed decision making process. Journal of Coastal Research 67(Special Issue):1–15.

Piazza, S. C., G. D. Steyer, K. F. Cretini, C. E. Sasser, J. M. Visser, G. O. Holm Jr., L. A. Sharp, D. E. Evers, and J. R. Meriwether. 2011. Geomorphic and ecological effects of hurricane Katrina and Rita on coastal Louisiana Marsh communities. OFR 2011-1094, U.S. Geological Survey, Reston, Virginia, USA.

Rivera-Monroy, V. H., B. Branoff, E. A. Meselhe, A. McCorquodale, M. Dortch, G. D. Steyer, J. Visser, and H. Wang. 2013. Landscape-level estimation of nitrogen loss in coastal Louisiana wetlands: potential sinks under different restoration scenarios. Journal of Coastal Research 67(Special Issue):75–87.

Sasser, C. E., J. M. Visser, E. Mouton, and J. Linscombe. 2008. Vegetation types in coastal Louisiana in 2007. USGS Open-File Report 2008-1224. http://pubs.usgs.gov/of/2008/1224/

Snedden, G. A., K. Cretini, and B. Patton. 2015. Inundation and salinity impacts to above- and belowground productivity in *Spartina patens and Spartina alterniflora* in the Mississippi River deltaic plain: implications for using river diversions as restoration tools. Ecological Engineering 81:133–139.

Stagg, C. L., D. R. Schoolmaster, S. C. Piazza, G. Snedden, G. D. Steyer, C. J. Fischenich, and R. W. McComas. 2016. A landscape-scale assessment of above- and belowground primary production in coastal wetlands: implications for climate change-induced community shifts. Estuaries and Coasts. https://doi.org/10.1007/s12237-016-0177-y

Steyer, G. D., B. R. Couvillion, H. Wang, W. Sleavin, J. Rybczyk, N. Trahan, H. Beck, C. Fischenich, R. Boustany, and Y. Allen. 2012. Louisiana's 2012 Coastal Master Plan: Wetland Morphology Model Technical Report. Louisiana Coastal Protection and Restoration Authority (CPRA), Baton Rouge, Louisiana, USA. http://coastal.la.gov/our-plan/2012-coastal-masterplan/cmp-appendices/

Twilley, R. R., and A. Nyman. 2005. The role of biogeochemical processes in marsh restoration: implications to freshwater diversions. Final Report to Louisiana Department of Natural Resources, Baton Rouge, Louisiana, USA.

VanZomeren, C. M., J. R. White, and R. D. DeLaune. 2013. Ammonification and denitrification rates in coastal Louisiana bayou sediment and marsh soil: implications for Mississippi River diversion management. Ecological Engineering 54:77–81.

Visser, J. M., S. M. Duke-Sylvester, J. Carter, and W. P. Broussard. 2013. A computer model to forecast wetland vegetation changes resulting from restoration and protection in coastal Louisiana. Journal of Coastal Research 67(Special Issue):51–59.

Visser, J. M., D. Reed, G. D. Steyer, J. Callaway, E. M. Swenson, G. M. Suir, and J. Suhayda. 2003. Wetland nourishment module. *In* R. R. Twilley, editor. Coastal Louisiana Ecosystem Assessment and Restoration (CLEAR) Model of Louisiana Coastal Area (LCA) Comprehensive Ecosystem Restoration Plan. Volume I: tasks 1–8. Final Report to Department of Natural Resources, Coastal Restoration Division, Baton Rouge, Louisiana, USA.

Wang, H., S. C. Piazza, L. A. Sharp, C. L. Stagg, B. R. Couvillion, G. D. Steyer, and T. E. McGinnis. 2016. Determining the spatial variability of wetland soil bulk density, organic matter, and the conversion factor between organic matter and organic carbon across coastal Louisiana, U.S.A. Journal of Coastal Research. https://doi.org/10.2112/jcoastres-d-16-00014.1

Wang, H., R. Wang, Y. Yu, M. J. Mitchell, and L. Zhang. 2011. Soil organic carbon of degraded wetlands treated with freshwater in the Yellow River Delta, China. Journal of Environmental Management 92:2628–2633.

Wang, H., et al. 2014. Forecasting landscape effects of Mississippi River diversions on elevation and accretion in Louisiana deltaic wetlands under future environmental uncertainty scenarios. Estuarine, Coastal and Shelf Science 138:57–68.

Williams, S. J. 2013. Sea-level rise implications for coastal regions. Journal of Coastal Research 63 (Special Issue):184–196.

8

Structural heterogeneity affects raptor assemblages and niche characteristics in mixed-grass ecosystems

Fidelis A. Atuo† and Timothy J. O'Connell

Department of Natural Resource Ecology and Management, Oklahoma State University, Stillwater, Oklahoma 74078 USA

Abstract. Landscape function is largely determined by landscape structure. Evidence suggests that landscape heterogeneity often benefits generalists over sympatric specialists. The scale at which animals perceive that heterogeneity likely varies across species and could play an important role in determining the species assemblages supported by different landscapes. In this study, we investigated the relative importance of landscape heterogeneity in the structuring of raptor assemblages in mixed-grass prairie. We compared niche breadth and overlap of 14 species in two landscapes that differed in heterogeneity. We used a modified line transect approach and distance sampling to determine approximate locations of raptors and mapped these locations in a GIS. We then compared land cover in buffered areas around detection locations to randomly selected locations at different scales. We calculated Outlying Mean Index to quantify niche breadth and compare among species. Species composition was similar in both landscapes, but niche characteristics differed according to heterogeneity. Raptor assemblages were better predicted by land cover composition in heterogeneous landscapes than by land cover configuration in homogenous landscapes. Outlying Mean Index analysis suggested that six of the 14 species occupied marginal niches and narrower niche breadths in homogeneous landscapes. None of these species demonstrated marginality but occupied broader niche breadths in heterogeneous landscapes. Our study provides evidence that heterogeneity is important in providing conditions necessary to support multiple, sympatric birds of prey in mixed-grass prairie. Species overlapped broadly under heterogeneity but less in landscapes with larger patches of distinct land cover types.

Key words: heterogeneity; landscape composition; landscape configuration; niche breadth; niche characteristics; raptor communities.

† **E-mail:** fidelis.atuo@okstate.edu

Introduction

The uneven distribution of species is often mediated by the mosaic of land cover patches in the landscapes in which they occur (Pickett and Cadenasso 1995, Stein et al. 2014). A structurally complex landscape is presumed to offer diverse opportunities for species to exploit resources in such a way that reduces interspecific competition, thereby promoting species diversity (Holt 1984, Wiens 1995, Tews et al. 2004). Empirical support for this relationship, however, is mixed: High land cover diversity (i.e., heterogeneity) has been associated with both increases and decreases in species diversity (August 1983, Hovick et al. 2015). For example, both compositional and configurational heterogeneity have been shown to increase the diversity of butterflies and carabid beetles in agricultural landscapes (Flick et al. 2012, Slancarova et al. 2014). Furthermore, landscapes that are characterized by high patch densities and small patch sizes have been shown to benefit generalist species to the detriment of habitat specialists (Devictor et al. 2008). In these landscapes and depending on the species, structural attributes are more likely to be perceived as fragmentation rather than heterogeneity (Bertrand et al. 2016). In a Mesoamerican tropical rainforest ecosystem, the α-diversity (number of species within a habitat) of generalist bird species showed positive correlation with landscapes that had lower forest cover and higher number of forest patches (Carrara et al. 2015). The same study also demonstrated that α-diversity of specialist species that depend on large patches of distinct cover types was positively associated with large patches of old-growth forest cover and was negatively related to the percentage of secondary forest cover with higher patch density (PD). These studies point to the fact that anthropogenic disturbances that alter the structure of a landscape are likely to negatively affect specialists by placing them in competitive proximity with generalist species thus increasing their risk of extinction or extirpation. This is because interspecific competition of co-occurring species in patchy landscapes is likely to result in spatial avoidance and a reduction or loss of foraging opportunities for less competitive species though in a scale-dependent way (Turner et al. 1997, Fuhlendorf et al. 2002, Boyce et al. 2003).

Several landscape features including but not limited to vegetation matrix, food availability, and predator abundance play important roles in driving community interactions, co-occurrence, and distribution of populations (Matter and Mannan 2005, Monteith et al. 2014). Predators offer ample opportunities for understating the role of habitat mosaics on species habitat selection (Gorini et al. 2012). Given their important contribution in determining community structure of co-occurring prey species, understanding predator distribution, community composition, abundances, and niche relationships is often important in setting landscape-level conservation goals (Sánchez-Zapata and Calvo 1999, Krüger and Lindström 2001, Penteriani et al. 2002). The structural landscape mosaic has been shown to affect the way predators compete for resources and the attendant prey vulnerability risk (Gorini et al. 2012). Predator–predator competition may reduce hunting success among predators by limiting the use of habitats that provide good hunting grounds (Kauffman et al. 2007). On the other hand, spatial configuration of the landscape may reduce the amount of refugia available to sympatric prey species in a way that might increase the vulnerability of less competitive individuals to predation (Koivunen et al. 1998).

Vegetation cover, structure, food resource, and availability of nesting and perching substrate are often responsible for the differential distribution and composition of raptor species in many landscapes (Tapia et al. 2007). For example, Anderson (2001) demonstrated that landscape heterogeneity was the most significant predictor of direct increases in raptor diversity, density, and richness in the tropical moist forest of the Río Plitáno watershed in northeastern Honduras.

Information on the predator composition, abundance, distribution in space and time, and preferred habitats (i.e., organism's use of a habitat relative to its availability; Aarts et al. 2008, Beyer et al. 2010) is essential in making informed decisions on how best to manage for prey species, especially in landscapes where predators are believed to limit prey population substantially (Thirgood et al. 2000, Valkama et al. 2005). This is practically relevant in the mixed-grass prairie landscapes of the southern Great Plains where drastic decline of game bird species has been partly attributed to avian predation pressure (Cox et al. 2004, Hagen et al. 2007, Wolfe et al. 2007). In this study, our objectives were (1) to describe the abundance and structure of raptor communities at two mixed-grass prairie; (2) to estimate the relative importance of landscape composition and landscape configuration in the structuring of raptor community assemblages in mixed-grass prairie ecosystems; and (3) to describe the niche characteristics of raptor communities and the role of habitat heterogeneity on species specialization.

Methods

Study areas

Our study was conducted within two Wildlife Management Areas (WMAs) located in northwestern Oklahoma. Packsaddle WMA covers ~6475 ha of mixed-grass prairie with an elevation that ranges from approximately 579 to 762 m above sea level. The average annual precipitation is ~53 cm, with the majority occurring during spring and summer (DeMaso et al. 1997). Detailed vegetation and landscape characteristics of the area are described in DeMaso et al. (1997) and Hall (2015). Beaver River WMA is ~7163 ha in area, consisting of a mixture of upland grassland, floodplain, and river bottom with a mean annual precipitation of ~7.6 cm. Vegetation in the uplands is predominantly sagebrush (*Artemisia filifolia*) and buffalograss (*Bouteloua dactyloides*) interspersed with sand plum (*Prunus angustifolia*) thickets and gently rolling sandhills. The floodplain portion of the WMA is comprised mostly of grasses mixed with cottonwood (*Populus deltoides*), hackberry (*Celtis occidentalis*), and American elm (*Ulmus americana*). The river bottom is dominated by woody vegetation consisting of sand plum thickets and invasive salt cedar (*Tamarix* spp.). The Oklahoma Department of Wildlife Conservation manages the two WMAs mainly for hunting and cattle grazing. Most of the management practices on the WMAs are intended to increase the population of the Northern Bobwhite and other game species. A combination of prescribed grazing by cattle and prescribed burns is conducted (mostly at Packsaddle) to produce promote the growth of native forbs. There is also significant oil exploration and extraction at Packsaddle WMA.

Raptor survey

To quantify raptor occurrences, we used a modified distance sampling protocol based on line transect surveys (Bibby et al. 2000, Buckland et al. 2001). We established a total of 14 line transects at Packsaddle and 16 at Beaver River WMA. Twenty-six of the 28 transects were established on existing roads and trails that were accessible in a 4 × 4 truck (Fuller and Mosher 1987). To access a roadless area at Packsaddle WMA, we included two transects that required survey on foot. All 28 transects measured 2–5 km in length and were surveyed at least twice each month between January 2013 and December 2015.

Given that raptors usually occur in low densities, we adjusted counts to account for detectability following the methods described in Thompson (2002) and Buckland et al. (2015). To reduce the chance of counting an individual more than once, we placed transect lines along trails separated by a distance ≥900 m irrespective of whether they were on existing trails or freshly cut trails. Surveys along existing trails were conducted from a 4 × 4 truck, with the observer usually driving but sometimes accompanied by a driver. The observer scanned a distance of approximately 400 m on either side of the transect line with the truck driven at a speed of 20–30 km/h (Andersen 2007). We used the same sampling methods for the two transects that were accessed on foot, with the exception that survey speed was slower (e.g., 2–5 km/h). Birds that were flying high and fast across were considered to be migrating through and were not included in our surveys.

When a raptor was detected, the observer paused to record estimated distance to the bird using a laser rangefinder (Nikon 8398 ACULON, Nikon Inc., Melville, New York, USA) and approximate deviation from the transect line in azimuth degrees using a magnetic compass. We georeferenced each raptor detection at the point of observation on the transect line using a Garmin Montana 650TM GPS unit.

To develop a spatially explicit model of raptor–habitat association, we plotted each georeferenced point to the point of actual occurrence in time using the "Bearing Distance to Line" tool in ArcGIS 10.2 (Atuo 2017, Atuo and O'Connell 2017). This tool created a new feature class containing a geodetic line for each raptor occurrence point based on the values in the *x*- and *y*-coordinate fields, angle of observation, and detection distance. Next, we used the "Feature Vertices to Points" tool to create a feature class containing georeferenced points at the end of each geodetic line feature. These new points containing *x*, *y*-coordinates represented the actual location of the individual bird at the time it was sighted. Thus, we developed a new layer of offset points that represented the actual locations of individual sightings collected during surveys, rather than simply recording the location along a transect from where we observed a raptor. To account for inconsistency in performance of rangefinders and the highly mobile behavior of many raptors, we summarized estimated raptor use locations at a coarse resolution.

Compositional variables

We collected vegetation variables from 10-m-resolution imagery acquired through the National Agriculture Imagery Program. Raw imagery acquired in 2015 was pre-processed and classified by the Oklahoma Department of Wildlife Conservation (http://www.wildlifedepartment.com/facts_maps/ecoregions.htm). First, we clipped the land cover map to our study extent in order to reduce the number of land cover classes and processing time. The original vegetation map included a layer of 31 land cover classes for Packsaddle and 24 classes for Beaver River. Hence, we reclassified each layer by combining similar vegetation classes to create a new layer containing nine dominant vegetation cover types (landscape composition variables) for Packsaddle and eight classes for Beaver River (Table 1). To estimate landscape-level raptor–habitat association, we stratified each study site into squared polygons of 1000 × 1000 m. Within each polygon, we extracted landscape composition (vegetation) variables within an 800 m radius buffer centered at a centrally placed point location in the polygon. Next, we overlaid our offset raptor locations over the stratified maps and summarized raptor abundances within each polygon.

Table 1. Landscape composition and configuration variables calculated for each 1000 × 1000 m cell.

Variable category	Variables
Configurational variables	
Class level	Length (m) of riparian woody cover
	Length (m) of upland woody cover
	Length (m) of riparian shrub cover
	Number of isolated upland woodland patches
	Number of isolated riparian woodland patches
	Number of isolated riparian shrubland patches
	Number of isolated upland shrub patches
Landscape level	Landscape shape index
	Percentage of like adjacencies
	Slope
	Elevation
Composition variables	
Woodland	Area (m^2) of riparian woody cover
	Area (m^2) of upland woody cover
Shrubland	Area (m^2) of riparian shrub cover
	Area (m^2) of upland shrub cover/shinnery motte
Grassland	Area (m^2) of grass/prairie cover
Pasture land	Area (m^2) of agricultural land
Open water	Area (m^2) open water including dams, rivers, lakes
Barren/bare areas	Area (m^2) of completely bare land
Sandhill shinnery	Area (m^2) of dispersed sandhill shinnery cover

Configurational variables

To obtain landscape configuration variables at each site, we used the raster-split tool in ArcMap to split our land cover rasters layers into 50 smaller raster segments (cells) of 1000 × 1000 m for Packsaddle and 45 cells for Beaver River. Using these smaller segments, we selected several landscape-level metrics to quantify the structure of the landscape mosaic available in each cell. We also quantified class-level metrics focusing on the area and configuration of dominant vegetation cover types at Packsaddle (sandhill shinnery, upland woodland, and riparian woodland) and Beaver River (mixed-grass prairie, upland land shrub, and riparian woodland). For landscape-level metrics, we quantified indices of (1) Shannon diversity (SHDI), which measures the diversity of the landscape mosaic based on information on the number of patches (NP) and extent of each patch type; (2) edge length (EL), which measures the total edge (TE) segments of all patch types without contrast weighting; (3) the perimeter–area fractal dimension, which provides a global measure of landscape complexity; (4) PD, which measures the total density of individual patches in the landscape; and (5) landscape shape index, which provides a standardized measure of TE or edge density that adjusts for the size of the landscape. At the class level, we calculated (1) percentage of like adjacencies, which measures the degree of aggregation of focal patch types; (2) TE, which is an absolute measure of total EL of a particular patch type; and (3) NP, which measures the number of distinct patches in each cell (McGarigal and Ene 2015). In addition to the landscape configuration matrices listed above, we also calculated elevation and slope from the Digital Elevation Model (DEM) in ArcGIS 10.2.2. DEM data were collected from the United States Geological Survey data portal at 1/3 arc-seconds (10 m) resolution. For each landscape cell, we extracted elevation and slope values based on a centrally place point in the cell.

Data analysis

At each study site, we estimated a distance detection function for each species by computing detection probabilities using the multiple-covariate distance sampling approach (Marques et al. 2007; Buckland et al. 2015). In each case, we compared a suite of a priori candidate models including half-normal and hazard-rate function keys with cosine adjustment terms. To increase the explanatory power of our models, we included covariates time of the day (tod), month of survey (month), observer ID (obs), and their interactions. We treated all covariates as factors including time of day which was categorized into four time intervals: morning (07:00–10:00), late morning (10:01–12:00), afternoon (12:01–15:00), and later afternoon (>15:00 Central Standard/Daylight Time).

We ranked models using the Akaike Information Criterion (AIC) and collected detection probabilities based on the best competing models within a ΔAIC values <2. For all species with sufficient encounters (at least 20), we estimated detection-corrected density and abundance based on the best model for the detection function. To account for the effects of multiple visits on density estimates, we computed survey efforts as the number of survey events for each transect multiplied by the transect length (Buckland et al. 2015). We performed all distance analyses using program *Distance 6.2* (Thomas et al. 2010).

We assessed raptor species richness and sample coverage (a measure of sample completeness expressed as the proportion of total number of individuals in a community that belong to the species represented in the sample) using the extrapolation and rarefaction methods (Chao et al. 2014). To construct an estimated variance and the associated confidence interval for sample coverage, we completed 100 bootstrap runs for each sample and extrapolated the number of species to twice the number of individuals in each sample using the iNEXT package v2.0.5 (Hsieh et al. 2016) in program R. Given that coverage-based rarefaction is known to provide minimal biases of community species richness by keeping the ratio of species richness instead of compressing it (Chao and Jost 2012), we plotted both individual-based species accumulation curves and sample coverage curves for each site (Fig. 1).

We used multivariate analyses to demonstrate the relationship between multiple raptor species and the environmental gradients. We quantified the impact of landscape heterogeneity on raptor communities by performing direct gradient analyses and variation partitioning in Canoco 5.03 (Ter Braak and Smilauer 2012). For all analyses, we log-transformed ($\log(x + 1)$) our abundance data and downweighted rare species to dampen the effects of extreme samples (Ter Braak and Smilauer 2012, Šmilauer and Lepš 2014). Prior to performing analyses, we log-transformed a selection of the landscape composition variables to maximize the linearity of their relation and ensure that the ecological importance of all the land cover types was considered (Šmilauer and Lepš 2014, Neumann et al. 2016).

To understand the effects of landscape composition and landscape configuration on raptor assemblages, we performed separate partial canonical correspondence analyses (pCCAs) for each variable set. For each analysis, we selected variables that best summarized the effects of landscape composition or landscape configuration on raptor assemblages using partial interactive forward selection (pIFS). This allowed the removal of collinear and redundant variables from the pCCA model (Ter Braak and Smilauer 2012, Šmilauer and Lepš 2014). For each pIFS, we included significant variables that were collected over a global Monte Carlo permutation test with 999 iterations (Šmilauer and Lepš 2014). In each model, we determined the significance of explanatory variables using a P-value < 0.05 (Ter Braak and Smilauer 2012). We applied the false

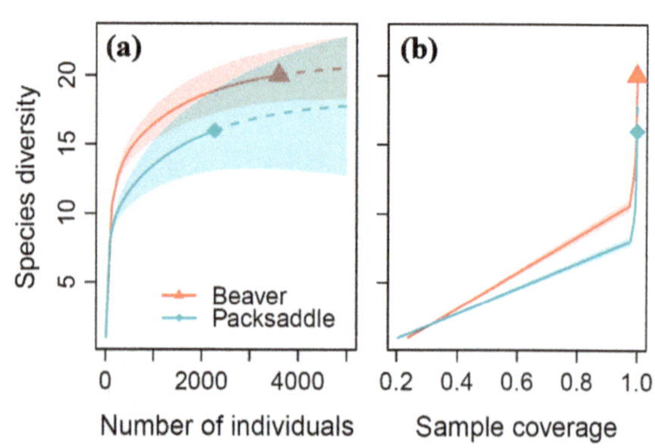

Fig. 1. (Right) Individual-based integrated extrapolation–rarefaction curves and (left) species richness as a function of sample coverage for Beaver River and Packsaddle Wildlife Management Areas, Oklahoma, USA. Solid lines indicate interpolations, and dashed lines indicate extrapolated data. Shaded areas represent 95% confidence intervals.

discovery rate method to adjust the significance values for potential inflated family-wise type 1 errors, thus reducing the likelihoods of false significance (Ter Braak and Smilauer 2012, Šmilauer and Lepš 2014). To further determine the unique and shared contributions of the landscape composition and configuration variables in explaining raptor species composition, we performed a variance partitioning of the important variables identified by the pIFS models for each site. The variance partitioning procedure quantifies whether and to what degree the variability in raptor assemblages at each study site is explained by landscape composition variables only, landscape configuration variables only, or whether the overlap of the two variable sets together has an additive explanatory effect (Legendre 2008, Šmilauer and Lepš 2014).

To estimate the ecological niche filled by each raptor species within each community, we performed an Outlying Mean Index (OMI) analysis (Dolédec et al. 2000) using the same set of landscape composition and landscape configuration variables as in the pCCA models. The OMI or species marginality analysis is a multivariate analysis technique (based on principal component analysis) that measures the distance between the mean habitat conditions used by a species (species centroid) and the mean habitat conditions that exist in the study landscape. An OMI analysis places each species along a habitat gradient (niche hyperspace) based on its habitat requirements. A measure of how far a species occurs away from the origin of the niche hyperspace is termed its marginality. Hence, the marginality of a species under investigation depends on its deviation from the origin of the niche hyperspace. Species with higher marginality scores select unique environmental conditions different from the mean available conditions in the landscape. The OMI analysis also calculates a species' tolerance (i.e., species niche breadth). Species with higher tolerance values can occupy varying habitat conditions (generalist species), while those with low tolerance values are limited in their habitat use (specialists). The OMI technique also addresses some of the limitations of canonical correspondence analyses (CCA) and redundancy analysis (RDA), which are best suited for data with unimodal distribution (with respect to CCA) or linear distribution (with respect to RDA; Dolédec et al. 2000). Unlike the CCA which tends to over-emphasize rare species in an ordination output, the OMI technique makes no assumptions about species responses to environmental variables, thus assigning equal weights to all sample units irrespective of their species abundance scores (Dolédec et al. 2000). We performed the OMI analysis using the ade4 package (Chessel et al. 2016) in program R and determined significance of the OMI based on a Monte Carlo simulation of 999 random permutation values of species marginalities. Finally, we used a two-sample Student's *t*-test to obtain the difference in landscape heterogeneity by comparing patch densities collected within each 1000 × 1000 m cell between our study systems.

Results

We recorded a total of 3230 raptor detections belonging to 19 species at the Beaver River WMA and 2348 detections of 15 species at the Packsaddle WMA (Table 2). At both sites, the hazard-rate key function provided the strongest support for detection. The best detection models for raptors varied depending on species or study site (Table 2). Nonetheless, time of day and observer were major covariates that influenced raptor detections at both sites (Table 2). Estimated densities and abundances also varied depending on species and study site (Table 2).

Red-tailed Hawk and Northern Harrier were the most abundant predatory raptors, respectively constituting 36% and 31% of detected species at Beaver River, and 35% and 13% at Packsaddle. Under-represented species with <15 detections were excluded from our ordination analysis. At both study sites, overall raptor abundance was higher during winter than during summer. Shannon diversity was higher during the winter (6.13 ± 0.19) than during the summer (4.49 ± 0.12) at Beaver River but comparable between seasons (5.07 ± 0.15 during winter and 5.03 ± 0.14 during summer) at Packsaddle. Overall species richness was higher at Beaver River than at Packsaddle WMA (Fig. 1).

Confidence intervals of rarefaction–extrapolation curves overlapped (Fig. 1), indicating that the two sites shared similarities in species richness both in the number of species detected and when extrapolated to twice the number of detected individuals. For both sites, the rarefaction–extrapolation curves

Table 2. Summary details for top detection models fitted to individual raptors at Beaver River and Packsaddle Wildlife Movement Areas, Oklahoma, USA, 2013–2015.

Species	Code	Scientific names	Model	Key	Delta AIC	ESW	D/100 ha	D LCL	D UCL	N	P.D.
(a) Beaver River											
Turkey Vulture	TUVU	*Cathartes aura*	obs	HR	0	195.7	0.57	0.49	0.66	4	0.56
Northern Harrier	NOHA	*Circus cyaneus*	tod	HN	0	183.72	0.79	0.73	0.86	6	0.68
Mississippi Kite	MIKI	*Ictinia mississippiensis*	month	HN	0	104.27	0.17	0.06	0.39	2	0.52
Sharp-shinned Hawk	SSHA	*Accipiter striatus*	tod × obs	HN	0	37.76	0.10	0.07	0.16	6	0.53
Cooper's Hawk	COHA	*Accipiter cooperii*	obs × tod	HR	0	105.39	0.18	0.15	0.39	6	0.59
Swainson's Hawk	SWHA	*Buteo swainsoni*	month	HN	0	171.25	0.54	0.44	0.67	4	0.57
Red-tailed Hawk	RTHA	*Buteo jamaicensis*	obs	HR	0	226.67	1.33	1.15	1.53	9	0.76
Ferruginous Hawk	FEHA	*Buteo regalis*	tod	HN	0	173.72	0.47	0.35	0.62	3	0.58
American Kestrel	AMKE	*Falco sparverius*	obs	HN	0	94.6	0.54	0.51	0.57	4	0.63
Prairie Falcon	PRFA	*Falco mexicanus*	tod × obs	HN	0	31.92	0.08	0.01	0.1	2	0.28
Great Horned Owl	GHOW	*Bubo virginianus*	tod	HN	0	111.53	0.88	0.53	1.47	6	0.37
(b) Packsaddle											
Turkey Vulture	TUVU	*Ca. aura*	tod	HN	0	180.6	0.64	0.64	0.71	4	0.6
Northern Harrier	NOHA	*Ci. cyaneus*	obs × tod	HN	0	175.55	0.81	0.81	0.96	9	0.65
Mississippi Kite	MIKI	*I. mississippiensis*	obs	HN	0	127.2	1.68	1.68	1.84	11	0.64
Sharp-shinned Hawk	SSHA	*A. striatus*	month	HR	0	103.31	0.16	0.06	0.3	2	0.4
Cooper's Hawk	COHA	*A. cooperii*	month	HN	0	103.91	0.19	0.05	0.28	2	0.49
Swainson's Hawk	SWHA	*But. swainsoni*	tod	HR	0	185.6	0.60	0.30	0.77	4	0.62
Red-tailed Hawk	RTHA	*But. jamaicensis*	obs	HN	0	195.97	1.08	0.83	1.83	7	0.72
Ferruginous Hawk	FEHA	*But. regalis*	obs	HR	0	249.87	0.13	0.02	0.34	3	0.28
American Kestrel	AMKE	*F. sparverius*	obs	HN	0	195.97	0.71	0.71	0.77	5	0.63
Prairie Falcon	PRFA	*F. mexicanus*	tod	HN	0	167.96	0.06	0.01	0.18	4	0.37
Great Horned Owl	GHOW	*Bub. virginianus*	month	HR	0	186.67	0.1	0.01	0.25	7	0.42
Eastern Screech Owl**	ESOW	*Megascops asio*	–	–	–	–	–	–	–	–	–
Broad-winged Hawk**	BWHA	*Buteo platypterus*	–	–	–	–	–	–	–	–	–
Rough-legged Hawk**	RLHA	*Buteo lagopus*	–	–	–	–	–	–	–	–	–
Golden Eagle*	GOEA	*Aquila chrysaetos*	–	–	–	–	–	–	–	–	–
Bald Eagle*	BAEA	*Haliaeetus leucocephalus*	–	–	–	–	–	–	–	–	–
Merlin**	MERL	*Falco columbarius*	–	–	–	–	–	–	–	–	–
Burrowing Owl**	BUOW	*Athene cunicularia*	–	–	–	–	–	–	–	–	–
Osprey*	OSPR	*Pandion haliaetus*	–	–	–	–	–	–	–	–	–
Merlin*	MERL	*F. columbarius*	–	–	–	–	–	–	–	–	–
Peregrine Falcon*	PEFA	*Falco peregrinus*	–	–	–	–	–	–	–	–	–
Short-eared Owl*	SEOW	*Asio flammeus*	–	–	–	–	–	–	–	–	–
Barred Owl*	BAOW	*Strix varia*	–	–	–	–	–	–	–	–	–

Notes: AIC, Akaike Information Criterion. Each model represents the top out of a suite of models for each species. Detection functions were defined using distance key functions including half-normal (HN) and hazard-rate (HR). Also shown are effective detection width (ESW), estimated density (D/100 ha), estimated abundance (N), and probability of detection (P.D.). Lower confidence limit (LCL) and upper confidence limits (UCL) of estimated D are also presented. Model covariates were abbreviated as obs (observer), tod (time of day), and month (month of survey). Species with an asterisk (*) had <20 detections and were not included in distance sampling analyses. Species with double asterisks (**) had <10 detections and were included in neither distance sampling nor ordination analysis.

approached asymptotes (Fig. 1a) and the sample coverages were at ~100% (Fig. 1b), indicating that raptor assemblages were adequately sampled and that most of the species in these landscapes were detected.

Landscape composition and raptor communities

Partial canonical correspondence analysis for compositional environmental variables explained 32.3% of the total variation in raptor assemblages at Beaver River WMA ($F = 2.9$, $P = 0.003$). At

Packsaddle WMA, pCCA ordination based on landscape composition variables alone explained only 3.5% ($F = 4.0$, $P = 0.36$) of the total variation in raptor assemblages. The pIFS analysis identified three explanatory variables (prairie cover, upland woodland, and riparian woodland) at Beaver River and four variables (sandhill shinnery, upland woodland, riparian woodland, and prairie cover) at Packsaddle as variables that best summarized the effects of landscape composition on raptor assemblages (Fig. 3). Raptor species that typically hunt from perches, for example, Red-tailed Hawk (*Buteo jamaicensis*), Cooper's Hawk (*Accipiter cooperii*), and Sharp-shinned Hawk (*Accipiter striatus*), were associated with compositional variables that provided perches for hunting (Fig. 2). Northern Harrier, which does not typically hunt from perches, was more likely to be associated with open prairie vegetation (Fig. 2). Swainson's Hawk (*Buteo swainsoni*) and Ferruginous Hawk (*Buteo regalis*) were associated with prairie vegetation at Packsaddle WMA (Fig. 2a) but were mostly generalist in their habitat selection at Beaver River WMA (Fig. 2b).

Landscape configuration and raptor communities

Forward selection of configurational variables suggested that the EL of riparian woodland, elevation, and the EL of upland shrub cover were the most significant ($P < 0.05$) variables influencing raptor distribution at the Beaver River WMA. The three variables together explained 17.1% of the total variation in raptor assemblages at this landscape. A global permutation test with 999 Monte Carlo permutations indicated that this association was significant (pCCA, $F = 5.0$, $P = 0.01$; Fig. 3a). The EL of riparian woodland was the most important variable accounting for 6.6% of the total variation and 32.6% of the explained variation in raptor assemblages. The two other variables, elevation and EL of upland shrub, respectively accounted for 5.8% and 4.7% of the total variation and 28.2% and 22.9% of the explained variation in raptor assemblages. At Packsaddle WMA, the interactive forward selection identified total EL of upland shrubland, SHDI index, total EL of riparian woodland, and elevation as variables that best summarized the effects of landscape configuration on raptor assemblages. Landscape configuration variables explained 13.1% of the total variation in raptor assemblages. Our global permutation test over 999 Monte Carlo permutations for raptor association with landscape configuration variables at Packsaddle was also significant (pCCA, $F = 5.2$, $P = 0.031$). Elevation ($P = 0.03$) was the most

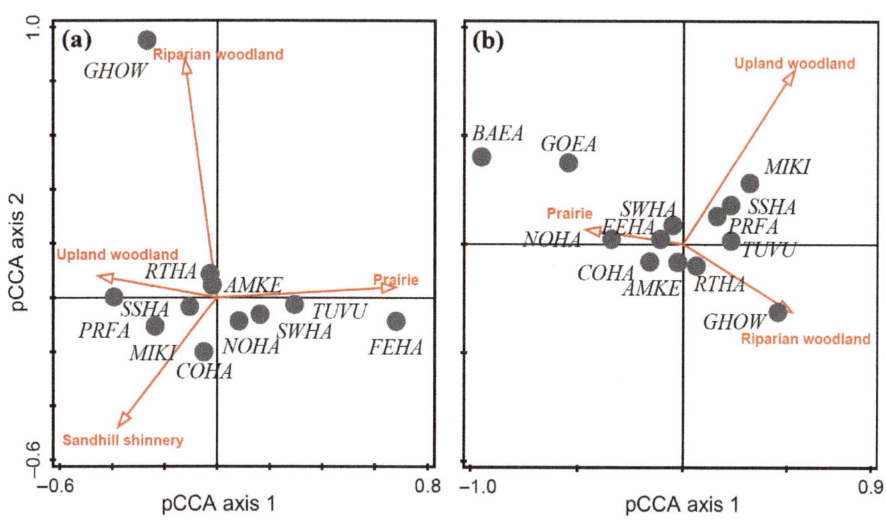

Fig. 2. The first two axes of partial canonical correspondence analysis (pCCA) biplots illustrating significant landscape composition variables (as identified by partial interactive forward selection) that explain raptor assemblages at Packsaddle (a) and Beaver River (b) Wildlife Management Areas, Oklahoma, USA. Refer to Table 1 for a description of landscape composition variables and Table 2 for an explanation of species codes.

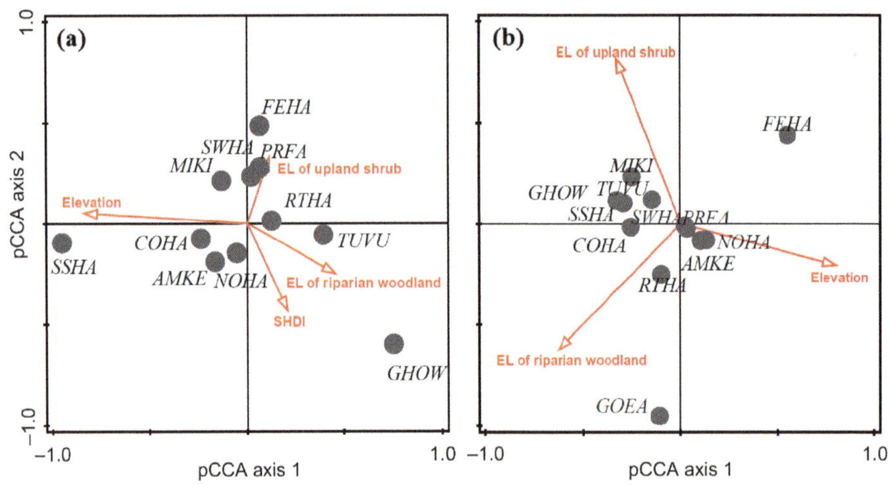

Fig. 3. The first two axes of partial canonical correspondence analysis (pCCA) biplots illustrating significant landscape configuration variables (as identified by partial interactive forward selection) that explain raptor assemblages at Packsaddle (a) and Beaver River (b) Wildlife Management Areas, Oklahoma, USA. Refer to Table 1 for a description of landscape configuration variables and Table 2 for an explanation of species codes. EL, edge length.

important variable accounting for 6.2% of the total variation and 37% of the explained variation in raptor assemblages. Shannon diversity index ($P = 0.74$), total EL of riparian woodland ($P = 0.69$), and total EL upland shrubland ($P = 0.75$) respectively accounted for 2.1%, 2.3%, and 2.0% of the total variation and 37%, 12.5%, and 13.4% of the explained variation in raptor assemblages. Accipiters at Packsaddle (Cooper's and Sharp-shinned hawks) were associated with higher elevation (Fig. 3a). Ferruginous Hawk, Mississippi Kite, Swainson's Hawk, and Prairie Falcon were more associated with upland woodland edge (Fig. 3a). Red-tailed Hawk, Northern Harrier, and American Kestrel showed no strong association with any of the configuration variables (Fig. 3a). Turkey Vulture and Great Horned Owl were more associated with landscape heterogeneity and riparian woodland edge (Fig. 3b). At Beaver River, most of the "perch and hunt" raptors were associated with edge of riparian woodland and upland shrub vegetation (Fig. 3b).

Variation partitioning

The total amount of the variation explained by both landscape composition and landscape configuration variables shows that at each study site, the separate effects of landscape composition and configuration were significant at $P < 0.05$ (Monte Carlo tests with 999 permutations). The total variation captured by both configuration and composition variables was 31.6% at Beaver River WMA (Fig. 4a) and 33.5% at Packsaddle WMA (Fig. 4b). Landscape composition explained a greater proportion (20.2%, $P = 0.001$) of the variation at Beaver River WMA, whereas the reverse was the case at Packsaddle where a greater proportion of the variation (25.1%, $P < 0.001$) was explained by configurational variables. Landscape configuration and composition respectively explained 9.5% ($P = 0.005$) and 5.7% ($P = <0.001$) of the variation at Beaver River and Packsaddle WMAs. The joint effect at both study sites (1.9% at Beaver River and 2.7% at Packsaddle) indicated some degree of explanatory overlap between landscape composition and landscape configuration variables at these two study systems.

Raptor communities and niche characteristics

The OMI analysis shows that six of the 14 raptors (43%) included in our analysis for the Beaver River WMA demonstrated significant deviation (i.e., marginal niches) from the mean compositional landscape conditions available on the landscape. Also, five (36%) of the 14 species demonstrated marginal niches from the mean landscape configurational conditions available at Beaver River (Table 3). Contrary to Beaver River

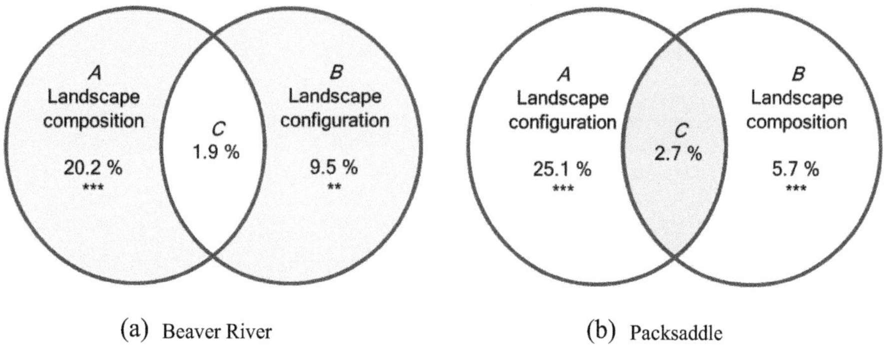

Fig. 4. Results of variation partitioning for the total raptor community composition in terms of fractions of variation explained by landscape composition and configuration variables at Beaver River (a) and Packsaddle (b) Wildlife Management Areas, Oklahoma, USA. *A* and *B* represent unique effects of landscape composition and configuration variables, while the overlapping area (C) indicates their joint effects.

WMA, none of the species examined at the Packsaddle WMA demonstrated any marginality with the mean landscape composition or landscape configuration conditions available on the landscape (Table 3). Fifty-five percent of the raptor species that we examined at Beaver River WMA and all of the ones we examined at Packsaddle WMA had low OMI values and high residual tolerance index values, both indicating low to no niche specialization for these species (Table 3).

Landscape heterogeneity

Patch density was significantly higher ($P < 0.001$) at Packsaddle WMA than at Beaver River WMA, indicating a higher degree of landscape heterogeneity at Packsaddle WMA. Mean PD within sampled cells was 19.44 ± 0.89 SE for Packsaddle WMA and 9.87 ± 0.49 SE for Beaver River WMA.

Discussion

Our study underscores the significance of landscape heterogeneity as an underlying condition for species distribution and community coexistence in mixed-grassed prairie ecosystems. Specifically, we focused on understanding the influence of landscape composition and landscape configuration on the distribution of ecologically sensitive species (raptors) at scales that are relevant to both individual species and wildlife populations (Storch 1997) in mixed-grass prairie ecosystems.

Raptor assemblages at our two study sites were similar in terms of community composition but differed in their niche characteristics. At Packsaddle WMA, where heterogeneity was high, raptor species assemblages were better explained by landscape configuration than by landscape composition. Most of the species at Packsaddle behaved as habitat generalists, making use of major vegetation cover types in proportion to their availability (Fig. 2a). Notwithstanding, few species such as the Prairie Falcon, Sharp-shinned Hawk, Ferruginous Hawk, and Great Horned Owl showed proven associations with vegetation cover types that provided perches. In terms of configurational heterogeneity, elevation, EL of upland shrub, and the EL of riparian woodland were the most important variables that drove species assemblages at Packsaddle.

Dominant upland vegetation at Packsaddle WMA is shinnery oak mixed with grass (DeMaso et al. 1997). Although the height of shinnery stems is typically <0.5 m, individuals that have hybridized with post oaks (*Quercus stellata*) regularly form mottes with stems of 3–4 m in height (Hall 2015). The presence of these hybrid mottes thus provides elevated vantage points for raptors that typically hunt from perches and might otherwise be restricted to riparian woodland where trees are more consistently available. Upland grassland at Packsaddle also provided habitat for grassland specialists such as Northern Harrier that do not typically hunt from perches. The

Table 3. Results of the OMI analysis showing the relationships between raptor species, landscape composition, and landscape configuration variables at Beaver River and Packsaddle Wildlife Management Areas.

Site	Species	Inertia	OMI	Tol	RTol	*omi*	*tol*	*rtol*	*P*-value
Beaver River, landscape composition	AMKE	7.32	0.10	1.25	5.97	1.40	17.10	81.50	0.25
	BAEA	5.05	0.60	0.31	4.14	11.80	6.10	82.10	0.73
	BUOW	2.10	1.71	0.12	0.27	81.30	5.90	12.80	0.63
	COHA	**6.57**	**0.89**	**0.99**	**4.70**	**13.50**	**15.10**	**71.40**	**<0.001**
	FEHA	4.97	0.77	0.31	3.90	15.40	6.20	78.40	0.07
	GHOW	**7.71**	**1.36**	**0.67**	**5.67**	**17.70**	**8.70**	**73.60**	**0.02**
	GOEA	4.47	0.63	0.88	2.96	14.10	19.70	66.20	0.79
	MIKI	**8.93**	**1.27**	**1.52**	**6.15**	**14.20**	**17.00**	**68.80**	**<0.001**
	NOHA	**7.45**	**0.15**	**1.82**	**5.48**	**2.00**	**24.40**	**73.60**	**0.05**
	PRFA	8.22	0.63	1.58	6.02	7.60	19.20	73.20	0.27
	RTHA	**7.85**	**0.16**	**1.76**	**5.94**	**2.10**	**22.40**	**75.60**	**<0.001**
	SSHA	9.65	1.20	1.16	7.29	12.40	12.00	75.60	0.06
	SWHA	**7.09**	**0.18**	**1.49**	**5.42**	**2.60**	**21.00**	**76.40**	**0.04**
	TUVU	**6.82**	**0.57**	**1.55**	**4.70**	**8.40**	**22.80**	**68.90**	**<0.001**
Beaver River, landscape configuration	AMKE	5.23	0.07	0.70	4.47	1.20	13.40	85.40	0.39
	BAEA	4.14	0.58	0.58	2.98	14.10	14.00	71.90	0.62
	BUOW	2.18	1.92	0.00	0.26	87.90	0.20	11.90	0.44
	COHA	4.96	0.53	0.85	3.58	10.60	17.20	72.20	0.07
	FEHA	**5.82**	**0.84**	**0.41**	**4.57**	**14.40**	**7.00**	**78.60**	**0.01**
	GHOW	5.63	0.58	1.38	3.66	10.40	24.50	65.10	0.18
	GOEA	5.34	0.85	1.29	3.20	15.90	24.20	59.90	0.53
	MIKI	**6.32**	**0.70**	**1.24**	**4.38**	**11.10**	**19.60**	**69.20**	**0.02**
	NOHA	5.48	0.03	0.81	4.63	0.60	14.90	84.50	0.52
	PRFA	6.38	0.27	1.57	4.54	4.20	24.60	71.30	0.59
	RTHA	5.82	0.08	1.46	4.28	1.40	25.00	73.50	0.07
	SSHA	**5.44**	**0.91**	**0.64**	**3.88**	**16.80**	**11.80**	**71.40**	**0.05**
	SWHA	**5.46**	**0.09**	**0.78**	**4.59**	**1.70**	**14.30**	**84.00**	**0.20**
Packsaddle River, landscape composition	**TUVU**	**5.43**	**0.32**	**1.12**	**3.99**	**5.80**	**20.70**	**73.50**	**0.01**
	COHA	5.85	0.62	1.05	4.18	10.60	18.00	71.40	0.25
	FEHA	3.69	0.94	0.30	2.45	25.50	8.10	66.40	0.55
	GHOW	7.96	1.07	0.47	6.42	13.40	5.90	80.70	0.46
	MIKI	5.21	0.38	0.87	3.96	7.40	16.80	75.90	0.44
	NOHA	5.84	0.31	1.35	4.18	5.30	23.20	71.60	0.51
	PRFA	8.29	0.68	0.88	6.72	8.30	10.70	81.10	0.85
	RTHA	6.15	0.28	1.17	4.70	4.60	19.00	76.40	0.15
	SSHA	6.90	0.43	0.94	5.53	6.20	13.70	80.20	0.47
	SWHA	5.80	0.20	1.64	3.96	3.40	28.30	68.30	0.59
	TUVU	6.36	0.18	0.91	5.27	2.90	14.40	82.80	0.59
Packsaddle River, landscape configuration	**TUVU**	8.38	0.29	3.62	4.46	3.50	43.20	53.30	0.70
	COHA	7.90	0.63	1.63	5.63	8.00	20.70	71.30	0.45
	FEHA	8.43	1.84	0.37	6.22	21.80	4.40	73.80	0.30
	GHOW	6.80	0.64	0.85	5.31	9.40	12.50	78.10	0.84
	MIKI	8.45	0.55	1.94	5.96	6.50	22.90	70.60	0.36
	NOHA	7.86	0.51	2.05	5.31	6.50	26.00	67.50	0.34
	PRFA	10.88	1.81	5.74	3.33	16.70	52.80	30.60	0.49
	RTHA	7.77	0.07	1.03	6.68	0.80	13.30	85.90	0.93
	SSHA	7.43	0.23	1.16	6.04	3.10	15.60	81.30	0.88
	SWHA	8.42	0.32	2.55	5.55	3.80	30.30	65.90	0.45
	TUVU	8.65	0.44	1.78	6.44	5.00	20.50	74.40	0.21

Notes: inertia = variance or weighted sum of squared distances to the origin of the environmental axes; OMI = Outlying Mean Index (marginality); Tol = tolerance index, which represents niche breadth of the two predators across the measured environmental variables; RTol = residual tolerance. Italicized terms represent the percentages of variability corresponding to a specific statistic. P = frequency based on number of random permutations (out of 10,000) that yielded a higher value (highlighted in bold) than the observed OMI ($P \leq 0.05$ indicates a significant influence of the environmental variables for a species). Species expansions are present in Table 2.

abundance of woody vegetation in upland grassland areas thus offered ample opportunities for multiple raptor species to co-occur at Packsaddle.

Compositional heterogeneity was comparatively low at Beaver River WMA with mean PD approximately half that at Packsaddle WMA. Unlike Packsaddle WMA, raptor assemblages at Beaver River were better summarized by landscape composition than they were by landscape configuration. Compared to Packsaddle, Beaver River was less heterogeneous with large patches of distinct vegetation cover types (Atuo and O'Connell 2017). Raptors that typically hunt from perches were associated with riparian areas that offered trees for perching. Grassland specialists either showed generalist behavior (in terms of using the most abundant cover type) or were associated with large patches of grass cover.

Based on an OMI analysis, none of the raptor species at Packsaddle WMA showed a significant ($P < 0.05$) deviation from the mean environmental conditions available in the landscape (Table 3). Overlap in habitat use by multiple species of raptors was comparatively high at Packsaddle. At Beaver River WMA with its larger patches of homogenous vegetation cover, niche overlap was low, suggesting segregation in habitat use. For example, half of the raptor species at Beaver River demonstrated significant ($P < 0.05$) deviation from the mean compositional heterogeneity of the landscape. Five species also showed significant deviation from the mean landscape configuration (Table 3).

These results emphasize the effect of spatial heterogeneity on the dynamics of species interaction and structure of raptor communities in these mixed-grass prairie ecosystems. At Beaver River WMA with lower PD, raptors formed identifiable associations with vegetation cover types that best supported their hunting strategies. For example, Red-tailed Hawk that typically hunts from perches was more associated with riparian woodland that provided perching substrate. This cover type also provided trees large enough to support their relatively large nests. In contrast, Northern Harrier does not typically require perches for hunting nor trees for nesting and was strongly associated with upland grass cover. Because riparian woodland and upland grassland were spatially discrete, so too were these abundant raptors. At Packsaddle WMA, these same species showed no specific or strong associations with any of the vegetation cover types available in the landscape, presumably because nearly every spot on the landscape provided accessible patches of both grassland and woodland.

While competition is the most frequently documented interaction among sympatric species that rely on similar resources, our results supported studies that have demonstrated that species interactions are more frequently mediated by the structural mosaic of the landscape (Danielson 1991, Cornulier and Bretagnolle 2006, Gorini et al. 2012). Increases in configurational heterogeneity reduce large patches of homogeneous vegetation into smaller ones (Hargis et al. 1998, Fahrig et al. 2011). If the reduction in minimum patch sizes for specialist species forces individuals to make use of multiple cover types, including suboptimal cover, they could be at a competitive disadvantage to generalists (Hanski 1995). Spatial heterogeneity is therefore key to competitive ability and coexistence.

While landscape heterogeneity enhances the competitive abilities of generalists, specialists are frequently placed at competitive disadvantage (Tattersall et al. 2002, Fahrig et al. 2011, Bertrand et al. 2016). Poor competitors (specialists) with limited dispersal abilities are either displaced to poorer habitats or outcompeted by generalist species (Amarasekare 2008, Hanski 2008, Lin et al. 2013). For example, Palmer (2003) demonstrated that spatial heterogeneity associated with Odontotermes mounds strongly influenced the dynamics of competition and community structure in acacia ants. Competitively dominant ants are disproportionately successful in displacing more subordinate species in high heterogeneous patches, while competitively subordinate species were more successful in less heterogeneous patches. Similarly, Kurki et al. (1998) showed that increased landscape complexity resulting from habitat fragmentation altered the competitive behavior of Red Fox (*Vulpes vulpes*) and Pine Marten (*Martes martes*) in forested landscapes. Habitat fragmentation increased PD and reduced patch size which favored the more generalist fox to the disadvantage marten.

As sensitive species (Bildstein 2001, Sergio et al. 2008), avian predators offer ample opportunities for understanding the role of the habitat mosaic on intraspecific and interspecific species interactions.

Also, given the important contribution of raptors in determining community structures of co-occurring prey species, knowledge of interspecific interactions among raptors can be relevant for conservation planning. Both of our study systems are managed by the Oklahoma Department of Wildlife Conservation for the recovery of the globally "near threatened" Northern Bobwhite (*Colinus virginianus*). While habitat loss and fragmentation are considered ultimate factors driving Northern Bobwhite decline in North America, several lines of evidence have demonstrated that predation could act as a major proximate factor in population declines (Cox et al. 2004, Rolland et al. 2010, Hernández et al. 2013). Knowledge of competitive interactions within a predator guild mediated by landscape heterogeneity could offer important insights for management to reduce the impact of predation on prey populations of conservation concern. For example, the co-occurrence and competition between multiple predator species are likely to put a prey population under intense pressure prior to density-dependent prey switching. Prey species under these systems need anti-predation strategies to avoid attack from multiple generalist predators with different hunting strategies. Spatial heterogeneity has been shown to greatly enhance the rate at which predators reduce prey populations by increasing prey vulnerability and predator hunting success (Matter and Mannan 2005, Gorini et al. 2012).

Overall, our study provides new and valuable insights into the role of landscape heterogeneity in structuring interspecific habitat partitioning by raptor communities in mixed-grass prairie landscapes of varying heterogeneity gradients. Our study supports the growing hypothesis that landscape heterogeneity plays a major role in shaping the co-occurrence of sympatric species. Raptor communities responded differently to changes in landscape heterogeneity by increasing overlap in more heterogeneous landscapes while partitioning their habitats in a less heterogeneous landscape. Information of this nature can be relevant in structuring conservation and land management decisions at the landscape scale. It may seem obvious that the greater the diversity of cover types in a landscape, the more species that landscape is able to accommodate. Nonetheless, it is equally important to take into account the relationship between patch diversity and patch size. For any given area in the landscape, an increase in the number of cover types will result in a decrease in the amount of each (Fahrig et al. 2011). This will negatively affect habitat specialists that often rely on large patches of distinct cover types for population persistence. Maintaining large patches of distinct cover types will benefit specialist species while reducing competition among generalist species. At fine scales, this dynamics will reduce competition and predation pressure for grassland specialists (e.g., Northern Bobwhite) while promoting increased diversity at broad scale.

Our analyses were based on a broad scale relevant for land management and conservation actions. Further studies that examine effects of different spatial scales will be important in disentangling the effects of scale on composition and configuration.

Acknowledgments

Funding for this study was provided by the Oklahoma Department of Wildlife Conservation grant no. F11AF00069, to TJO. The Oklahoma Cooperative Fish and Wildlife Research Unit administered funding and provided logistical and administrative support. Further support was provided by the Oklahoma Agricultural Experiment Station through the USDA National Institute of Food and Agriculture, McIntire-Stennis project at Oklahoma State University. We are grateful to the staff of Packsaddle and Beaver River Wildlife Management Areas for study area access and vital logistic support during fieldwork. Emily Sinnott, Jonathan Harris, and several field technicians provided valuable field assistance. For their assistance in study design and analytical advice, we thank David M. Leslie, Jr., Craig Davis, Dwayne Elmore, and Mona Papes. Two anonymous reviewers provided suggestions that greatly improved an earlier version of this manuscript.

Literature Cited

Aarts, G., M. MacKenzie, B. McConnell, M. Fedak, and J. Matthiopoulos. 2008. Estimating space-use and habitat preference from wildlife telemetry data. Ecography 31:140–160.

Amarasekare, P. 2008. Spatial dynamics of foodwebs. Annual Review of Ecology, Evolution, and Systematics 39:479–500.

Andersen, D. E. 2007. Survey techniques. Pages 89–100 *in* D. M. Bird and K. L. Bildstein, editors. Raptor research and management techniques manual.

Raptor Research Foundation, Hancock House Publishers, Surrey, British Columbia, Canada.

Anderson, D. L. 2001. Landscape heterogeneity and diurnal raptor diversity in Honduras: the role of indigenous shifting cultivation. Biotropica 33:511–519.

Atuo, F. A. 2017. Spatial distribution of aerial predators: influences on usable space for quail. Dissertation. Oklahoma State University, Stillwater, Oklahoma, USA.

Atuo, F. A., and T. J. O'Connell. 2017. The landscape of fear as an emergent property of heterogeneity: contrasting patterns of predation risk in grassland ecosystems. Ecology and Evolution. https://doi.org/10.1002/ece3.3021

August, P. V. 1983. The role of habitat complexity and heterogeneity in structuring tropical mammal communities. Ecology 64:1495–1507.

Bertrand, C., F. Burel, and J. Baudry. 2016. Spatial and temporal heterogeneity of the crop mosaic influences carabid beetles in agricultural landscapes. Landscape Ecology 31:451–466.

Beyer, H. L., D. T. Haydon, J. M. Morales, J. L. Frair, M. Hebblewhite, M. Mitchell, and J. Matthiopoulos. 2010. The interpretation of habitat preference metrics under use–availability designs. Philosophical Transactions of the Royal Society of London B: Biological Sciences 365:2245–2254.

Bibby, C. J., N. D. Burgess, D. A. Hill, and S. Mustoe. 2000. Bird census techniques. Second edition. Academic press, London, UK.

Bildstein, K. L. 2001. Why migratory birds of prey make great biological indicators. Pages 169–179 *in* K. L. Bildstein and D. Klem, editors. Hawkwatching in the Americas. Hawk Migration Association of North America, North Wales, Pennsylvania, USA.

Boyce, M. S., J. S. Mao, E. H. Merrill, D. Fortin, M. G. Turner, J. Fryxell, and P. Turchin. 2003. Scale and heterogeneity in habitat selection by elk in Yellowstone National Park. Ecoscience 10:421–431.

Buckland, S. T., D. R. Anderson, K. P. Burnham, J. L. Laake, D. Borchers, and L. Thomas. 2001. Introduction to distance sampling estimating abundance of biological populations. Oxford University Press, New York, New York, USA.

Buckland, S. T., E. A. Rexstad, T. A. Marques, and C. S. Oedekoven. 2015. Distance sampling: methods and applications. Springer, Cham, Switzerland.

Carrara, E., V. Arroyo-Rodríguez, J. H. Vega-Rivera, J. E. Schondube, S. M. de Freitas, and L. Fahrig. 2015. Impact of landscape composition and configuration on forest specialist and generalist bird species in the fragmented Lacandona rainforest, Mexico. Biological Conservation 184:117–126.

Chao, A., N. J. Gotelli, T. Hsieh, E. L. Sander, K. Ma, R. K. Colwell, and A. M. Ellison. 2014. Rarefaction and extrapolation with Hill numbers: a framework for sampling and estimation in species diversity studies. Ecological Monographs 84:45–67.

Chao, A., and L. Jost. 2012. Coverage-based rarefaction and extrapolation: standardizing samples by completeness rather than size. Ecology 93:2533–2547.

Chessel, D., A. Dufour, and S. Dray. 2016. Analysis of ecological data: exploratory and Euclidean methods in environmental sciences (ade4). R Package version 1.7–4. https://pbil.univ-lyon1.fr/ade4/ade4-html/00Index.html

Cornulier, T., and V. Bretagnolle. 2006. Assessing the influence of environmental heterogeneity on bird spacing patterns: a case study with two raptors. Ecography 29:240–250.

Cox, S. A., A. D. Peoples, S. J. DeMaso, J. J. Lusk, and F. S. Guthery. 2004. Survival and cause-specific mortality of Northern Bobwhites in western Oklahoma. Journal of Wildlife Management 68:663–671.

Danielson, B. J. 1991. Communities in a landscape: the influence of habitat heterogeneity on the interactions between species. American Naturalist 138:1105–1120.

DeMaso, S. J., A. D. Peoples, S. A. Cox, and E. S. Parry. 1997. Survival of northern bobwhite chicks in western Oklahoma. Journal of Wildlife Management 61:846–853.

Devictor, V., R. Julliard, and F. Jiguet. 2008. Distribution of specialist and generalist species along spatial gradients of habitat disturbance and fragmentation. Oikos 117:507–514.

Dolédec, S., D. Chessel, and C. Gimaret-Carpentier. 2000. Niche separation in community analysis: a new method. Ecology 81:2914–2927.

Fahrig, L., J. Baudry, L. Brotons, F. G. Burel, T. O. Crist, R. J. Fuller, C. Sirami, G. M. Siriwardena, and J. L. Martin. 2011. Functional landscape heterogeneity and animal biodiversity in agricultural landscapes. Ecology Letters 14:101–112.

Flick, T., S. Feagan, and L. Fahrig. 2012. Effects of landscape structure on butterfly species richness and abundance in agricultural landscapes in eastern Ontario, Canada. Agriculture, Ecosystems & Environment 156:123–133.

Fuhlendorf, S. D., A. J. Woodward, D. M. Leslie, and J. S. Shackford. 2002. Multi-scale effects of habitat loss and fragmentation on lesser prairie-chicken populations of the US Southern Great Plains. Landscape Ecology 17:617–628.

Fuller, M. R., and J. A. Mosher. 1987. Raptor survey techniques. Pages 37–65 *in* B. A. G. Pendleton, B. A. Millsap, K. W. Cline, and D. M. Bird, editors.

Raptor management techniques manual. National Wildlife Federation, Washington, D.C., USA.

Gorini, L., J. D. Linnell, R. May, M. Panzacchi, L. Boitani, M. Odden, and E. Nilsen. 2012. Habitat heterogeneity and mammalian predator–prey interactions. Mammal Review 42:55–77.

Hagen, C. A., J. C. Pitman, B. K. Sandercock, R. J. Robel, and R. D. Applegate. 2007. Age-specific survival and probable causes of mortality in female Lesser Prairie-Chickens. Journal of Wildlife Management 71:518–525.

Hall, S. 2015. Vegetation communities and land cover in the mixedgrass prairie: an analysis of contemporary and historic data sources. Dissertation. University of Oklahoma, Norman, Oklahoma, USA.

Hanski, I. 1995. Effects of landscape pattern on competitive interactions. Pages 203–224 *in* L. Hansson, L. Fahrig, and G. Merriam, editors. Mosaic landscapes and ecological processes. Chapman & Hall, London, UK.

Hanski, I. 2008. Spatial patterns of coexistence of competing species in patchy habitat. Theoretical Ecology 1:29–43.

Hargis, C. D., J. A. Bissonette, and J. L. David. 1998. The behavior of landscape metrics commonly used in the study of habitat fragmentation. Landscape Ecology 13:167–186.

Hernández, F., L. A. Brennan, S. J. DeMaso, J. P. Sands, and D. B. Wester. 2013. On reversing the Northern Bobwhite population decline: 20 years later. Wildlife Society Bulletin 37:177–188.

Holt, R. D. 1984. Spatial heterogeneity, indirect interactions, and the coexistence of prey species. American Naturalist 124:377–406.

Hovick, T. J., R. D. Elmore, S. D. Fuhlendorf, D. M. Engle, and R. G. Hamilton. 2015. Spatial heterogeneity increases diversity and stability in grassland bird communities. Ecological Applications 25:662–672.

Hsieh, T. C., K. H. Ma, and A. Chao. 2016. An R package for rarefaction and extrapolation of species diversity (Hill numbers). Methods in Ecology and Evolution 7:1451–1456.

Kauffman, M. J., N. Varley, D. W. Smith, D. R. Stahler, D. R. MacNulty, and M. S. Boyce. 2007. Landscape heterogeneity shapes predation in a newly restored predator–prey system. Ecology Letters 10:690–700.

Koivunen, V., E. Korpimäki, and H. Hakkarainen. 1998. Refuge sites of voles under owl predation risk: Priority of dominant individuals? Behavioral Ecology 9:261–266.

Krüger, O., and J. Lindström. 2001. Habitat heterogeneity affects population growth in goshawk (*Accipiter gentilis*). Journal of Animal Ecology 70:173–181.

Kurki, S., A. Nikula, P. Helle, and H. Linden. 1998. Abundances of red fox and pine marten in relation to the composition of boreal forest landscapes. Journal of Animal Ecology 67:874–886.

Legendre, P. 2008. Studying beta diversity: ecological variation partitioning by multiple regression and canonical analysis. Journal of Plant Ecology 1:3–8.

Lin, W.-T., C.-H. Hsieh, and T. Miki. 2013. Difference in adaptive dispersal ability can promote species coexistence in fluctuating environments. PLoS ONE 8:e55218.

Marques, T. A., L. Thomas, S. G. Fancy, B. M. Pierce, S. T. Buckland, and C. Handel. 2007. Improving estimates of bird density using multiple-covariate distance sampling. The Auk 124:1229–1243.

Matter, W. J., and R. W. Mannan. 2005. How do prey persist? Journal of Wildlife Management 69: 1315–1320.

McGarigal, K., and E. Ene. 2015. Fragstats: spatial pattern analysis program for categorical maps. Version 4.2, computer software program produced by the authors at the University of Massachusetts, Amherst. http://www.umass.edu/landeco/research/fragstats/fragstats.html

Monteith, K. L., V. C. Bleich, T. R. Stephenson, B. M. Pierce, M. M. Conner, J. G. Kie, and R. T. Bowyer. 2014. Life-history characteristics of mule deer: effects of nutrition in a variable environment. Wildlife Monographs 186:1–62.

Neumann, J. L., G. H. Griffiths, A. Hoodless, and G. J. Holloway. 2016. The compositional and configurational heterogeneity of matrix habitats shape woodland carabid communities in wooded-agricultural landscapes. Landscape Ecology 31:301–315.

Palmer, T. M. 2003. Spatial habitat heterogeneity influences competition and coexistence in an African acacia ant guild. Ecology 84:2843–2855.

Penteriani, V., M. Gallardo, and P. Roche. 2002. Landscape structure and food supply affect Eagle Owl (*Bubo bubo*) density and breeding performance: a case of intra-population heterogeneity. Journal of Zoology 257:365–372.

Pickett, S. T., and M. L. Cadenasso. 1995. Landscape ecology: spatial heterogeneity in ecological systems. Science 269:331.

Rolland, V., J. A. Hostetler, T. C. Hines, H. F. Percival, and M. K. Oli. 2010. Impact of harvest on survival of a heavily hunted game bird population. Wildlife Research 37:392–400.

Sánchez-Zapata, J., and J. F. Calvo. 1999. Raptor distribution in relation to landscape composition in semi-arid Mediterranean habitats. Journal of Applied Ecology 36:254–262.

Sergio, F., T. Caro, D. Brown, B. Clucas, J. Hunter, J. Ketchum, K. McHugh, and F. Hiraldo. 2008. Top predators as conservation tools: ecological rationale, assumptions, and efficacy. Annual Review of Ecology, Evolution, and Systematics 39:1–19.

Slancarova, J., J. Benes, M. Kristynek, P. Kepka, and M. Konvicka. 2014. Does the surrounding landscape heterogeneity affect the butterflies of insular grassland reserves? A contrast between composition and configuration. Journal of Insect Conservation 18:1–12.

Šmilauer, P., and J. Lepš. 2014. Multivariate analysis of ecological data using CANOCO 5. Cambridge University Press, UK.

Stein, A., K. Gerstner, and H. Kreft. 2014. Environmental heterogeneity as a universal driver of species richness across taxa, biomes and spatial scales. Ecology Letters 17:866–880.

Storch, I. 1997. The importance of scale in habitat conservation for an endangered species: the capercaillie in Central Europe. Pages 310–330 *in* J. A. Bissonette, editor. Wildlife and landscape ecology. Springer, New York, New York, USA.

Tapia, L., P. L. Kennedy, and R. W. Mannan. 2007. Habitat sampling. Pages 153–169 *in* D. M. Bird and K. L. Bildstein, editors. Raptor research and management techniques manual. Raptor Research Foundation, Hancock House Publishers, Surrey, British Columbia, Canada.

Tattersall, F., D. Macdonald, B. Hart, P. Johnson, W. Manley, and R. Feber. 2002. Is habitat linearity important for small mammal communities on farmland? Journal of Applied Ecology 39:643–652.

Ter Braak, C. J., and P. Smilauer. 2012. Canoco reference manual and user's guide: software for ordination, version 5.0. Cambridge University Press, Cambridge, UK.

Tews, J., U. Brose, V. Grimm, K. Tielbörger, M. Wichmann, M. Schwager, and F. Jeltsch. 2004. Animal species diversity driven by habitat heterogeneity/diversity: the importance of keystone structures. Journal of Biogeography 31:79–92.

Thirgood, S. J., S. M. Redpath, P. Rothery, and N. J. Aebischer. 2000. Raptor predation and population limitation in red grouse. Journal of Animal Ecology 69:504–516.

Thomas, L., S. T. Buckland, E. A. Rexstad, J. L. Laake, S. Strindberg, S. L. Hedley, J. R. Bishop, T. A. Marques, and K. P. Burnham. 2010. Distance software: design and analysis of distance sampling surveys for estimating population size. Journal of Applied Ecology 47:5–14.

Thompson, W. L. 2002. Towards reliable bird surveys: accounting for individuals present but not detected. Auk 119:18–25.

Turner, M. G., S. M. Pearson, W. H. Romme, and L. L. Wallace. 1997. Landscape heterogeneity and ungulate dynamics: What spatial scales are important? Pages 331–348 *in* J. A. Bissonette, editor. Wildlife and landscape ecology. Springer, New York, New York, USA.

Valkama, J., E. Korpimäki, B. Arroyo, P. Beja, V. Bretagnolle, E. Bro, R. Kenward, S. Manosa, S. M. Redpath, and S. Thirgood. 2005. Birds of prey as limiting factors of gamebird populations in Europe: a review. Biological Reviews 80:171–203.

Wiens, J. A. 1995. Landscape mosaics and ecological theory. Pages 1–26 *in* L. Hansson, L. Fahrig, and G. Merriam, editors. Mosaic landscapes and ecological processes. Chapman & Hall, London, UK.

Wolfe, D. H., M. A. Patten, E. Shochat, C. L. Pruett, and S. K. Sherrod. 2007. Causes and patterns of mortality in Lesser Prairie-chickens (*Tympanuchus pallidicinctus*) and implications for management. Wildlife Biology 13:95–104.

9

Comparison of two photographic methodologies for collecting and analyzing the condition of coral reef ecosystems

D. E. P. Bryant,[1,2,†] A. Rodriguez-Ramirez,[1] S. Phinn,[2] M. González-Rivero,[1] K. T. Brown,[1] B. P. Neal,[1,3] O. Hoegh-Guldberg,[1] and S. Dove[1]

[1]*The Global Change Institute, ARC Centre for Excellence in Coral Reef Studies, and School of Biological Sciences, The University of Queensland, St Lucia, Queensland 4072 Australia*
[2]*Remote Sensing Research Centre, School of Geography Planning and Environmental Management, The University of Queensland, St Lucia, Queensland 4072 Australia*
[3]*Bigelow Laboratory for Ocean Sciences, 60 Bigelow Drive, East Boothbay, Maine 04544 USA*

Abstract. Coral reefs are declining rapidly in response to unprecedented rates of environmental change. Rapid and scalable measurements of how benthic ecosystems are responding to these changes are critically important. Recent technological developments associated with the XL Catlin Seaview Survey have begun to provide high-resolution 1-m^2 photographic quadrats of ~1.8–2 km of coral reefs at 10 m depth by using a semi-autonomous image collection SeaView II camera system (SVII). The rapid collection of images by SVII can result in images being taken at a variable distance (1–2 m) from the substrate as well as having natural light variability between images captured. This variability can affect the quality of taxonomic resolution archived from photographic quadrats captured by SVII. Conventional approaches for taking photographic quadrats of coral reefs involve taking images from a fixed distance with the use of artificial light. These methods often provide images that enable high taxonomic resolution, but are typically used to cover areas of 50–150 m. Here, we select key metrics associated with coral reef condition from a functional perspective to contrast how much is lost in terms of association and agreement from image annotations using SVII images (1-m^2 photoquadrats) compared to conventional methods, capturing 0.5-m^2 photoquadrats using a digital single lens reflex camera from a fixed height of 0.5 m above the benthos. Comparisons were made over the same 50-m linear transects at 15 sites in the Maldives. Photoquadrats were manually analyzed using an online image annotation tool and image repository called coral net with 25 randomly placed dots per 0.5 m^2 of a quadrat. Our results reveal high levels of correlation and agreement between methods when measuring the abundance of hard corals as a functional group and individual labels, which make up hard corals as a functional group. These results demonstrate the two methods are comparable when measuring functional groups and coral communities on coral reefs. Therefore, involving rapid semi-automated technologies that maximize data collection for monitoring coral reefs does not necessarily imply that taxonomic resolution is compromised. This insight has important ramifications for detecting important changes in coral reef condition, such as a decline in coral cover or shifts in benthic community composition.

Key words: benthic assessment; coral reefs; environmental monitoring; image analysis; Maldives; photoquadrats; underwater photography.

† **E-mail:** dominic.bryant@uq.edu.au

Introduction

Coral reefs are undergoing rapid reduction in hard coral cover (Gardner et al. 2003, Pandolfi et al. 2003, Bruno and Selig 2007, Edmunds and Elahi 2007, De'ath et al. 2012) and shifts in community structure (Gleason 1993, Glynn et al. 2001, Pratchett 2010, Pratchett et al. 2011) in many regions globally. These changes are occurring as a result of climate change (Hoegh-Guldberg et al. 2007) as well as local anthropogenic stressors (Mumby 2006, Williams et al. 2015). The increased frequency and severity of disturbances such as cyclones, coral bleaching events, *Acanthaster planci* (crown-of-thorns starfish, COTS) outbreaks, as well as other impacts from anthropogenic influences, is heralding the need for monitoring these changes in community ecology of coral reefs rapidly and at larger scales (Edmunds and Bruno 1996, Knowlton and Jackson 2008, De'ath et al. 2012, González-Rivero et al. 2014). The XL Catlin Seaview Survey (XL CSS) was launched in 2012 with the aim of developing a broad-scale assessment of the state of key coral reefs around the world (González-Rivero et al. 2014, 2016) and providing a much-needed standardized global baseline (Knowlton and Jackson 2008, Trygonis and Sini 2012) that is rapid, accurate, and rigorous.

Previous attempts to assess the condition of benthic communities on coral reefs vary in spatial scale and taxonomic resolution (Fig. 1; Andréfouët et al. 2002, Mumby and Steneck 2008, Hedley et al. 2012, González-Rivero et al. 2014). Satellite imagery and airborne photography (Fig. 1d, e) provide remote sensing at scales of 1–1000 km of reef surveyed enabling the construction of coarse habitat boundary maps that provide information on percent cover for simple categories of benthos (e.g., living vs. dead coral cover, macroalgae, and soft sediment; Mumby et al. 1997, Hochberg and Atkinson 2003, Mumby et al. 2004). Furthermore, in-water surveys must be performed in order to allow for object-based image analysis within the photographs collected (Phinn 1998, Mumby et al. 1999, Roelfsema et al. 2010, Phinn et al. 2012). This reduction in detail and taxonomic resolution means scientists and decision makers miss out on useful information such as changes in the diversity and community structure of coral reefs. Being able to measure the frequency and

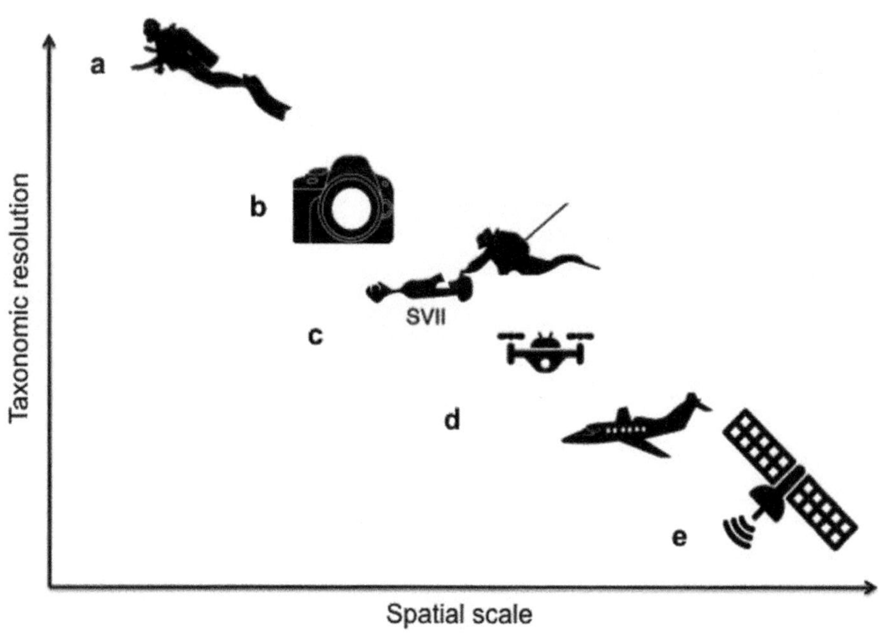

Fig. 1. Coral reef survey technique differences in spatial scale and taxonomic resolution. (a) In situ diver assessment, (b) high-resolution photoquadrats (0.25–1 m^2), (c) SVII with diver propulsion vehicle, (d) airborne photography (aircraft and drone), and (e) satellite imagery.

abundance of the key organisms is essential if we are to broaden our understanding of the drivers and resulting changes to the world's coral reefs (Knowlton and Jackson 2008). On the other hand, in-water diver assessment techniques (Fig. 1a) offer the highest level of taxonomic resolution. The time, cost, and effort for in-water diver assessment techniques, however, reduce the ability to cover large spatial scales (i.e., 50- to 200-m SCUBA dive; Mumby et al. 1999, Hill and Wilkinson 2004). Therefore, remote sensing techniques (Fig. 1d, e), or diver surveys based on the rapid collection of imagery (e.g., video or semi-autonomous photography; Fig. 1b, c), offer a cost-effective monitoring technique for rapidly measuring the state of coral reefs at larger scales (González-Rivero et al. 2014).

Recent advances in photography, computer learning, and image classification have increased the scale over which measurements with a high degree of precision and accuracy can be made. The XL CSS utilizes an innovative camera assemblage (SeaView II or SVII camera) which is made up of a high-resolution 360° panoramic camera system connected to a diver propulsion vehicle (DPV). This technology has opened up new possibilities for understanding reef ecosystems at multiple kilometer scales, at a depth of 8–12 m (González-Rivero et al. 2014). The SVII has the ability to collect and analyze over a thousand images per 1.5- to 2-km transect, which would otherwise take a large amount of time and cost using conventional human-based methods. Furthermore, surveys performed without a DPV become nearly impossible in areas with strong currents (González-Rivero et al. 2014, 2016).

Semi- and fully automated underwater vehicles (AUVs) are increasingly being used for the collection of large image sets that archive taxonomic information of benthic marine habitats (Durrant and Dunbabin 2011, Williams et al. 2012, González-Rivero et al. 2014, Roelfsema et al. 2015). Although AUVs offer advantages in terms of the amount of data collected, there are limitations when using certain types of uncontrolled photography that can introduce error into the quality of the data collected due to a reduction in image quality within the image set. This error can be attributed to factors such as changes in light availability, changes in the distance an image is taken from the benthos, and changes in attitude (i.e., pitch, yaw, and roll) that an image was taken (González-Rivero et al. 2014).

The research described here aimed to select key metrics associated with coral reef condition from a functional perspective and contrast how much is lost in terms of association and agreement between image annotations using a semi-autonomous image collection method (SVII camera system) and a more conventional method, which offers more consistent image quality at the cost of spatial scale. Exploring the capabilities of a semi-autonomous AUV methodology to provide similar or better association and agreement at larger spatial scales is an important step in developing the capacity to make broad-scale interpretations regarding coral reef conditions during unprecedented rates of decline (Edmunds and Bruno 1996, Balmford et al. 2005, Edmunds and Elahi 2007, Knowlton and Jackson 2008, De'ath et al. 2012, González-Rivero et al. 2014).

Materials and Methods

Image collection

Images were collected during the 2015 XL Catlin SS Maldives expedition from the 30th of March to the 19th of April 2015. Fifteen of the sites from the expedition were chosen to explore the agreement and performance of the two methods (Fig. 2). A 50-m transect tape was laid out at a depth of 10 m at the starting point of each of the 1.5- to 2-km SVII transects (Fig. 4e).

SVII method

The SVII camera system (Fig. 4a) is a customized high-resolution imaging system (González-Rivero et al. 2014) approximately 2 m in length and 50 kg in weight. The SVII houses three remotely triggered digital single lens reflex Canon 5d MkII cameras (Canon Inc., Tokyo, Japan; maximum resolution = 5613 × 3744 mpixels), which have been aligned in a housing designed to capture a 360° panoramic of the surrounding environment every three seconds. The camera housing is attached to a DPV, allowing the diver to cover up to 2 km per dive. A fisheye lens with the aperture set to f8 is connected to the camera, and ambient light is used in order to preserve battery power. The SVII diver can change the ISO and shutter speed manually via a Samsung

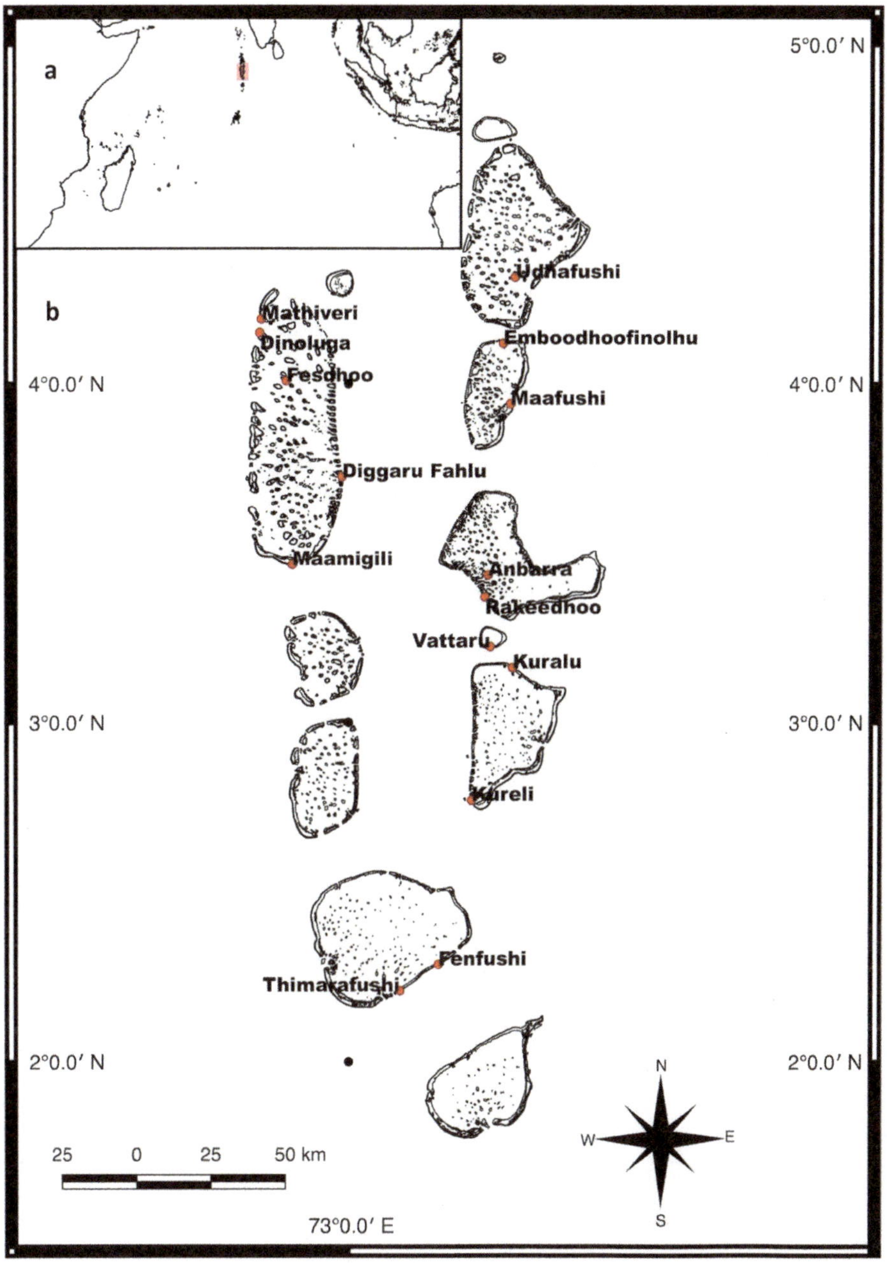

Fig. 2. Map of survey sites for comparison of SVII and conventional photoquadrat methods in the Maldives taken during the March–April–2015 XL Catlin Seaview Survey expedition. (a) Map of the Indian Ocean with the survey area of the Central Maldives highlighted in red. (b) Survey sites in the Central Maldives.

Galaxy tablet connected to the top of SVII. The diver has the ability to view each image on the Samsung Galaxy tablet screen and will make adjustments to ISO according to light availability. The shutter speed is maintained at 1/320 due to the propulsion of the SVII while taking images. For more details regarding SVII and its use, refer to González-Rivero et al. (2014).

SVII captured images over a 50-m transect at 10 m depth to compare with conventional methods in ~1–2 minutes depending on current conditions. Photoquadrats of 1 m^2, with an average resolution of 20 and 10 pixels/cm, were color-corrected and then cropped from the SVII images by using a time-stamped altitude measurement taken from a Micron Echosounder altimeter by Tritech (Tritech International, Westhill, Aberdeenshire, UK), attached to the bottom of the DPV body on SVII. A pressure sensor was also used to record the depth each image was captured. Furthermore, a global positioning system, which has been time synchronized with the cameras and sensors, was attached to a surface marker tethered to the SVII diver. Therefore, every image is spatially georeferenced (González-Rivero et al. 2014).

Fig. 3. (a) Cropped and color-corrected SVII 1 × 1 m photoquadrat from Emboodhoofinolhu reef with a full resolution of 1171 × 1171 pixels. (b) Un-color-corrected conventional 0.5 × 0.5 m photoquadrat using two strobes taken at Emboodhoofinolhu reef with a full resolution of 4608 × 3456 pixels.

Between 20 and 40 1-m^2 SVII photographic quadrats (Fig. 3a) were selected from the SVII images at each site (Table 1). A 50-m transect tape running through the center of the image was used to identify which 1-m^2 SVII photographic quadrats were selected for image analysis. SVII photographic quadrats (1 m^2) were assessed to ensure there was no overlap of area between quadrats before being manually annotated on coral net. Differences in the number of 1-m^2 SVII photographic quadrats are caused by the speed the DPV is traveling. Differences in DPV speed can be due to the current at the time and the diver driving the DPV.

Conventional method

Divers used a 0.5 × 0.5 m quadrat attached to Olympus mirrorless EM-5 digital camera (Olympus Corporation, Tokyo, Japan; maximum resolution = 4608 × 3456) at a fixed distance of 0.5 m away from the substrate to take pictures (Fig. 3b) during 60-minute SCUBA dive. Two Sea and Sea YS-D2 strobes were also attached provided artificial light for each image. Between 50 and 90 photographic quadrats (Fig. 3b) were taken using the conventional method on a single 50 × 1 m linear transect (Table 1). On occasion, strong currents and equipment failures caused dives using conventional methods to be abandoned before the 50 m could be completed.

Image annotation and analysis

Images were manually annotated by an expert annotator using CoralNet (Beijbom et al. 2012), an online repository for analyzing coral reef images; 100 randomly allocated points were used for the analysis of each 1-m^2 photoquadrat captured by SVII (Fig. 4b) and 25 points were used for the 0.5-m^2 photoquadrats collected employing conventional methods (Fig. 4d). The number of points was chosen in order to assess relatively the same amount of space inside the photoquadrats from both methods (i.e., 25 points per 0.5 m^2 of photoquadrat). Correlation and agreement between the two methods was explored when estimating label categories made up of (1) broad benthic parameters at a functional level (i.e., dead hard coral, hard coral, macroalgae, and other invertebrates); (2) morphological groups within the hard coral

Table 1. Area covered in m^2 by SVII and conventional methods over the same 50-m linear transects at 15 sites in the Maldives, surveyed during the 2015 XL Catlin Seaview Survey expedition (www.catlinseaviewsurvey.com).

Sites	Latitude/Longitude	No. SVII 1-m^2 quadrats = area covered (m^2)	No. conventional 0.5-m^2 quadrats × 0.25 = area covered (m^2)
Anbarra	3′26.11325 N/73′25.32778 E	30 × 1 = 30	90 × 0.25 = 22.50
Diggaru Fahlu	3′42.80441 N/72′58.88254 E	31 × 1 = 31	58 × 0.25 = 14.50
Dinoluga	4′9.04307 N/72′43.87115 E	30 × 1 = 30	54 × 0.25 = 13.50
Emboodhoofinolhu	4′7.12197 N/73′28.06809 E	30 × 1 = 30	74 × 0.25 = 18.50
Fenfushi	2′17.28663 N/73′16.41937 E	30 × 1 = 30	89 × 0.25 = 22.25
Fesdhoo	4′0.50388 N/72′48.68643 E	30 × 1 = 30	80 × 0.25 = 20.00
Kuralu	3′9.75164 N/73′29.72463 E	39 × 1 = 39	89 × 0.25 = 22.25
Kureli	2′47.02307 N/73′22.10777 E	30 × 1 = 30	87 × 0.25 = 21.75
Maafushi	3′56.73654 N/73′30.00041 E	30 × 1 = 30	90 × 0.25 = 22.50
Maamigili	3′28.00391 N/72′50.1526 E	30 × 1 = 30	80 × 0.25 = 20.00
Mathiveri	4′11.30840 N/72′48.68643 E	22 × 1 = 22	90 × 0.25 = 22.50
Rakeedhoo	3′18.63745 N/73.28.07505 E	30 × 1 = 30	84 × 0.25 = 21.00
Thimarafushi	2′12.60078 N/73′9.45339 E	30 × 1 = 30	89 × 0.25 = 22.25
Udhafushi	4′18.80890 N/73′30.21989 E	30 × 1 = 30	90 × 0.25 = 22.50
Vattaru	3′13.47848 N/73′25.69829 E	36 × 1 = 36	50 × 0.25 = 12.50
Mean (±standard error)		31 × 1 = 31 (0.92)	80 × 0.25 = 20 (0.95)

Fig. 4. Methods being compared over 50-m linear transect at 15 sites in the Maldives during the March–April 2015 XL Catlin Seaview Survey expedition. (a) SVII surveying a Maldivian reef slope at 10 m. (b) Cropped and color-corrected 1-m^2 photoquadrat taken from Maafushi using SVII (a) with 100 randomly allocated points using coral net (https://coralnet.ucsd.edu/). (c) Conventional 0.5-m^2 photoquadrats being captured with a 0.5-m stand off the substrate over a 50-m linear transect on a Maldivian reef slope. (d) 0.5-m^2 photoquadrat taken Maafushi using conventional methods (c) with 25 randomly allocated points using coral net (https://coralnet.ucsd.edu/). (e) Illustration of photoquadrat placement over 50-m linear transects in the Maldives, blue squares indicate 1 m^2 from SVII (a), and red squares indicate 0.5-m^2 photoquadrats from conventional methods (b). Images a and c were used with permission from the copyright owners of the Ocean Agency/Christophe Bailhache.

functional group along with genera easily identified based on color and texture from a 2D image (i.e., branching tabulate/corymbose *Acropora*, branching arborescent *Acropora*, branching *Pocillopora*, branching *Porites*, and branching other, massive/submassive/encrusting [MASE] Porites, MASE small invisible polyps, MASE large round polyps, MASE meandering, and thin/foliose/plating [TFP]); and (3) other invertebrates such as soft corals, tunicates, and sponges (Table 2).

Statistical analysis

The percent cover for each quadrat was determined, and then, the average cover per site for each label and functional group was calculated. A D'Agostino and Pearson normality test was performed to check whether the percent cover of each label and functional group for both methods was normally distributed across all 15 sites. If the percent cover for each site was normally distributed, then a Pearson correlation coefficient was performed to measure the strength of relationship between the percent cover estimates of the two methodologies for that specific label or functional group. A Spearman's rank correlation coefficient was determined using GraphPad Prism version 7 (Mac OS X, GraphPad Software, Inc., La Jolla, California, USA) on labels and functional groups that did not pass the D'Agostino and Pearson normality test.

Bland–Altman method comparison plots, commonly used to compare estimation in agreement between two methods of measurement where the true value remains unknown (Bland and Altman 1986, 2003, Giavarina 2015), were constructed to assess agreement between SVII and conventional methods. These methods also give a visualization if one method was biased at estimating coverage of one label over the other. Bland–Altman plot was done by plotting the mean percent cover between the two methods at each site on the x-axis and the ratio between the SVII and conventional methods for each site on the y-axis. If the ratio bias (average ratio across all sites) was more than one, the methods were observed as more biased toward SVII in estimates of benthic cover associated with a specific label or functional group. Ratio bias of less than one meant there were more biases toward conventional methods. Furthermore, an equal proportion of sites scattered on either side of one, and with a ratio closer to one is deemed as suggestive of

Table 2. Label categories and functional groups used for the image annotation of photoquadrats taken using SVII and conventional methodologies on Maldivian reef slopes during the March–April 2015 XL Catlin Seaview Survey expedition (www.catlinseaviewsurvey.com).

Functional group	Label and description
Dead hard coral (DHC)	DHC covered by epilithic algal matrix (EAM)
	Rubble covered by epilithic algal matrix
	DHC covered by crustose coralline algae (CCA)
	Rubble covered by crustose coralline algae (CCA)
	DHC covered by cyanobacteria
	Rubble covered by cyanobacteria
Hard coral	Tabulate/corymbose/plating *Acropora* spp.
	Digitate *Acropora* spp.
	Arborescent *Acropora* spp.
	Branching *Pocillopora* spp.
	Branching *Porites* spp.
	Branching other
	Massive/submassive/encrusting (MASE) *Porites* spp.
	Massive/submassive/encrusting small polyps
	Massive/submassive/encrusting large round polyps
	Massive/submassive/encrusting meandering *Lobophyllia* spp.
	Massive/submassive/encrusting meandering other
	Massive/submassive/encrusting bleached
	Thin/foliose/plating (TFP) with visible relief structures
	Thin/foliose/plating with visible round corallites
Macroalgae	Filamentous macroalgae >1 cm in height
	Leathery macrophytes >1 cm in height
	Halimeda spp. >1 cm in height
Other	Fish
	Transect hardware
	Anthropogenic marine debris
	Unclear (i.e., shadow, overexposed, blurry)
Other invertebrates	*Alcyonacea* spp.
	Crinoids
	Hydroids
	Individual tunicates (*Didemnum molle*)
	Massive or encrusting sponges
	Branching or rope-forming sponges
	Gorgonacea spp. (sea fans and sea whips)
	Mobile invertebrates (echinoderms)
	Other invertebrates (colonial ascidians and hexacorallia)
Soft sediment	Sand

greater method agreement, compared with a lopsided scatter of points more removed from one.

RESULTS

Correlation and agreement of functional groups

Percent cover across all six functional groups surveyed by SVII methods and conventional methods over the 50-m transects at the 15 sites surveyed was significantly correlated. The Bland–Altman method comparison plots depict different ratio bias and agreement between the two methods for percent cover of different functional groups (Fig. 5).

The amount of dead hard coral (DHC) measured by the two methodologies was significantly correlated ($r = 0.9401$, $n = 15$, $P < 0.0001$; Fig. 5a) with a ratio bias of 1.17 and all sites exhibiting a ratio greater than one (Fig. 5b). Estimates of the amount of hard coral cover from the two methodologies were also significantly correlated (Fig. 5c; $r = 0.9527$, $n = 15$, $P < 0.0001$), with a ratio bias of 0.906, with 80% of sites having a ratio of less than one (Fig. 5d).

Macroalgae were significantly correlated, when using a Spearman's rank correlation coefficient (r_s; $r_s = 0.6739$, $n = 15$, $P = 0.0073$; Fig. 5e), and had a ratio bias of 0.58 with 73% of sites having a ratio of less than one (Fig. 5f). Three of the 15 sites (Dinoluga, Feshdhoo, and Vattaru) did not pick up macroalgae using the SVII method, while one site (Diggaru Fahlu) did not find macroalgae using conventional methods. Seventy-three percent of sites had a ratio of less than one. Other (mostly unclear) labels were significantly correlated ($r = 0.641$, $n = 15$, $P = 0.0101$) and had a ratio bias of 1.3 with 66% of the sites having a ratio >1 (Fig. 5h). The benthic category, other

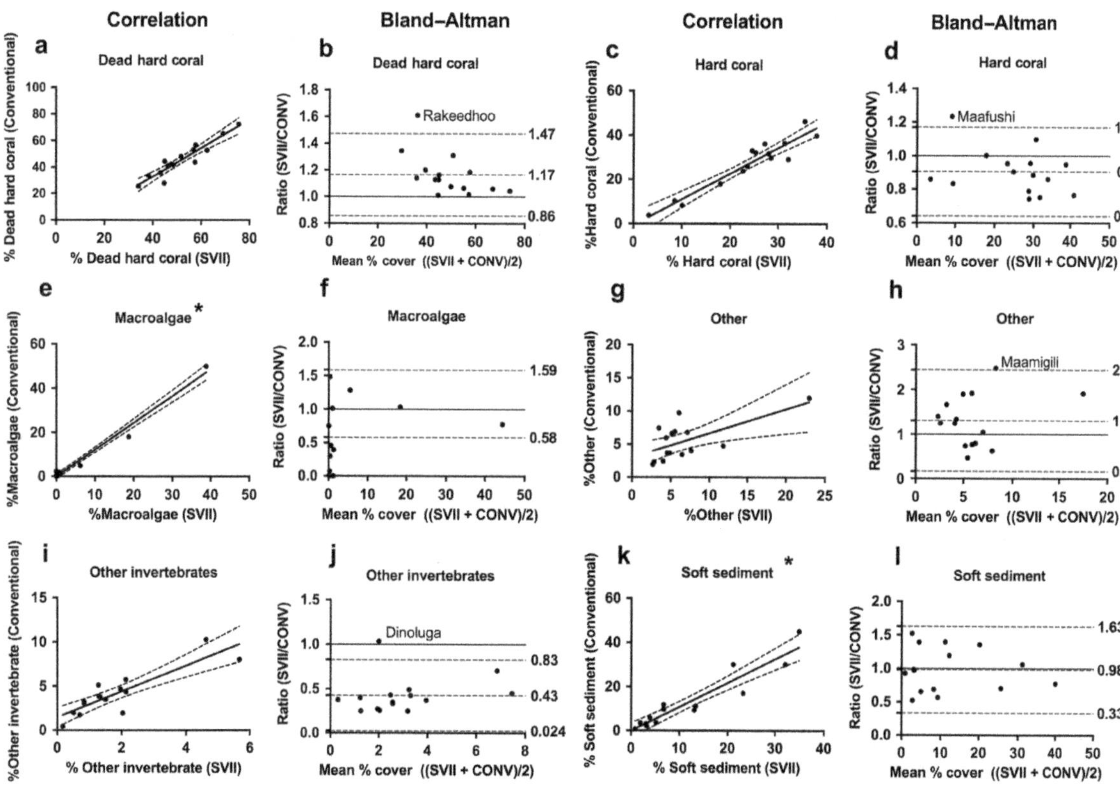

Fig. 5. Correlation (with 95% confidence bands) and Bland–Altman method agreement plots for percent cover of functional groups surveyed by SVII and conventional (CONV methods across 50-m linear transects at 15 coral reef sites from the XL Catlin Seaview Survey Maldives expedition from the 29th of March until the 17th of April 2015). * Indicates Spearman's rank correlation performed for correlation coefficient. Red line indicates 95% limit agreements, and blue line represents ratio bias between SVII and conventional methods for Bland–Altman plots.

invertebrates, was significantly correlated between the two methods ($r = 0.762$, $n = 15$, $P < 0.0001$) with a ratio bias of 0.43 (Fig. 5j) and 93% of the sites having a ratio of less than one. Similarly, the percent cover of sand was significantly correlated ($r_s = 0.9429$, $n = 15$, $P < 0.0001$; Fig. 5k) and had a ratio bias of 0.98 with 53% of sites being under 1 (Fig. 5l).

Correlation and agreement within hard coral functional groups

The percent cover of tabulate/corymbose corals was correlated between methods ($r = 0.9325$, $n = 15$, $P < 0.0001$; Fig. 6a). The ratio bias for tabulate/corymbose corals was 0.91 with 60% of the sites having a ratio of less than one (Fig. 6b). Arborescent *Acropora* was correlated ($r = 0.9004$, $n = 13$, $P < 0.0001$) with a ratio bias of 0.91; however, only 30.7% of the sites had a ratio below one (Fig. 6d). Digitate *Acropora* coral cover was correlated ($r = 0.9063$, $n = 15$, $P < 0.0001$; Fig. 6e) and had a ratio bias of 1.736 (Fig. 6f). A ratio of above one was found in 73.3% of sites (Fig. 7f). Percent *Pocillopora* cover was correlated between the two methods ($r = 0.954$, $n = 14$, $P < 0.0001$; Fig. 6g) with a ratio bias of 0.9229 (Fig. 6h). A ratio of less than one was found in 71.4% of the sites surveyed (Fig. 6h).

The abundance of massive *Porites* was correlated between sites ($r = 0.9312$, $n = 15$, $P < 0.0001$; Fig. 7a) and had a ratio bias of 0.9033 (Fig. 7b). A ratio of less than one was found in 60% of the sites. Massive small/invisible polyps were correlated ($r = 0.7126$, $n = 15$, $P = 0.0029$; Fig. 7c) and had a ratio bias of 1.067. At least 53% of sites had a ratio of less than one between massive small/invisible polyp labels (Fig. 7d). Massive large round polyps corals were correlated ($r_s = 0.9455$, $n = 15$, $P < 0.0001$; Fig. 7e) and had a ratio bias of 0.594 (Fig. 7f). Eighty percent of sites had a ratio of less than one between massive large round polyp labels (Fig. 7f). Massive, meandering corals were correlated ($r = 0.6979$, $n = 15$, $P = 0.0046$; Fig. 8g) and had a ratio bias of 0.8119 (Fig. 7h). A ratio of less than one was found in 66.7% of sites between massive, meandering coral labels (Fig. 7h).

Thin/foliose/plating label (Fig. 8) was analyzed by adding the total percentages of TFP ridged, round polyps and *Porites* labels because the sums of each label alone were too small to conduct an analysis. Estimations of TFP corals between methods were correlated ($r_s = 0.787$, $n = 15$, $P = 0.0009$; Fig. 8a) and had a ratio bias of 0.81 with 63% of sites ratio scattered below one (Fig. 8b).

Correlation and agreement among other invertebrates found in the Maldives

There were three labels from the functional group of "other invertebrates" (i.e., soft corals, solitary tunicates, and massive encrusting sponges) that were identified during annotation and analyzed for the amount of correlation and agreement (Fig. 9). Estimations of soft corals from the *Alcyonaceans* were correlated ($r_s = 0.787$, $n = 15$, $P = 0.009$; Fig. 9a) and had a ratio bias of 1.09 with 55.5% of the sites ratio scattered below one (Fig. 9b). Individual solitary tunicates (i.e., *Didemnum molle*) were correlated ($r_s = 0.7688$, $n = 15$, $P = 0.0013$; Fig. 9), with a ratio bias of 0.34 and 92.3% of the sites with a ratio of less than one (Fig. 9d). Finally, estimation of massive encrusting sponges was not correlated ($r = 0.5231$, $n = 15$, $P = 0.0549$; Fig. 9e) and had a ratio bias of 0.68 with 100% of the sites having a ratio difference of less than one (Fig. 9f).

Discussion

The results presented here provide strong evidence that the two methods explored here are interchangeable. Broad benthic parameters and labels within the hard coral functional group were correlated between SVII and conventional methodologies. These results demonstrate that SVII is a useful tool for rapidly interpreting coral reef conditions and composition (González-Rivero et al. 2014). Although SVII methods cause a slight reduction in taxonomic resolution, by reducing the ability to see organisms <5 cm. Identifying different types of coral categories mentioned in Table 2 are therefore not impeded when using SVII images and it can be done more rapidly and at much larger spatial scale (i.e., 1.5–2 km^2) compared with more conventional in situ survey methods (González-Rivero et al. 2016).

Broad benthic parameters

Functional groups were correlated between methods, indicating that the two methods are estimating similar estimates of the percent cover of broad benthic parameters on coral reefs. However, the SVII method consistently estimated a

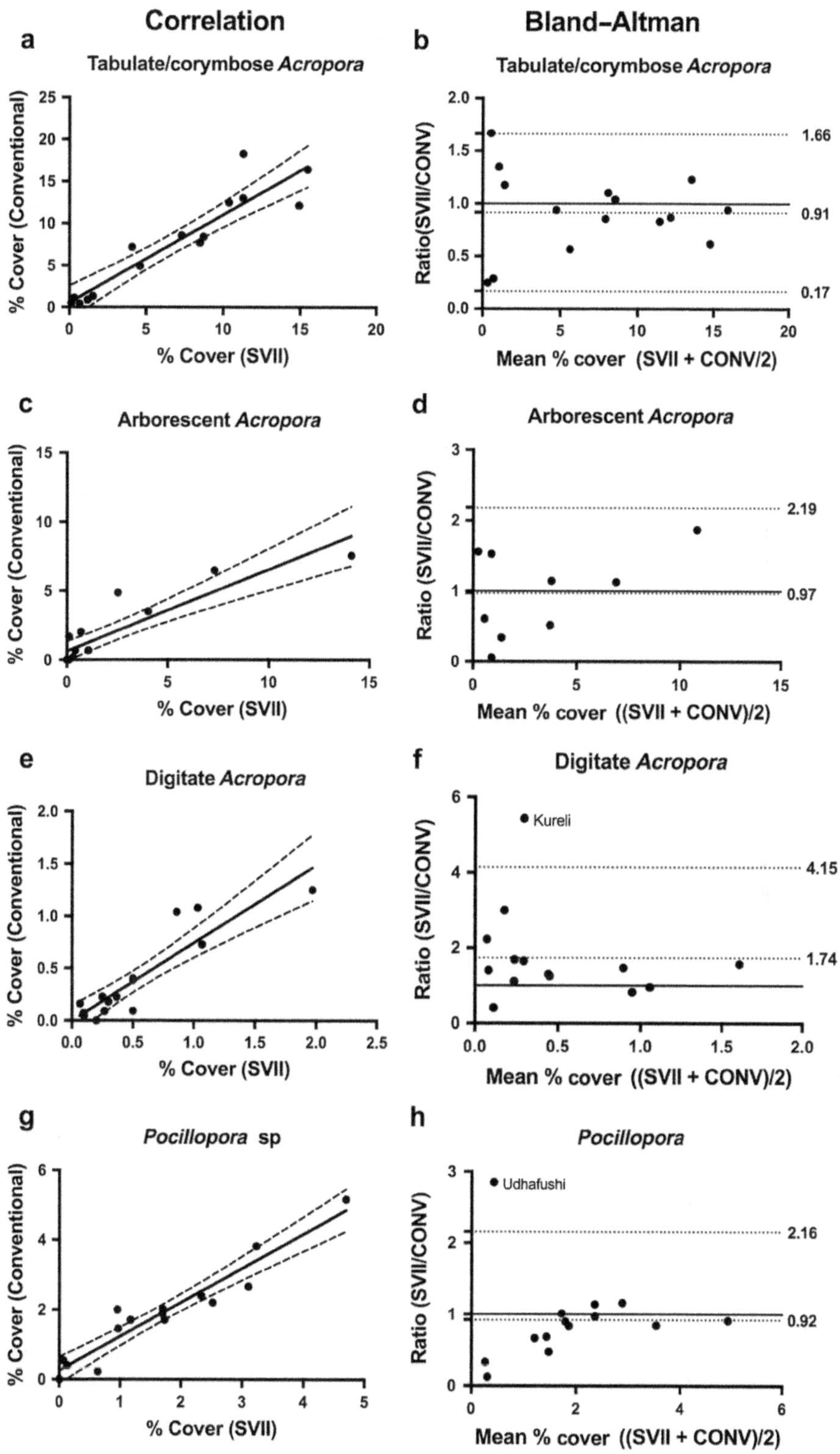

Fig. 6. Correlation (with 95% confidence bands) and Bland–Altman method agreement plots for percent cover of *Acropora* labels surveyed by SVII and conventional (CONV) methods across 50-m linear transects at 15 coral

(Fig. 6. *Continued*)
reef sites from the XL Catlin Seaview Survey Maldives expedition from the 29th of March until the 17th of April 2015. * Indicates Spearman's rank correlation performed for correlation coefficient. Red line indicates 95% limit agreements, and blue line represents ratio bias between SVII and conventional methods for Bland–Altman plots.

higher percent cover of DHC and a consistently lower percent cover of other invertebrates compared with conventional methods. This can be attributed to the difference in image resolution between images by SVII after post-processing (1031 × 1031 pixels) and the conventional photo-quadrats (4608 × 3456 pixels). The lowered image resolution limits the annotator's ability to identify organisms, which are too small or have smaller distinguishing features such as individual and colonial tunicates, small encrusting sponges, and hydrozoa. Therefore, these organisms are often missed with lowered image resolution and labeled as one of the labels within the DHC functional group. Most likely DHC was covered by epilithic algae matrix (EAM), which is usually defined as a multispecific assemblage of up to 1 cm in height (Connell et al. 2014, Beijbom et al. 2015). Small invertebrates such as the solitary tunicate (*Didemnum molle*) are known to only take up a small amount of benthic cover on coral reefs (<3%). However, *D. molle* was seen to reach nearly 12% cover at one site in the Maldives following the 1998 coral bleaching event (McClanahan 2000). *Didemnum molle* can quickly colonize empty spaces after disturbances and have been found to be settled on settlement plates in the Maldives after just two months (Clark 2000).

The hard coral functional group

Overall, the two methods were in significant agreement when identifying the different types of coral within the label set (Table 2), emphasizing the ability of the SVII to assess coral community composition. The two methods recorded very similar amounts of overall percent hard coral. Agreement between the two methods improved to different extents when more defined labels were assessed (e.g., tabulate/corymbose *Acropora*, arborescent *Acropora*, massive *Porites*, massive small and invisible polyps, massive meandering, and TFP labels). The correlation and agreement within coral labels is an important result as the potential loss of major reef-building coral such as *Acropora* on coral reefs can lead to declines in spatial complexity and therefore eliminate habitat for many reef-dwelling organisms such as reef fish and invertebrates (Bellwood et al. 2004, Alvarez-Filip et al. 2009).

Image resolution may have influenced the ability of the annotator to pick out the definition of massive large round polyps (i.e., *Favia* sp., *Favites* sp., *Diploastrea heliopora, Goniopora* sp.). Therefore, massive corals with large round polyps may have been identified as massive corals with small or invisible polyps, further explaining why SVII estimated a higher percent cover of massive small/invisible polyp colonies over conventional methods. However, being able to differentiate between different types of massive corals can potentially provide greater insights into the diversity of coral within the community composition. The life history traits of many massive corals are thought to be similar (e.g., stress tolerant and slow growing) (Darling et al. 2012). Shifts in community assemblages toward more "stress-tolerant" and "weedy" (e.g., *Pocillopora* sp.) species have been observed after bleaching events and in areas with high fishing impacts (Darling et al. 2013).

Quantifying the coral community composition on coral reefs is extremely important and more valuable to management than knowing hard coral cover alone (Bellwood et al. 2004). Disturbance events can affect some coral morphologies more than others. Cyclones, for example, can often lead to a much greater reduction in branching *Acropora* than other massive or encrusting corals (Brown 1997). Furthermore, it is important to identify shifts in coral community composition through time, after major disturbance events, in order to explore the reef recovery process. Frequent disturbance events in Moorea, French Polynesia, between 1979 and 2003 caused taxonomic shifts in coral composition from tabulate *Acropora*- and *Montipora*-dominated reefs to reefs with a relative abundance of *Pocillopora* and massive *Porites* (Pratchett et al. 2011).

The Maldives went through an extreme underwater heatwave in April 2016, which leads to the mass die-off of reef-building corals and COTS (*Ancanthaster planci*) in some sites, which drove a major decline in *Acropora* at

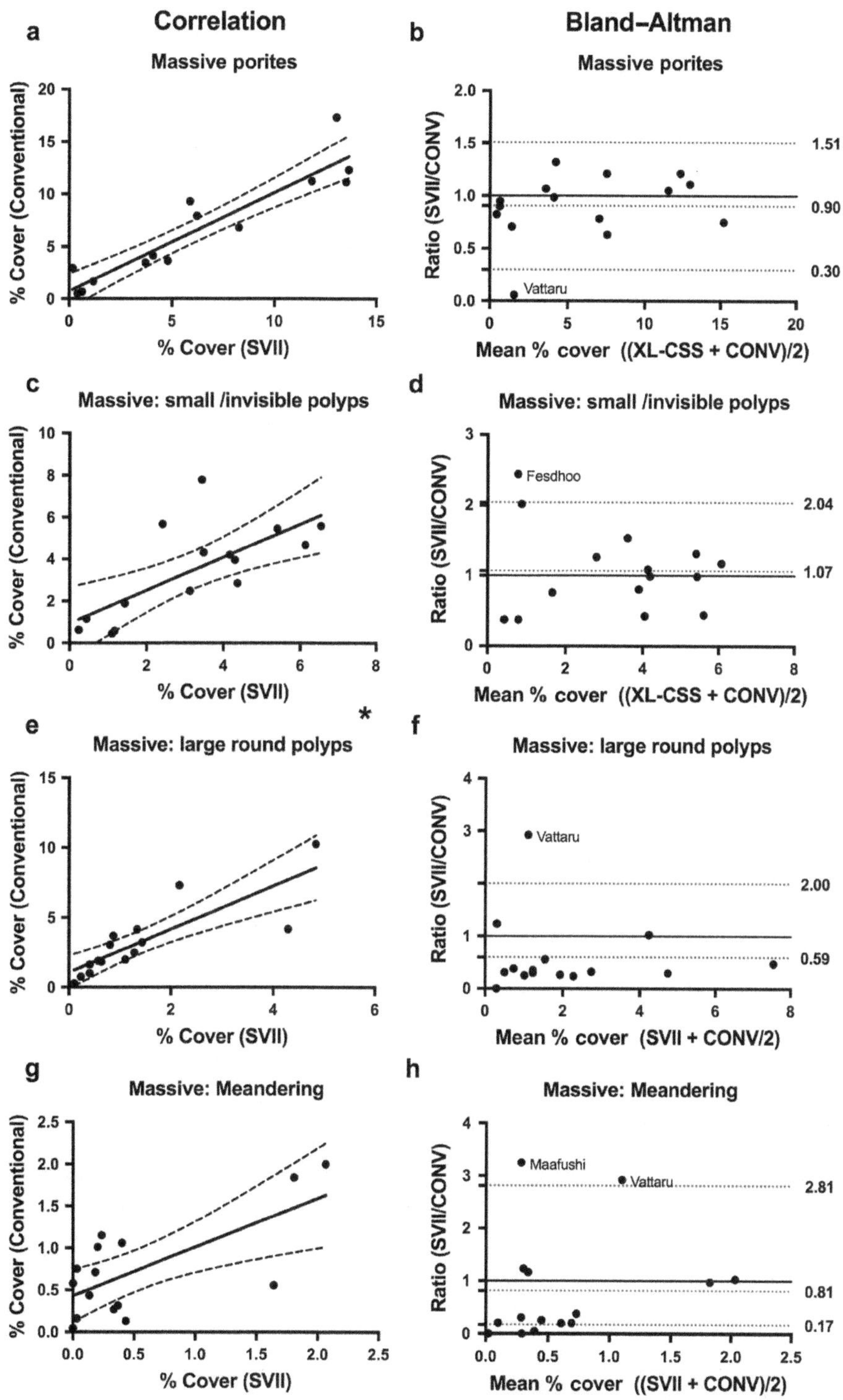

Fig. 7. Correlation (with 95% confidence bands) and Bland–Altman method agreement plots for percent cover of massive coral labels surveyed by SVII and conventional (CONV) methods across 50-m linear transects at 15 coral reef sites from the XL Catlin Seaview Survey Maldives expedition from the 29th of March until the 17th of April 2015. * Indicates Spearman's rank correlation performed for correlation coefficient. Red line indicates 95% limit agreements, and blue line represents ratio bias between SVII and conventional methods for Bland–Altman plots.

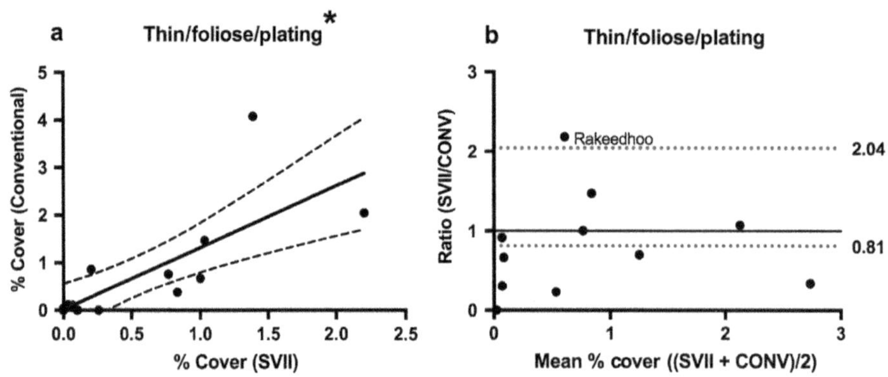

Fig. 8. Correlation (with 95% confidence bands) and Bland–Altman method agreement plots for percent cover of thin/foliose/plating corals surveyed by SVII and conventional (CONV) methods across 50-m linear transects at 15 coral reef sites from the XL Catlin Seaview Survey Maldives expedition from the 29th of March until the 17th of April 2015. * Indicates Spearman's rank correlation performed for correlation coefficient. Red line indicates 95% limit agreements, and blue line represents ratio bias between SVII and conventional methods for Bland–Altman plots.

various sites in 2015 (Pisapia et al. 2016). Being able to detect a change in coral composition at larger scales on coral reefs will provide valuable information in terms of initial mortality and recovery from both the coral bleaching event and *A. planci* outbreaks.

Other invertebrates

There was significant correlation and relative agreement between the two methods when identifying soft corals (*Alcyonaceans*). This is important to know because soft corals in the family *Alcyonacea* are a major sessile invertebrate on Indo-Pacific reefs, often dominating disturbed reefs when they are present (Dai 1991). However, they can also be affected by disturbances such as bleaching events (Wilkinson 1999). Also, *D. molle* (solitary tunicate) estimations were correlated between methods, but were consistently showing a higher percent cover in the conventional photo-quadrats. As mentioned above, this is most probably due to image resolution making it difficult to identify small solitary tunicates (<2 cm) that are not dominating to compositions of the reef. Massive encrusting sponges were not correlated between methods. This is most likely due to image resolution and rarity of encrusting sponges found at the sites visited in the Maldives during the 2015 XL CSS expedition, or the cryptic nature of encrusting species in general. Encrusting sponges have key roles on coral reefs such as bio-erosion and competition for space (Diaz and Rützler 2001, Bell 2008, González-Rivero et al. 2013). Detecting changes in rare or exotic species requires different survey techniques than the two used for this study, such as those used to detect marine bioinvasions (Campbell et al. 2007).

Conclusions

Our results demonstrate that the SVII provides an important tool for rapidly, accurately, and rigorously surveying coral reefs at multi-kilometer scales while maintaining taxonomic resolution that is comparable to conventional methods at much greater scales. The larger spatial scale (1.5–2 km per dive) that the SVII was able to achieve provides an important bridge to remote sensing techniques (Mumby et al. 2004, Roelfsema and Phinn 2010, Phinn et al. 2012). Furthermore, the scale provided by SVII also provides a link with satellite imagery that is used to measure and relate biophysical parameters (e.g., sea surface temperature (SST), chlorophyll-*a*, wind/wave exposure, and light attenuation) to the spatial patterns observed using the SVII (Mumby et al. 2011, Chollett and Mumby 2012, Yamano 2013, González-Rivero et al. 2016). These linkages are likely to become of increasing importance as large-scale changes to the environment from climate change impact reefs more and more. This type of information will be critical as the scientific

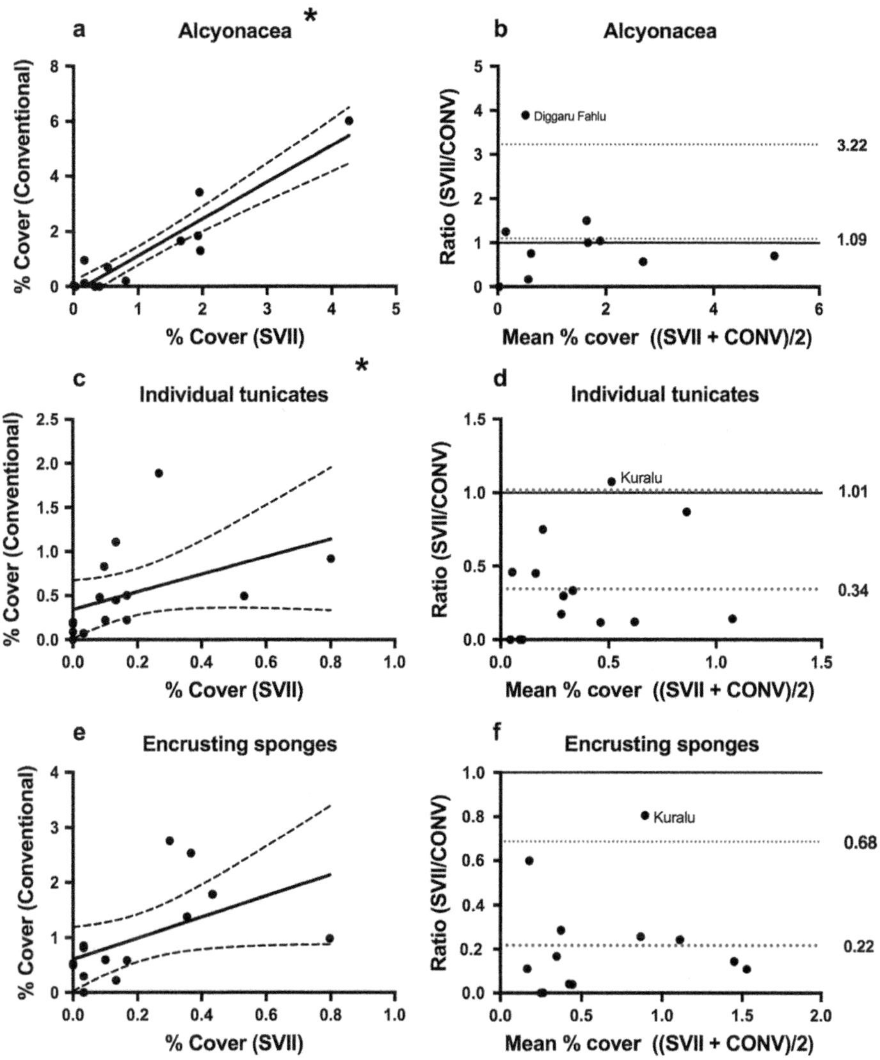

Fig. 9. Correlation (with 95% confidence bands) and Bland–Altman method agreement plots for percent cover of other invertebrates surveyed by SVII and conventional methods across 50-m linear transects at 15 coral reef sites from the XL Catlin Seaview Survey Maldives expedition from the 29th of March until the 17th of April 2015. *Indicates Spearman's rank correlation performed for correlation coefficient. Red line indicates 95% limit agreements, and blue line represents ratio bias between SVII and conventional methods for Bland–Altman plots.

and management communities seek to understand the resilience of coral reefs in a changing world. In the latter case, technologies such as the SVII are likely to offer the ability to rapidly detect more robust (or indeed, less robust) sections of coral reefs for conservation action. Technologies such as the SVII have the potential for providing a global series of baseline measurements estimating the state of coral reefs, which is essential for understanding how coral reefs are likely to change over the coming decade and century.

Acknowledgments

This study was undertaken as part of the XL Catlin Seaview Survey, designed and undertaken by the Global Change Institute, and funded by XL Catlin Pty Ltd in partnership with The Ocean Agency and The University of Queensland. DEPB was supported by The Australian Government Research Training Program Scholarship through The University of Queensland, and OHG was supported by an ARC Laureate Fellowship. The SVII camera system was invented by Richard Ververs, Aaron Spence, and Christophe

Bailhache. The authors would like to also acknowledge the XL Catlin Seaview Survey project management and field team at the Global Change Institute for their tireless efforts in arranging all associated field activities, and also acknowledge the technical assistance with image processing from Panedia. We would also like to thank the Maldives government and members of the Marine Research Centre, Ministry of Fisheries and Agriculture, and Environmental Protection Agency for their assistance with permitting and in-country knowledge. Finally, we would like to recognize the excellent efforts of boat crew and administration team from MV Emperor Voyager for proving an excellent research vessel with no delays occurring during our expedition.

Literature Cited

Alvarez-Filip, L., N. K. Dulvy, J. A. Gill, I. M. Côté, and A. R. Watkinson. 2009. Flattening of Caribbean coral reefs: region-wide declines in architectural complexity. Proceedings. Biological Sciences 276: 3019–3025.

Andréfouët, S., R. Berkelmans, L. Odriozola, T. Done, J. Oliver, and F. Müller-Karger. 2002. Choosing the appropriate spatial resolution for monitoring coral bleaching events using remote sensing. Coral Reefs 21:147–154.

Balmford, A., P. Crane, A. Dobson, R. E. Green, and G. M. Mace. 2005. The 2010 challenge: data availability, information needs and extraterrestrial insights. Philosophical Transactions of the Royal Society of London. Series B, Biological Sciences 360:221–228.

Beijbom, O., P. J. Edmunds, D. I. Kline, B. G. Mitchell, and D. Kriegman. 2012. Automated annotation of coral reef survey images. Pages 1170–1177 *in* IEEE Conference on Computer Vision and Pattern Recognition (CVPR). Institute of Electrical and Electronics Engineers, Providence, USA.

Beijbom, O., P. J. Edmunds, C. Roelfsema, J. Smith, D. I. Kline, B. P. Neal, M. J. Dunlap, V. Moriarty, T.-Y. Fan, and C.-J. Tan. 2015. Towards automated annotation of benthic survey images: variability of human experts and operational modes of automation. PLoS ONE 10:e0130312.

Bell, J. J. 2008. The functional roles of marine sponges. Estuarine, Coastal and Shelf Science 79:341–353.

Bellwood, D. R., T. P. Hughes, C. Folke, and M. Nystrom. 2004. Confronting the coral reef crisis. Nature 429:827–833.

Bland, J. M., and D. Altman. 1986. Statistical methods for assessing agreement between two methods of clinical measurement. Lancet 327:307–310.

Bland, J. M., and D. G. Altman. 2003. Applying the right statistics: analyses of measurement studies. Ultrasound in Obstetrics & Gynecology 22:85–93.

Brown, B. E. 1997. Disturbances to reefs in recent times. Pages 354–379 *in* C. Birkeland, editor. Life and death of coral reefs. Chapman & Hall, New York, New York, USA.

Bruno, J. F., and E. R. Selig. 2007. Regional decline of coral cover in the Indo-Pacific: timing, extent, and subregional comparisons. PLoS ONE 2:e711.

Campbell, M. L., B. Gould, and C. L. Hewitt. 2007. Survey evaluations to assess marine bioinvasions. Marine Pollution Bulletin 55:360–378.

Chollett, I., and P. Mumby. 2012. Predicting the distribution of Montastraea reefs using wave exposure. Coral Reefs 31:493–503.

Clark, S. 2000. Evaluation of succession and coral recruitment in Maldives. Pages 169–175 *in* D. Souter, D. Obura, and O. Linden, editors. Coral reef degradation in the Indian Ocean. (CORDIO) status report 2000. SAREC Marine Science Program, Department of Zoology, Stockholm University, Stockholm, Sweden.

Connell, S., M. Foster, and L. Airoldi. 2014. What are algal turfs? Towards a better description of turfs. Marine Ecology Progress Series 495:299–307.

Dai, C.-F. 1991. Distribution and adaptive strategies of alcyonacean corals in Nanwan Bay, Taiwan. Hydrobiologia 216:241–246.

Darling, E. S., L. Alvarez-Filip, T. A. Oliver, T. R. McClanahan, and I. M. Cote. 2012. Evaluating life-history strategies of reef corals from species traits. Ecology Letters 15:1378–1386.

Darling, E. S., T. R. McClanahan, and I. M. Cote. 2013. Life histories predict coral community disassembly under multiple stressors. Global Change Biology 19:1930–1940.

De'ath, G., K. E. Fabricius, H. Sweatman, and M. Puotinen. 2012. The 27-year decline of coral cover on the Great Barrier Reef and its causes. Proceedings of the National Academy of Sciences of the USA 109: 17995–17999.

Diaz, M. C., and K. Rützler. 2001. Sponges: an essential component of Caribbean coral reefs. Bulletin of Marine Science 69:535–546.

Durrant, A., and M. Dunbabin. 2011. Real-time image classification for adaptive mission planning using an autonomous underwater vehicle. Pages 1–6 *in* Proceedings of Oceans 2011. IEEE, Kona, Hawaii, USA.

Edmunds, P. J., and J. F. Bruno. 1996. The importance of sampling scale in ecology: kilometer-wide variation in coral reef communities. Marine Ecology Progress Series 143:165–171.

Edmunds, P. J., and R. Elahi. 2007. The demographics a 15-year decline in cover of the Caribbean reef coral Montastrea annularis. Ecological Monographs 77:3–18.

Gardner, T. A., I. M. Cote, J. A. Gill, A. Grant, and A. R. Watkinson. 2003. Long-term region-wide declines in Caribbean corals. Science 301:958–960.

Giavarina, D. 2015. Understanding bland Altman analysis. Biochemia Medica 25:141–151.

Gleason, M. 1993. Effects of disturbance on coral communities: bleaching in Moorea, French Polynesia. Coral Reefs 12:193–201.

Glynn, P. W., J. L. Maté, A. C. Baker, and M. O. Calderón. 2001. Coral bleaching and mortality in Panama and Ecuador during the 1997–1998 El Niño-Southern Oscillation event: spatial/temporal patterns and comparisons with the 1982–1983 event. Bulletin of Marine Science 69:79–109.

González-Rivero, M., O. Beijbom, A. Rodriguez-Ramirez, T. Holtrop, Y. González-Marrero, A. Ganase, C. Roelfsema, S. Phinn, and O. Hoegh-Guldberg. 2016. Scaling up ecological measurements of coral reefs using semi-automated field image collection and analysis. Remote Sensing 8:30.

González-Rivero, M., P. Bongaerts, O. Beijbom, O. Pizarro, A. Friedman, A. Rodriguez-Ramirez, B. Upcroft, D. Laffoley, D. Kline, and C. Bailhache. 2014. The Catlin Seaview Survey–kilometre-scale seascape assessment, and monitoring of coral reef ecosystems. Aquatic Conservation: Marine and Freshwater Ecosystems 24:184–198.

González-Rivero, M., A. V. Ereskovsky, C. H. Schönberg, R. Ferrari, J. Fromont, and P. J. Mumby. 2013. Life-history traits of a common Caribbean coral-excavating sponge, *Cliona tenuis* (Porifera: Hadromerida). Journal of Natural History 47:2815–2834.

Hedley, J. D., C. M. Roelfsema, S. R. Phinn, and P. J. Mumby. 2012. Environmental and sensor limitations in optical remote sensing of coral reefs: implications for monitoring and sensor design. Remote Sensing 4:271–302.

Hill, J., and C. Wilkinson. 2004. Methods for ecological monitoring of coral reefs. Australian Institute of Marine Science, Townsville, Queensland, Australia.

Hochberg, E. J., and M. J. Atkinson. 2003. Capabilities of remote sensors to classify coral, algae, and sand as pure and mixed spectra. Remote Sensing of Environment 85:174–189.

Hoegh-Guldberg, O., et al. 2007. Coral reefs under rapid climate change and ocean acidification. Science 318:1737–1742.

Knowlton, N., and J. B. C. Jackson. 2008. Shifting baselines, local impacts, and global change on coral reefs. PLoS Biology 6:e54.

McClanahan, T. R. 2000. Bleaching damage and recovery potential of Maldivian coral reefs. Marine Pollution Bulletin 40:587–597.

Mumby, P. J. 2006. The impact of exploiting grazers (Scaridae) on the dynamics of Caribbean coral reefs. Ecological Applications 16:747–769.

Mumby, P. J., I. A. Elliott, C. M. Eakin, W. Skirving, C. B. Paris, H. J. Edwards, S. Enríquez, R. Iglesias-Prieto, L. M. Cherubin, and J. R. Stevens. 2011. Reserve design for uncertain responses of coral reefs to climate change. Ecology Letters 14:132–140.

Mumby, P., E. Green, A. Edwards, and C. Clark. 1997. Coral reef habitat mapping: How much detail can remote sensing provide? Marine Biology 130: 193–202.

Mumby, P. J., E. P. Green, A. J. Edwards, and C. D. Clark. 1999. The cost-effectiveness of remote sensing for tropical coastal resources assessment and management. Journal of Environmental Management 55:157–166.

Mumby, P. J., W. Skirving, A. E. Strong, J. T. Hardy, E. F. LeDrew, E. J. Hochberg, R. P. Stumpf, and L. T. David. 2004. Remote sensing of coral reefs and their physical environment. Marine Pollution Bulletin 48:219–228.

Mumby, P. J., and R. S. Steneck. 2008. Coral reef management and conservation in light of rapidly evolving ecological paradigms. Trends in Ecology & Evolution 23:555–563.

Pandolfi, J. M., et al. 2003. Global trajectories of the long-term decline of coral reef ecosystems. Science 301:955–958.

Phinn, S. R. 1998. A framework for selecting appropriate remotely sensed data dimensions for environmental monitoring and management. International Journal of Remote Sensing 19:3457–3463.

Phinn, S. R., C. M. Roelfsema, and P. J. Mumby. 2012. Multi-scale, object-based image analysis for mapping geomorphic and ecological zones on coral reefs. International Journal of Remote Sensing 33:3768–3797.

Pisapia, C., D. Burn, R. Yoosuf, A. Najeeb, K. Anderson, and M. Pratchett. 2016. Coral recovery in the central Maldives archipelago since the last major mass-bleaching, in 1998. Scientific Reports 6:34720.

Pratchett, M. 2010. Changes in coral assemblages during an outbreak of *Acanthaster planci* at Lizard Island, northern Great Barrier Reef (1995–1999). Coral Reefs 29:717–725.

Pratchett, M. S., M. Trapon, M. L. Berumen, and K. Chong-Seng. 2011. Recent disturbances augment community shifts in coral assemblages in Moorea, French Polynesia. Coral Reefs 30:183–193.

Roelfsema, C., M. Lyons, M. Dunbabin, E. M. Kovacs, and S. Phinn. 2015. Integrating field survey data with satellite image data to improve shallow water seagrass maps: The role of AUV and snorkeller surveys? Remote Sensing Letters 6:135–144.

Roelfsema, C., and S. Phinn. 2010. Integrating field data with high spatial resolution multispectral satellite imagery for calibration and validation of coral reef benthic community maps. Journal of Applied Remote Sensing 4:043527–043528.

Roelfsema, C., S. Phinn, S. Jupiter, J. Comley, M. Beger, and E. Paterson. 2010. The application of object based image analysis of high spatial resolution imagery for mapping large coral reef systems in the West Pacific at geomorphic and benthic community spatial scales. Pages 4346–4349 *in* 2010 IEEE International Geoscience and Remote Sensing Symposium, Honolulu, Hawaii, USA.

Trygonis, V., and M. Sini. 2012. photoQuad: a dedicated seabed image processing software, and a comparative error analysis of four photoquadrat methods. Journal of Experimental Marine Biology and Ecology 424:99–108.

Wilkinson, C. 1999. The 1997–1998 mass bleaching event around the world. Australian Institute of Marine Science, Townsville, Queensland, Australia.

Williams, G. J., J. M. Gove, Y. Eynaud, B. J. Zgliczynski, and S. A. Sandin. 2015. Local human impacts decouple natural biophysical relationships on Pacific coral reefs. Ecography 38:751–761.

Williams, S. B., O. R. Pizarro, M. V. Jakuba, C. R. Johnson, N. S. Barrett, R. C. Babcock, G. A. Kendrick, P. D. Steinberg, A. J. Heyward, and P. J. Doherty. 2012. Monitoring of benthic reference sites: using an autonomous underwater vehicle. IEEE Robotics & Automation Magazine 19:73–84.

Yamano, H. 2013. Multispectral applications. Pages 51–78 *in* J. Goodman, S. J. Purkis, and S. R. Phinn, editors. Coral reef remote sensing. Springer, Dordrecht, The Netherlands.

10

Seasonal and inter-annual variability in litter decomposition and nitrogen availability in a mid-Appalachian watershed

Jeff W. Atkins,[1,2,†] Howard E. Epstein,[1] and Daniel L. Welsch[3]

[1] *Department of Environmental Sciences, University of Virginia, 291 McCormick Ave, Charlottesville, Virginia 22902 USA*
[2] *Department of Biology, Virginia Commonwealth University, 1000 West Cary Street, Richmond, Virginia 23284 USA*
[3] *American Public University System, 111 W Congress St, Charles Town, West Virginia 25414 USA*

Abstract. The role that microclimates play on soil decomposition is poorly understood. Though litter decomposition is controlled by climate and substrate quality at coarse spatial scales, at the watershed scale, microclimates mediated by forest structure and landscape position can influence decomposition rates and in turn affect nitrogen cycling. To evaluate the effects of landscape position and vegetation heterogeneity on decomposition, we employed a two-year litterbag study (2011–2013) using yellow birch leaf litter across the Weimer Run watershed, a cool, humid watershed located near Davis, West Virginia. From our results, we created a spatially explicit empirical model that we tested against both a single-pool and three-pool decomposition model, each based on climate and derived from the Long-Term Intersite Decomposition Experiment Team. Initial litter decomposition varied by elevation, with greater rates of decomposition at locations lower in the watershed. Decomposition rates differed by elevation, except during the first winter of the study. No differences in decomposition rates were seen among elevation levels when snowfall was below average for the first winter (2011–2012). During the second winter (2012–2013), elevation levels showed separation in decomposition rates, with higher elevations exhibiting lower decomposition rate. This suggests important controls on decomposition exerted by the presence or absence of snow, inter-annual climate variability, and the interaction of both with topography. Our empirical model showed greater rates of decomposition during early stages of decomposition (<12 months), but converged with the three-pool decomposition model after 20 months. Plant available nitrogen differed by vegetation cover, largely driven by greater availability of nitrate (NO_3^-) beneath areas of canopy closure in the forest. Controls on decomposition and nitrogen cycling within the Weimer Run watershed vary spatially by elevation and vegetation cover and are subject to complex interactions and differ from standard models of decomposition. The effect of the inter-annual variance of snow depth on litter decomposition is of note and an important consideration moving forward. Climate-based models of decomposition greatly underestimate initial rates of decomposition, potentially leading to under-accounting and compounded uncertainty.

Key words: complex topography; inter-annual climate variability; litter decomposition; nitrogen cycling; snow.

† **E-mail:** jeffatkins@virginia.edu

Introduction

Decomposition is a key aspect of global nutrient and element cycling (Olson 1963, Raich and Schlesinger 1992, Vitousek 2004, Bonan et al. 2013) that integrates the chemical, physical, and biological processes involved in the breakdown of organic matter into labile, inorganic component forms (Berg and McClaugherty 2008) and more recalcitrant, complex organics (Chapin et al. 2002). At regional and global scales, decomposition is controlled by climate and substrate quality (Meentemeyer 1978, Aerts 1997, Silver and Miya 2001, Zhou et al. 2008) with broad-scale estimates of decomposition rates established from globally distributed, long-term decomposition studies (e.g., Long-term Intersite Decomposition Experiment Team [LIDET]; Gholz et al. 2000, Bonan et al. 2013). At these large scales, derived variables such as the climate decomposition index (CDI) have been shown to be strong predictors of decomposition rates (Adair et al. 2008). Climate decomposition index is calculated from air temperature, precipitation, and potential evapotranspiration (PET) and has been incorporated in ecosystem models (e.g., CENTURY). However, this broad-scale approach likely obscures the influence of intra-site ecosystem heterogeneities and climate variability, which may lead to unquantified uncertainty. These uncertainties are potentially quite high in ecosystems with strong seasonality or that are physically and biologically complex, such as mountainous watersheds.

At finer scales, decomposition is influenced by soil fauna composition and abundance (Singh and Gupta 1977) and interactions among varying leaf litter species (Hättenschwiler and Gasser 2005). However, the physical–biological structure and complexity of a system—from topographic heterogeneity to vegetation distribution—can profoundly influence decomposition by affecting the ground surface and altering microclimates. Litterfall amounts may decrease with increasing elevation over significant elevation gradients, but forest floor litter mass may increase due to slower turnover rates (Zhou et al. 2014). This effect can be compounded as elevation gradients often result in coniferous tree supplantation of deciduous trees at higher elevations (Reiners and Lang 1987). Across the mid-Appalachians, red spruce and balsam fir line the ridgetops. The additive effects of the more recalcitrant nature of this coniferous litter combined with decreased temperatures (Zhou et al. 2014) at altitude may account for increased forest floor mass as a function of lowered decomposition rates. Shanks and Olson (1961) showed differences in first-year decomposition rates of leaf litter from five different species of deciduous trees along an elevation gradient in the southern Appalachians. Decomposition rates were faster at lower elevations and faster underneath deciduous trees. Elevation effects can be seen across less extreme elevation gradients as well. Dwyer and Merriam (1981) showed that fine-scale forest floor topography (e.g., hummocks and hollows) was correlated with soil temperature along hillslopes in subalpine forests and affected leaf litter decomposition. Downslope winds may also result in the redistribution of fallen litter from higher to lower elevations, or increased heterogeneity in forest litter distribution due to aggregation around fallen logs, branches, or in depressions (Orndorff and Lang 1981).

Microclimates are determined by ecosystem structure—including forest heterogeneity and fragmentation (Hastwell and Morris 2013). The physical structure of plant communities can affect moisture and temperature regimes (Hastwell and Morris 2013, Atkins et al. 2015) through canopy-gap controls on wind, solar radiation, and rainfall interception and redirection (Belsky and Canham 1994, Breshears 2006). The effects of forest structure on litter decomposition rates can be significant (Hastwell and Morris 2013). There is a need to better constrain the high uncertainties in the spatial and temporal variation of decomposition rates, particularly in eastern North American forests such as those of the mid-Appalachians that may experience severe impacts from global climate change (Walter et al. 2017). Decreasing uncertainty can lead to improved understanding of the interactions among decomposition dynamics, nutrient cycling, and environmental heterogeneities.

In cold environments where there is frequent snowfall, carbon and nutrient cycling processes during the winter are driven by the physical properties and duration of the snowpack which acts as an insulative layer on the landscape (Saccone et al. 2013). A deep, well-formed snowpack limits the depth of soil freezing and can keep the soil surface at or above freezing temperatures during the winter. Climate-driven loss of snowpack may result in increased frequency of freeze–thaw cycles and harsher conditions during the winter for sub-nivean environments (Bokhorst et al. 2010, Saccone et al. 2013) affecting decomposition and nutrient cycling.

Nitrogen cycling is an intrinsic component of litter decomposition, the spatial variance of which is influenced by vegetation cover heterogeneity and forest structure (Reich et al. 2001).

Closed canopy areas of mature, conifer forests in the Upper Peninsula of Michigan have been shown to exhibit greater nitrogen mineralization rates than open, forest gaps (Schliemann and Bockheim 2014). And, while N mineralization rates increase with elevation in the mountains of the southern Appalachians (Garten and Van Miegroet 1994), at elevations above 700 m, there is increased variance among areas of differing vegetation, with significantly higher N mineralization rates occurring in areas where evergreen shrubs are absent—an indication that vegetation can differentially affect N mineralization in these ecosystems (Knoepp and Swank 1998).

N mineralization rates can also vary inter-annually; as an example, the watersheds at Coweeta Hydrologic Laboratory, located in the southern Appalachians, show evidence that inter-annual variance in N mineralization can be attributed to differences in rainfall rather than temperature, with increases in N mineralization rates during years of higher rainfall (Knoepp and Swank 1998). The temporal nature of N cycling can be affected by vegetation as well. The evergreen shrub layer of the southern and mid-Appalachians is dominated by mountain laurel (*Kalmia latifolia*) and rhododendron (*Rhododendron* spp.), species with highly recalcitrant litter that slows N cycling and increases N retention in the system—with N retention estimated to peak at 30 yr after shrub-stand establishment (Chastain et al. 2006).

Differences in N mineralization along elevation and vegetation gradients can be a function of the variance of soil temperature, soil moisture, and substrate quality (Powers 1990). While the forests and watersheds of the southern Appalachians share many similarities with the mid-Appalachians, there are important, ecologically relevant differences. Mid-Appalachian forests are often far less botanically diverse than their southern neighbors. Due to their more northerly latitude, mid-Appalachian forests are subject to colder winter temperatures, shorter growing seasons, increased snowfall, and more persistent snow coverage. These ecosystem factors would be expected to impact litter decomposition and nitrogen cycling.

Our understanding of the influence of the interactions between forest vegetation heterogeneity and elevation gradients on litter decomposition within cool, humid temperate watersheds is lacking. By discretizing the landscape within a watershed into both elevation and vegetation cover classes, we can attempt to parse out the controls exerted on decomposition and N availability by environmental heterogeneities. By employing a combined approach using litterbags in a factorial design across elevation and vegetation gradients coupled with measures of nitrogen availability, we ask the following questions:

1. What are the effects of vegetation cover on litter decomposition and N dynamics across an elevation gradient within a cool, humid watershed in West Virginia?
2. How does N availability during the growing season vary among vegetation cover classes across an elevation gradient within a cool, humid watershed in West Virginia?
3. How do decomposition dynamics respond to inter-annual variations in snow cover and snowpack depth?

Methods

Site description

The study was conducted in the Weimer Run watershed (374 ha), located in the Little Canaan Wildlife Management Area near Davis, West Virginia (39.1175–79.4430). The watershed has an elevation gradient of 940–1175 m with mixed northern hardwood–coniferous forest cover (Fortney 1975, Atkins et al. 2015). A prominent under-story vegetation layer exists in the watershed, populated by *Rhododendron maximum, Kalmia latifolia, Osmundastrum cinnamomeum,* and *Osmunda claytoniana* (Fortney, 1975). Mean annual precipitation is 1450 mm/yr (PRISM Climate Group). The mean daily maximum July temperature is 18.8°C, and the mean daily maximum January temperature is −3.9°C (National Climatic Data Center [NCDC], Davis, West Virginia, USA).

Elevation and vegetation cover classes

Three elevation classes were established along the northeastern aspect of the watershed to form an elevation gradient: LOW (975 m), MID (1050 m), and HIGH (1100 m). At each elevation level, three 2 × 2 m plots of three different vegetation cover classes were established—for a total of 27 plots across the entire watershed. The vegetation cover classes were defined as follows: CANOPY, closed canopy, forest interior with no shrub layer (canopy species included *Acer rubrum* and *Betula alleghaniensis*); SHRUB, closed canopy, forest interior, with dense shrub layer, dominated by *R. maximum*; and OPEN, forest gap with no canopy closure, within the forest interior (Atkins et al. 2015).

Litterbag experiment

To assess relative decomposition rates, five litterbags containing a common, senescent litter (*B. alleghaniensis*) were placed within each plot. The litterbags were constructed using a fine mesh nylon with openings of 0.92 mm^2 (0.96 × 0.96 mm; thickness of nylon = 0.1 mm). The edges of each bag were sealed using a ULINE H-163 Poly bag impulse heat sealer (U-Line; Pleasant Prairie, Wisconsin, USA). The impulse sealer allows for a non-reactive, weatherproof seal. The finished size of each bag was approximately 15 × 15 cm.

A common litter was used to control for species effects. Litter for each elevation level was collected at that level (e.g., litter used in litterbags at the LOW elevation level came from trees located near those plots). Litter was collected during September 2011, in litter traps. All senescent litter used was <7 d old, as traps were placed during the second week of September 2011, and all litter was collected the following week. Senescent litter was taken to the laboratory and sorted to ensure that only intact *B. alleghaniensis* leaves were used. Leaves were oven-dried for 72 h at 60°C. Each litterbag was filled with 2 ± 0.05 g of oven-dried litter, then individually weighed and tagged with a 1″ × 3–1/2″ (2.54 × 8.89 cm) all-weather, aluminum PermaTag (Forestry Suppliers; Jackson, Mississippi, USA). Each tag was marked with a unique identifying number corresponding to plot replicate, elevation level, and vegetation cover class.

Litterbags were attached to an aluminum wire via openings in the PermaTags on each bag. The aluminum wire was then staked to the ground using landscape pins (Forestry Suppliers). Litterbags were arranged in a circle to make sure that no bag was touching another and were placed on the soil surface, fully inside the assigned plot. Five litterbags were placed at each plot on 1 October 2011. The retrieval schedule was as follows: (1) November 2011 (1 month); (2) June 2012 (8 months); (3) October 2012 (12 months); (4) June 2013 (20 months); and (5) October 2013 (24 months).

Total leaf carbon and nitrogen were assessed for initial litter and for each litter collection as outlined above using a NA 2500 Elemental Analyzer (CE Instruments, Wigan, UK). Initial leaf litter C and N values were based on analysis of leaf litter of recently senesced leaves at each elevation level (e.g., initial values of C and N for leaves at the LOW elevation were based on C and N values derived from analysis of leaves collected at that level—as outlined previously).

Nitrogen availability

Supply rates of plant available nitrogen in the soil were estimated in the field using plant root simulator probes (PRS probes; Western Ag. Innovations, Saskatoon, Saskatchewan, Canada) which have been shown to be an effective and efficient tool at measuring available nitrogen (Johnson et al. 2004). Probes were incubated in the soils from 0 to 10 cm depth for three deployments in 2010: (1) between 14 June and 14 July (30 d); (2) 1 August–1 September (30 d); and (3) 1 September–6 October (36 d). Following incubations, they were collected, refrigerated, rinsed with deionized water, and shipped to Western Ag Innovations for analyses of NH_4^+ and NO_3^-.

Soils and the organic layer

Organic layer depth for each plot was determined at three points within the plot and then averaged for a plot mean organic layer depth during the fall of 2013. Organic layer soil samples were oven-dried at 105°C and sieved to 2 mm, with roots and gravel removed, and then analyzed for total soil carbon and nitrogen using a NA 2500 Elemental Analyzer.

Data analysis and model comparisons

Relative decomposition rates of leaf litter were calculated using a single-pool negative exponential model:

$$M_t = M_0 \exp(-kt) \tag{1}$$

where M_t is the mass of litter at time t, M_0 is the initial mass of litter, e is the base of the natural logarithm, and k is the decomposition constant.

Decomposition rates were not normally distributed and were analyzed using a Kruskal–Wallis test to compare differences by both elevation and vegetation classes.

Soil nutrient availability was compared across elevation, vegetation cover class, and elevation by vegetation cover class within each time period using analysis of variance (ANOVA). Differences among time periods were not reasonably comparable based on the method used to calculate NH_4^+ and NO_3^- by Western Ag.

Differences in organic layer depth (cm), organic layer C (%), organic layer N (%), and organic layer C:N were evaluated using ANOVA. Additionally, organic layer C, organic layer N, and organic layer C:N were further evaluated using analysis of covariance (ANCOVA) with organic layer depth as a covariate. All data analyses were conducted in R 3.4 (R Core Team 2017).

Empirically derived estimates of litter decomposition were compared to modeled estimates of expected decomposition based on CDI modeling approaches as outlined in Adair et al. (2008):

$$\mathrm{CDI}_i = F_t(T_i) \times F_w(\mathrm{PPT}_i, \mathrm{PET}_i) \quad (2)$$

$$F_t(T_i) = 0.5766 \times \exp\left[308.56 \times \left(\frac{1}{56.02} - \frac{1}{(273 + T_i) - 227.13}\right)\right] \quad (3)$$

$$F_w(\mathrm{PPT}_i, \mathrm{PET}_i) = \frac{1}{1 + 30 \times \exp\left(-8.5 \times \frac{\mathrm{PPT}_i}{\mathrm{PET}_i}\right)} \quad (4)$$

where T_i is mean monthly temperature (°C), PPT_i is monthly precipitation (mm), PET_i is monthly PET (mm), and CDI_i is the monthly CDI. A singular CDI value is derived for the site based on the average of monthly CDIs (CDI_i). Data from the NCDC (Station ID: GHCND:USC00462211) from 2010 to 2012 were used to calculate monthly means of temperature, precipitation, and PET. Potential evapotranspiration was calculated as a function of mean daily air temperature (T), saturation vapor pressure ($E_s(T)$ in kPa), and day length (D) (Hamon 1963):

$$\mathrm{PET} = 29.8 \times D \times \frac{e_s(T)}{T + 273.2} \times 1.2 \quad (5)$$

A correction coefficient of 1.2 is used to adjust Hamon's method (Sun et al. 2002). $e_s(T)$ is the saturation vapor pressure (kPa) from Campbell (1998):

$$e_s(T) = 0.611 \exp\left(\frac{17.502 \times T}{T + 240.97}\right) \quad (6)$$

Adair et al. (2008) suggest that a three-pool model that includes lignin and cellulose concentrations as secondary and tertiary pools in the model (the first pool being the remaining fraction) is the best model to predict decomposition. A standard single-pool CDI model was also tested for comparison:

Single-pool model.—

$$M_t = M_0 \exp(-k \times \mathrm{CDI} \times t) \quad (7)$$

Adair et al. (2008) derive a k value of 0.5278 for this model that is adjusted by the CDI. The CDI value for the Weimer Run watershed is 0.59.

Three-pool model.—

$$\begin{aligned} M_t = &M_1 \exp(-1.0669 \times \mathrm{CDI} \times t) \\ &+ M_2 \exp(-0.7676 \times \mathrm{CDI} \times t) \\ &+ M_3 \exp(0 \times \mathrm{CDI} \times t) \end{aligned} \quad (8)$$

where M_1 is defined as $100 - M_2 - M_3$, M_2 is the initial cellulose concentration (36.79%) and M_3 is the initial lignin concentration (20.94%). Initial lignin and cellulose concentrations were calculated as the mean of all *B. alleghaniensis* foliar chemistry data from the Northeastern Ecosystem Research Cooperative Foliar Chemistry Database (NERC 2017).

Both the single-pool CDI model and three-pool CDI model were then compared to a spatially explicit, empirically derived watershed model that considers landscape position:

$$\begin{aligned} M_t = &M_{\mathrm{LOW}} \exp(-k_1 \times t) + M_{\mathrm{MID}} \exp(-k_3 \times t) \\ &+ M_{\mathrm{HIGH}} \exp(-k_3 \times t) \end{aligned} \quad (9)$$

Results

Litterbag experiment

Litter decomposition rates were consistent with expectations with greater amounts of mass remaining at landscape positions lower in the watershed after two years (LOW 41.1%, MID 44.4%, and HIGH 47.8%). Results from beneath vegetation cover classes after two years were also as expected, with OPEN sites having the lowest values of total leaf litter mass remaining (41.7%), followed by CANOPY sites (44.2%) and SHRUB sites (47.7%; Figs. 1, 2).

Relative decomposition rates (*k*) and total mass remaining (%) among elevation classes were influenced strongly by season (Appendix S1: Table S1), with significant differences after 1 month (November, 2011; $P = 0.02$) and 12 months (October 2012; $P = 0.02$). Both mass loss and *k* show nearly significant differences at 20 months (June 2013; $P = 0.05$) and 24 months (October 2013; $P = 0.05$). At 8 months (June 2012), the greatest homogeneity in decomposition rates with regard to elevation was present. This is of note because the winter of 2011–2012 had below-average snowfall, with much of the watershed free of snow for the entire winter and soil temperatures well below average. In comparison, the winter of 2012–2013, which corresponds with collection after 20 and 24 months, was a year of average to above-average snowfall (Fig. 3). No statistically significant differences in decomposition or mass loss are seen by vegetation cover for the entire length of the experiment (Appendix S1: Table S2).

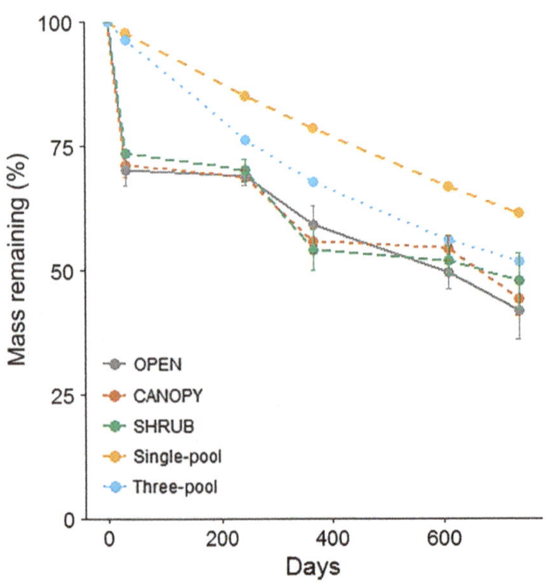

Fig. 1. Litterfall mass remaining (%) by vegetation cover class (OPEN, CANOPY, and SHRUB) with modeled decomposition estimates from both the single-pool (Eq. 7) and three-pool decay model (Eq. 8).

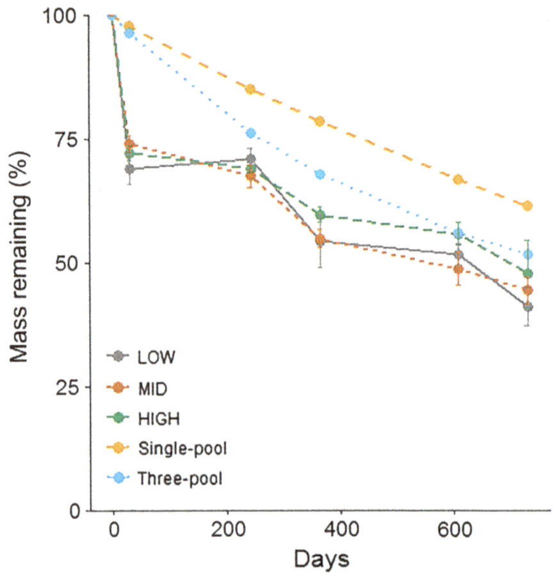

Fig. 2. Litterfall mass remaining (%) by elevation class (LOW, MID, and HIGH) with modeled decomposition estimates from both the single-pool (Eq. 7) and three-pool decay model (Eq. 8).

Initial values of litter C and N differed slightly by elevation class (carbon: LOW = 51.1 ± 0.3; MID = 52.7 ± 0.4; and HIGH = 53.3 ± 0.5; nitrogen: LOW = 2.2 ± 0.1; MID = 2.47 ± 0.0; and HIGH = 2.51 ± 0.1 with ± indicating standard deviation). Litter C:N ratio diverged underneath shrub plots from the other vegetation cover classes after 20 months (Fig. 4) driven by a decrease in N (Fig. 4). This collection period followed the second winter where both snowfall and snow depth were above the long-term mean (Fig. 3). Mean leaf litter C:N ratio for litterbag litter collected beneath shrubs was 24.32 during this collection and was much higher than at any other point. Elevated C:N ratios beneath shrubs were present across the elevation gradient. Overall decomposition rates were negatively correlated with final leaf N content ($r = -0.37$).

Nitrogen availability

Plant available soil nitrogen in the summer of 2010, measured with resin probes as NO_3^- and NH_4^+, was greater in plots beneath the forest canopy than other cover types (Table 1), an effect more pronounced at positions lower in the watershed. Plant available soil nitrogen was generally lower in August than in either June/July or

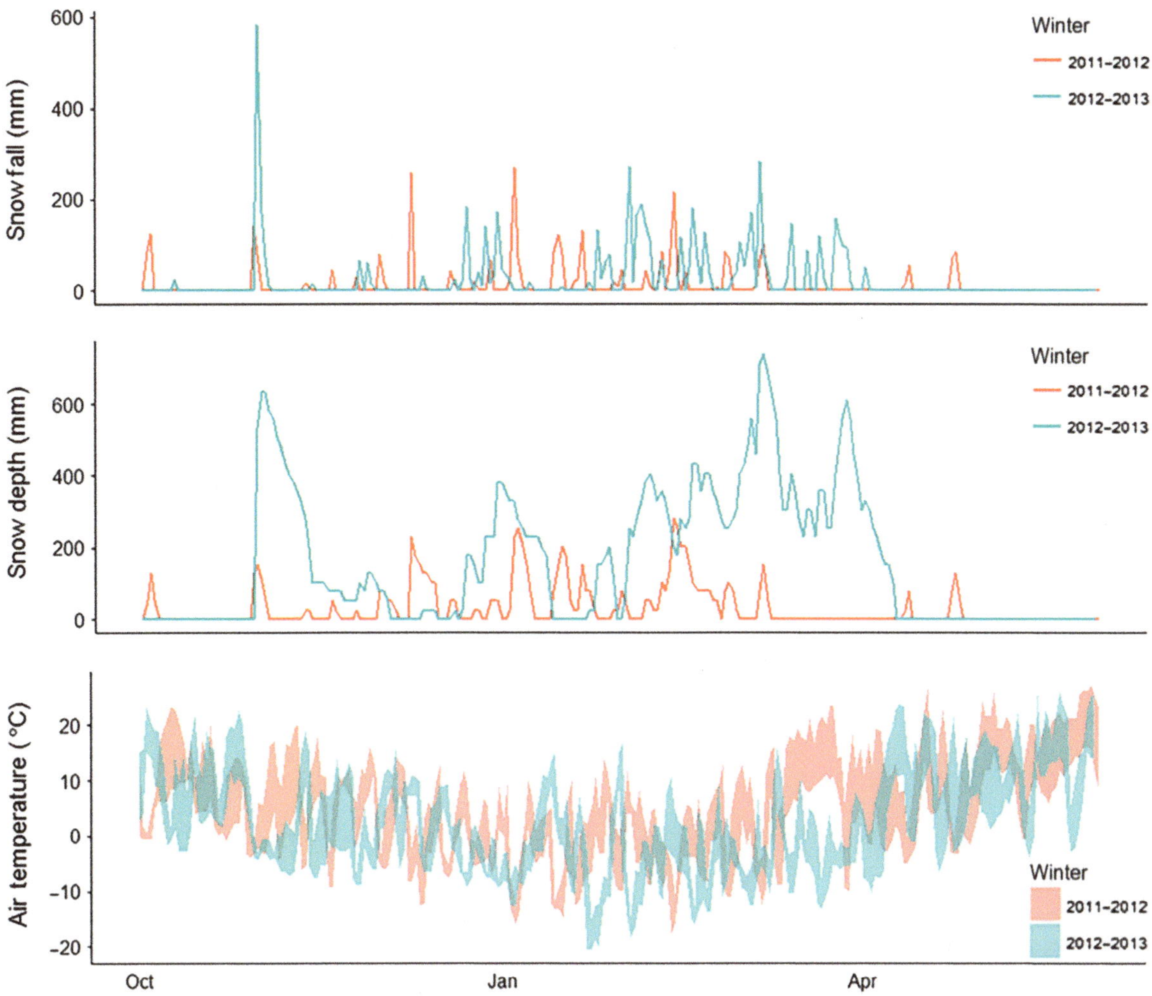

Fig. 3. Snow depth from nearby weather station CAAW2 (Canaan Valley Station 2, 39.05361, –79.44833). 2011–2012 total snowfall = 2984 mm; 2012–2013 total snowfall = 5684 mm.

September. While some plots show greater N in September, accurate comparisons between September and other months cannot be made due to the discrepancy in incubation time. However, by comparing plots within months, inferences can be made. Two-way ANOVA results comparing total inorganic N, NO_3^-, and NH_4^+ with respect to elevation, vegetation, and elevation × vegetation for June show differences in total inorganic N by vegetation cover (F = 3.62 P = 0.04), a pattern likely driven by statistically significant differences in NO_3^-, with greater NO_3^- availability in plots beneath the forest canopy (F = 4.29; P = 0.03; Appendix S1: Fig. S2) than other plots. No other significant relationships or differences were found in any other month (Table 1; Appendix S1: Table S2).

Organic layer

Two-way ANOVA results of organic layer depth show no differences among elevation and vegetation cover classes, though organic layer depth increases with elevation (Table 2; Appendix S1: Fig. S4). Organic layer C (%) does differ by vegetation (F = 3.88; P = 0.04), with shrub plots having significantly greater organic soil C (%). This in turn drives up the C:N ratios in shrub plot soils (F = 7.03; P < 0.01).

Model comparisons

The three-pool CDI decomposition model begins to converge with empirical measurements of decomposition at 20 months, but greatly underestimates the initial slope of decomposition

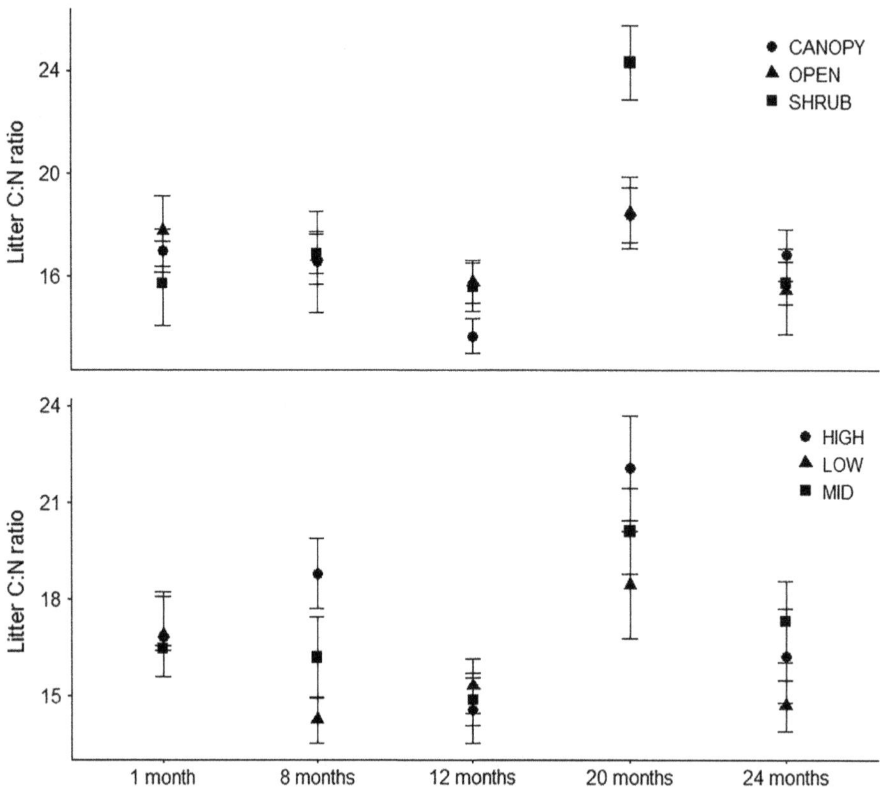

Fig. 4. Leaf litter C:N ratios following collection. Error bars indicate standard error. At top, by elevation class; at bottom, by vegetation.

Table 1. Analysis of variance results of nitrogen availability from Western Ag plant root simulator probes.

	June (30 d)		August (30 d)		September (36 d)
Species	*F*	*P*	*F*	*P*	*F*
Total N					
ELEV	1.357	0.286	0.193	0.827	0.789
VEG	3.618	**0.047**	1.447	0.271	1.609
ELEV:VEG	0.953	0.953	0.938	0.472	1.195
NO_3^-					
ELEV	0.271	0.765	2.006	0.174	0.467
VEG	4.293	**0.029**	2.586	0.113	1.444
ELEV:VEG	0.742	0.576	1.643	0.223	1.401
NH_4^+					
ELEV	2.467	0.113	0.208	0.815	0.818
VEG	2.008	0.163	0.869	0.442	1.729
ELEV:VEG	0.109	0.978	0.962	0.461	1.449

Notes: Collected during the summer of 2010. Statistically significant results shown in bold.

that occurs <8 months after litterfall (Figs. 1, 2). The single-pool CDI model fails to converge with empirical data during the two-year timeframe of this study.

Discussion

We find that spatial variance in initial decomposition of yellow birch leaf litter is primarily influenced by elevation, and secondarily by vegetation cover, with the greatest rates of decomposition found in open areas at lower elevations in the watershed. Elevation levels diverge in decomposition rates during the summer months, with greater decomposition at points lower in the watershed. Open forest gaps, lower in the watershed, tend to have greater light exposure, higher soil temperatures (OPEN = 15.6°C; SHRUB = 14.9°C; and CANOPY = 15.1°C), and greater

Table 2. Analysis of covariance (ANCOVA) results of organic layer C, N, and C:N.

Layer	ANCOVA results	
	F	*P*
Organic layer C		
Organic layer depth	0.49	0.491
ELEV	0.56	0.584
VEG	4.57	**0.029**
ELEV:VEG	0.30	0.869
Organic layer N		
Organic layer depth	2.16	0.162
ELEV	0.56	0.581
VEG	3.36	0.064
ELEV:VEG	0.74	0.579
Organic layer C:N		
Organic layer depth	0.04	0.841
ELEV	0.93	0.415
VEG	6.87	**0.008**
ELEV:VEG	0.86	0.507

Notes: Organic layer depth (cm) was used as the covariate. Statistically significant results shown in bold.

soil moisture (OPEN = 0.43 m^3/m^3; SHRUB = 0.32 m^3/m^3; and CANOPY = 0.41 m^3/m^3 measured as water-filled pore space; from Atkins et al. 2015). Given these environmental conditions, these results are expected. After one year of decomposition, similar significant results are seen (Appendix S1: Table S2). However, there are discrepancies during the winter months.

Seasonality and snowfall

Decomposition dynamics during and following the winter of 2011–2012 are an interesting anomaly in the findings. All other sampling times show significant or nearly significant differences in decomposition by elevation, yet the differences were less clear during this period (November 2011–June 2012). Snowfall during this winter totaled 2984 mm compared to the following winter (2012–2013) when snowfall totaled 5684 mm (Fig. 3). During the winter of 2011–2012, 65.9% of the time between 1 October and 31 May, snow depth was recorded at 0, whereas only 43.2% of the time during the winter of 2012–2013 was snow depth recorded at 0 (based on daily measurements of snow depth from the NCDC). Relatively low snowfall results in a spotty, underdeveloped snowpack that can lead to more prevalent and deeper soil freezing (Reinmann and Templer 2016), whereas soils beneath a well-developed snowpack remain at or above freezing during the course of the winter. Our results show support for decreased snowfall and the resultant absent or undeveloped snowpack leading to decreased decomposition and the minimization of expected altitudinal effects in this watershed. Altered snow regimes resulting in changing patterns in snow depth and inter-annual climatic variability affect water cycling which also controls carbon and nutrient cycling (e.g., nitrogen) with potential ecosystem cascades.

Observation and model comparison

The three-pool CDI model of decomposition, derived from the LIDET experiment, begins to converge with our empirical data at or near 20 months, while the single-pool model fails to converge over the two years of the study period (Figs. 1, 2). This is potentially an important implication for the use of these models. For longer-term studies, a three-pool CDI model would be appropriate. However, the initial underestimation of decomposition rates greatly affects carbon accounting. For example, litterfall mass at the Fernow Experimental Forest, located ~25 km to the east–southeast of the Weimer Run watershed, has an established range of 1.92–5.57 Mg $ha^{-1}\cdot yr^{-1}$ (Adams 2008), values in line with those of other central hardwood forests (Jenkins 2002). Fernow is similar to Weimer Run, but with a longer growing season. If we assume a conservative estimate of 2 Mg $ha^{-1}\cdot yr^{-1}$ of litterfall (~1 Mg $ha^{-1}\cdot yr^{-1}$ of C) for Weimer Run, we can compare the amount of C released from decomposition between an elevational composite, empirically based model (empirical model) of decomposition and the three-pool CDI model. Discretization of the Weimer Run watershed by elevation class shows that 31.7% of the watershed is in the LOW elevation, 41.3% in the MID elevation, and 26.9% classified as HIGH elevation (J. W. Atkins, H. E. Epstein, and D. L. Welsch, *unpublished manuscript*). After one month, the empirical model estimates 103.9 Mg C released, while the three-pool CDI model estimates only 7.4 Mg C. While estimates converge at 20 months as outlined previously (empirical model = 178.8 Mg C; three-pool CDI model = 173.9 Mg C), at sub-annual time scales C budgeting is substantially underestimated at the watershed scale by the CDI model.

Vegetation cover and C:N ratio

Differences in decomposition rates by vegetation cover classes may not be evident for several reasons. There may be fine-scale, localized differences in the microbial communities to which the lack of differences in decomposition could be attributed. These localized differences could be in part driven by microtopography. There is perhaps also variation in microbial composition and biomass associated with different vegetation-derived organic matter. Inferred increases in N mineralization seen beneath shrubs during periods of heavy snow indicate this as a possibility.

The relatively high N content of recently senesced yellow birch litter (overall mean initial senescent yellow birch litter: N% = 2.4; C% = 52.1; C:N = 21.7, n = 12), though within the range of past work (Schulz et al. 1982), may act as a top-down control on decomposition, whereby yellow birch litter decomposes readily despite bottom-up controls from differences in the microbial and fungal communities, along with different microclimates driven by the varying vegetation structure in this system. The absence of significant differences in decomposition rates by vegetation cover class is an unexpected result that necessitates future inquiry.

Elevated C:N ratios in leaf litter collected beneath shrub plots following the second winter in our study, where snowfall and snow depth were above long-term means, suggest that N mineralization rates increased beneath shrubs during the winter as this change in C:N is driven by decreased N (Fig. 5). A persistent, developed snowpack moderates soil temperature and moisture, creating conditions conducive to microbial activity, thus increasing decomposition and N mineralization. Conversely, the absence of snow could lead to decreased N mineralization in the system. Loss of snow driven by changes in climate and inter-annual variability could lead to colder soil temperatures and decreased plant available N within the system, which in turn could affect C sequestration, through limitations of plant growth from lack of N.

Elevated N may be observed in leaf litter over the course of decomposition, if this is a primarily bacterially dominated system, as soil bacteria typically have lower C:N ratios (5–10) than soil fungi (8–15) which leads to greater immobilization of N over the course of decomposition (Strickland and Rousk 2010). The C:N ratio of our litter remains rather low, below the estimated threshold for immobilization versus mineralization (~25:1; Chapin et al. 2002). The ratio of fungal to bacterial dominance is expected to decrease in areas of high N deposition or across increasing depositional gradients, as increased N deposition decreases fungal abundance (Lilleskov et al. 2002, Strickland and Rousk 2010). The Weimer Run watershed is

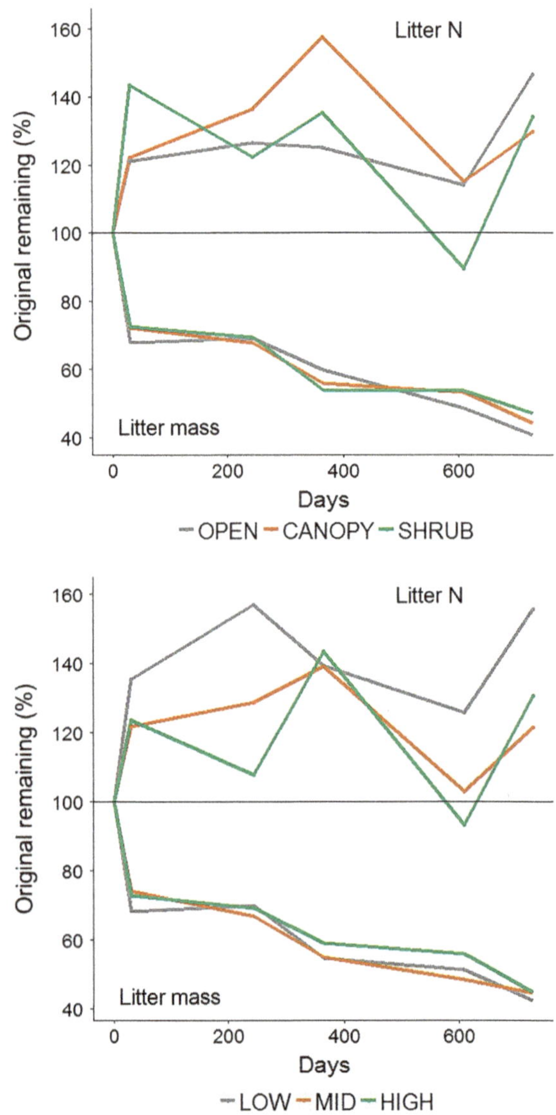

Fig. 5. Mass of litter remaining and percent nitrogen based on initial values (at top, by elevation class; at bottom, by vegetation class).

located in an area of West Virginia that receives high rates of both nitrate (12.09 kg/ha 2010–2012 mean) and ammonium deposition (3.11 kg/ha 2010–2012 mean; National Atmospheric Deposition Program [NADP]; Station WV18, Parsons, West Virginia, USA). However, the soils within the Weimer Run watershed are very acidic (below pH 4.3) and very wet (Atkins et al. 2015), conditions that do favor fungal over bacterial abundance (Rousk et al. 2009).

Given these caveats, it is possible that much of the effects of soil pH and nutrient addition examined in studies writ large are masked by methodologies that cannot accurately and conclusively determine fungal abundance (Strickland and Rousk 2010). There could also be substantial mosaicking of fungal and bacterial communities within the watershed. Our work points toward the need for further study, both broadly and within this watershed, to examine soil fungal and bacterial groups explicitly in order to tease out differences in composition and abundance along with their controls on decomposition across complex terrain.

Nitrogen availability

Plant available nitrogen, as assessed by Western Ag PRS probes, varies with vegetation, with more plant available N in areas beneath the forest canopy, early in the growing season (June). This pattern is driven by greater availability of NO_3^- which implies either greater rates of nitrification in those areas, low NO_3^- losses, or low NO_3^- uptake by vegetation—or some combination of all three processes. A comparison of the NH_4^+ to NO_3^- ratio by vegetation shows that more NO_3^- beneath shrubs and forest canopy areas is available relative to NH_4^+, implying greater rates of nitrification (Appendix S1: Fig. S4). However, the caveat should be made that direct rates of nitrification were not measured during this study. The environments beneath shrubs and canopy areas could be more conducive to nitrification, due to lower soil bulk density and higher soil porosity than open areas of the forest, despite being cooler and wetter than open areas.

Organic layer

Organic layer depth was notably deeper beneath shrubs and at higher elevations. Deeper organic soils at elevation imply lower rates of decomposition, which is in line with our findings. Higher organic layer C:N ratios beneath *Rhododendron* spp. shrubs, given their ericoid mycorrhizal associations, could lend credence to the idea of greater fungal dominance within the watershed. Conversely, nitrogen availability data coupled with implied spatial dynamics of nitrification driven by soil moisture heterogeneity (both spatial and temporal) might contradict these assertions. However, this is a very wet watershed, and lateral transfers and flows that distribute inorganic N and other nutrients more evenly are likely to be important, yet were not specifically assessed.

Threshold effects of elevation gradients

Litter decomposition has been shown to vary with elevation in the Coopers Rock State Forest in West Virginia, with higher decomposition rates along ridges and valleys and lower decomposition rates at mid-slope positions (Orndorff and Lang 1981). However, despite sharing similar climatology with Weimer Run, the elevation range tested in Coopers Rock State Forest (262–792 m) is much lower than the entire Weimer Run watershed (940–1175 m). The differences seen in decomposition along slopes at lower elevations than those tested in this study highlight the point that elevation in isolation is not a clear index of controls on decomposition in the Weimer Run watershed. This points toward a threshold effect of elevation on decomposition and element cycling that is not well understood. This possible explanation may be why differences in decomposition past one year along the elevation gradient are modulated by snowfall and the presence or absence of snowpack, and converge with broad-scale model estimates of decomposition rates.

There is a seasonal component to decomposition dynamics present in this mid-Appalachian watershed that is not evident in currently employed models of decomposition. This omission could lead to underestimating C fluxes at sub-annual time scales and lead to the propagation of uncertainty. The examination of the spatial distribution of fungal and bacterial functional groups in relation to landscape position and vegetation could further elucidate the controls on decomposition, organic layer dynamics, and nutrient cycling in this watershed. Landscape position should be considered not only as elevation, but also along determined flow paths within the watershed that could be defined via geospatial techniques (Seibert and McGlynn 2007).

Conclusions

By employing a litterbag study using a common leaf litter (yellow birch), this study showed that initial decomposition of leaf litter varies with elevation, with greater rates of decomposition at locations lower in the watershed. Initial rates of decomposition differ significantly from standard models of decomposition by greater that 20%. Elevational effects on decomposition are only apparent within the study period during the winter with greater snowfall and longer snowpack duration. However, there are no observed differences in decomposition rates by vegetation cover class either within or among elevation classes considered in this study. Organic layer depth also increases with increasing elevation, which agrees with observed lower rates of decomposition at higher elevations. Organic layer C:N ratios are higher in plots beneath shrubs at higher elevations in the watershed. Nitrogen availability, as measured using Western Ag PRS probes, differs seasonally and by vegetation cover. Both landscape position and vegetation cover influence decomposition, nutrient availability, and organic layer development within this watershed; however, there is a complex relationship with inter-annual climate variability that exerts strong controls on these ecosystem properties. Our results, viewed within the context of climate change, indicate that inter-annual climate variability, in this case snowpack dynamics, can alter the magnitude of carbon cycling processes. The terrestrial biosphere is a significant sink for anthropogenic carbon emissions (Müller et al. 2016), and understanding these complex relationships and how they may change in both the near and distant future is crucial.

Acknowledgments

We would like to acknowledge the Canaan Valley Institute, the Appalachian Stewardship Foundation, and the University Of Virginia Department Of Environmental Sciences for their funding and support of this research. We would like to sincerely thank Stesha Dunker, Jerrica Frazier, Virginia Mathurin, and Sang Mee Ko for their dedication and assistance in the laboratory and in the field. We would also like to thank Chris Gough and Todd Scanlon for suggestions that improved this manuscript and Margot Miller and Deb Lawrence for laboratory space and support. Thanks also to Lisa Turner, Ben Sagara, Ellen Stuart-Haentjens, and Cynthia Scheuermann for reading an early draft of this work.

Literature Cited

Adair, E. C., W. J. Parton, S. J. Del Grosso, W. L. Silver, M. E. Harmon, S. A. Hall, I. C. Burke, and S. C. Hart. 2008. Simple three-pool model accurately describes patterns of long-term litter decomposition in diverse climates. Global Change Biology 14:2636–2660.

Adams, M. B. (2008). Long-term leaf fall mass from three watersheds on the Fernow Experimental Forest, West Virginia, Proceedings, 16th Central Hardwood Forest Conference; 2008 April 8-9; West Lafayette, IN. Pages 179–186. Gen. Tech. Rep. NRS-P-24. U.S. Department of Agriculture, Forest Service, Northern Research Station, Newtown Square, Pennsylvania, USA.

Aerts, R. 1997. Nitrogen partitioning between resorption and decomposition pathways: A trade-off between nitrogen use efficiency and litter decomposibility? Oikos 80:603.

Atkins, J. W., H. E. Epstein, and D. L. Welsch. 2015. Vegetation and elevation influence the timing and magnitude of soil CO_2 efflux in a humid, topographically complex watershed. Biogeosciences 12:2975–2994.

Belsky, A. J., and C. D. Canham. 1994. Forest gaps and isolated savanna trees. BioScience 44:77–84.

Berg, B., and C. McClaugherty. 2008. Plant litter: decomposition, humus formation, carbon sequestration. Second edition. Springer, New York, New York, USA.

Bokhorst, S., J. W. Bjerke, J. Melillo, T. V. Callaghan, and G. K. Phoenix. 2010. Impacts of extreme winter warming events on litter decomposition in a sub-Arctic heathland. Soil Biology and Biochemistry 42:611–617.

Bonan, G. B., M. D. Hartman, W. J. Parton, and W. R. Wieder. 2013. Evaluating litter decomposition in earth system models with long-term litterbag experiments: an example using the Community Land Model version 4 (CLM4). Global Change Biology 19:957–974.

Breshears, D. D. 2006. The grassland-forest continuum: Trends in ecosystem properties for woody plant mosaics? Frontiers in Ecology and the Environment 4:96–104.

Campbell, G. S., and J. H. Norman. 1998. An introduction to environmental biophysics. Springer, New York, New York, USA.

Chapin, F. S., P. A. Matson, and H. A. Mooney. 2002. Principles of terrestrial ecosystem ecology. Springer, New York, New York, USA.

Chastain Jr., R. A., Currie, W. S., and Townsend, P. A. 2006. Carbon sequestration and nutrient cycling implications of the evergreen understory layer in Appalachian forests - ScienceDirect. Forest Ecology and Management 231:63–77.

Dwyer, L. M., and G. Merriam. 1981. Influence of topographic heterogeneity on deciduous litter decomposition. Oikos 37:228.

Fortney, R. H. 1975. The vegetation of Canaan Valley, West Virginia: a taxonomic and ecological study. Dissertation. West Virginia University, West Virginia, USA.

Garten Jr., C. T., and H. Van Miegroet. 1994. Relationships between soil nitrogen dynamics and natural ^{15}N abundance in plant foliage from Great Smoky Mountains National Park. Canadian Journal of Forest Research 24:1636–1645.

Gholz, H. L., D. A. Wedin, S. M. Smitherman, M. E. Harmon, and W. J. Parton. 2000. Long-term dynamics of pine and hardwood litter in contrasting environments: toward a global model of decomposition. Global Change Biology 6:751–765.

Hamon, W. R. 1963. Computation of Direct Runoff Amounts From Storm Rainfall. International Association of Hydrological Sciences 63:52–62.

Hastwell, G. T., and E. C. Morris. 2013. Structural features of fragmented woodland communities affect leaf litter decomposition rates. Basic and Applied Ecology 14:298–308.

Hättenschwiler, S., and P. Gasser. 2005. Soil animals alter plant litter diversity effects on decomposition. Proceedings of the National Academy of Sciences of the United States of America 102:1519–1524.

Jenkins, J. C., R. A. Birdsey, and Y. Pan. 2001. Biomass and NPP estimation for the Mid-Atlantic region (USA) using plot-level forest inventory data. Ecological Applications 11:1174–1193.

Johnson, D. W., P. S. J. Verburgh, and A. Arnoneb. 2004. Soil extraction, ion exchange resin, and ion exchange membrane measures of soil mineral nitrogen during incubation of a tallgrass prairie soil. Soil Science Society of America Journal 69:260–265.

Knoepp, J. D., and W. T. Swank. 1998. Rates of nitrogen mineralization across an elevation and vegetation gradient in the southern Appalachians. Plant and Soil 204:235–241.

Lilleskov, E. A., T. J. Fahey, T. R. Horton, and G. M. Lovett. 2002. Belowground ectomycorrhizal fungal community change over a nitrogen deposition gradient in Alaska. Ecology 83:104–115.

Meentemeyer, V. 1978. Macroclimate and lignin control of litter decomposition rates. Ecology 59:465–472.

Müller, C., E. Stehfest, J. G. van Minnen, B. Strengers, W. von Bloh, A. H. W. Beusen, S. Schaphoff, T. Kram, and W. Lucht. 2016. Drivers and patterns of land biosphere carbon balance reversal. Environmental Research Letters 11:44002.

Northeastern Ecosystem Research Cooperative (NERC). 2017. Compilation of soil chemistry data for the northeastern United States and southeastern Canada. NERC.13.5. http://www.nercscience.org

Olson, J. S. 1963. Energy storage and the balance of producers and decomposers in ecological systems. Ecology 44:322–331.

Orndorff, K. A., and G. E. Lang. 1981. Leaf litter redistribution in a West Virginia hardwood forest. Journal of Ecology 69:225–235.

Powers, R. F. 1990. Nitrogen mineralization along an altitudinal gradient: interactions of soil temperature, moisture, and substrate quality. Forest Ecology and Management 30:19–29.

R Core Team. 2017. R: a language and environment for statistical computing. R Foundation for Statistical Computing, Vienna, Austria.

Raich, J. W., and W. H. Schlesinger. 1992. The global carbon dioxide flux in soil respiration and its relationship to vegetation and climate. Tellus Series B 44:81–99.

Reich, P. B., et al. 2001. Plant diversity enhances ecosystem responses to elevated CO2 and nitrogen deposition. Nature 410:809–810.

Reiners, W. A., and G. E. Lang. 1987. Changes in litterfall along a gradient in altitude. Journal of Ecology 75:629.

Reinmann, A. B., and P. H. Templer. 2016. Reduced winter snowpack and greater soil frost reduce live root biomass and stimulate radial growth and stem respiration of red maple (*Acer rubrum*) trees in a mixed-hardwood forest. Ecosystems 19:129–141.

Rousk, J., P. C. Brookes, and E. Bååth. 2009. Contrasting soil pH effects on fungal and bacterial growth suggest functional redundancy in carbon mineralization. Applied and Environmental Microbiology 75:1589–1596.

Saccone, P., et al. 2013. The effects of snowpack properties and plant strategies on litter decomposition during winter in subalpine meadows. Plant and Soil 363:215–229.

Schliemann, S. A., and J. G. Bockheim. 2014. Influence of gap size on carbon and nitrogen biogeochemical cycling in Northern hardwood forests of the Upper Peninsula, Michigan. Plant and Soil 377: 323–335.

Schulz, J. C., P. J. Nothnagle, and I. T. Baldwin. 1982. Seasonal and individual variation in leaf quality of two northern hardwoods tree species. American Journal of Botany 69:753–759.

Seibert, J., and B. L. McGlynn. 2007. A new triangular multiple flow direction algorithm for computing upslope areas from gridded digital elevation models. Water Resources Research 43:1–8.

Shanks, R. E., and J. S. Olson. 1961. First-year breakdown of leaf litter in southern Appalachian forests. Science 134:194–195.

Silver, W. L., and R. K. Miya. 2001. Global patterns in root decomposition: comparisons of climate and litter quality effects. Oecologia 129:407–419.

Singh, J. S., and S. R. Gupta. 1977. Plant decomposition and soil respiration in terrestrial ecosystems. Botanical Review 43:449–528.

Strickland, M. S., and J. Rousk. 2010. Considering fungal:bacterial dominance in soils–Methods, controls, and ecosystem implications. Soil Biology and Biochemistry 42:1385–1395.

Sun, G., S. G. McNulty, D. M. Amatya, R. W. Skaggs, L. W. Swift, J. P. Shepard, and H. Riekerk. 2002. A comparison of the watershed hydrology of coastal forested wetlands and the mountainous uplands in the Southern US. Journal of Hydrology 263:92–104.

Vitousek, P. M. 2004. Nutrient cycling and limitation: Hawai'i as a model system. Princeton University Press, Princeton, New Jersey, USA.

Walter, J. A., J. C. Neblett, J. W. Atkins, and H. E. Epstein. 2017. Regional- and watershed-scale analysis of red spruce habitat in the southeastern United States: implications for future restoration efforts. Plant Ecology 218:305–316.

Zhou, G., L. Guan, X. Wei, X. Tang, S. Liu, J. Liu, D. Zhang, and J. Yan. 2008. Factors influencing leaf litter decomposition: an intersite decomposition experiment across China. Plant and Soil 311:61–72.

Zhou, Y., J. Su, I. A. Janssens, G. Zhou, and C. Xiao. 2014. Fine root and litterfall dynamics of three Korean pine (*Pinus koraiensis*) forests along an altitudinal gradient. Plant and soil 374:19–32.

11

The freshwater landscape: lake, wetland, and stream abundance and connectivity at macroscales

CAROL EMI FERGUS,[1,8,†] JEAN-FRANÇOIS LAPIERRE,[3,4] SAMANTHA K. OLIVER,[5] NICHOLAS K. SKAFF,[1] KENDRA S. CHERUVELIL,[1,6] KATHERINE WEBSTER,[1,7] CAREN SCOTT,[2] AND PATRICIA SORANNO[1]

[1]*Department of Fisheries and Wildlife, Michigan State University, East Lansing, Michigan 48824 USA*
[2]*NEON, 1685, 38th Street, Suite 100, Boulder, Colorado 80301 USA*
[3]*Département de Sciences Biologiques, Université de Montréal, Pavillon Marie-Victorin, 90, Avenue Vincent-d'Indy, Montreal, Québec H2V 2S9 Canada*
[4]*GRIL Université de Montréal, Pavillon Marie-Victorin, Département de Sciences Biologiques, C.P. 6128, succ. Centre-ville, Montreal, Québec H3C 3J7 Canada*
[5]*Center for Limnology, University of Wisconsin Madison, Madison, Wisconsin 53706 USA*
[6]*Lyman Briggs College, Michigan State University, East Lansing, Michigan 48825 USA*
[7]*School of Natural Sciences, Trinity College Dublin, The University of Dublin, College Green, Dublin 2, Ireland*

Abstract. Inland water bodies and their surface hydrologic connections are active components in the landscape, influencing multiple ecological processes that can propagate to broad-scale phenomena such as regional nutrient and carbon cycles and metapopulation dynamics. However, while lake, wetland, and stream abundance has been estimated at regional and global extents, less attention has been paid to freshwater connectivity attributes among aquatic systems at macroscales. Thus, regional to continental patterns of freshwater abundance and connectivity are poorly understood. We measured lake, wetland, and stream abundance and surface connectivity attributes (i.e., landscape position within stream networks) at a subcontinental extent in the Midwest and Northeast United States to characterize macroscale spatial patterns of the freshwater landscape (i.e., abundance and connectivity attributes of lakes, wetlands, and streams). We found that lake and wetland abundance exhibited opposite spatial patterns from stream density that generally followed glaciation extent boundaries—lake and wetland abundance was high north of the glaciation boundary, whereas stream density was high south of the glaciation boundary. Freshwater connectivity attributes exhibited distinct spatial patterns as defined by our integrated freshwater clusters and revealed a layer of complexity not captured by abundance measures. Patterns of freshwater abundance and connectivity in the study extent were associated primarily with glaciation and secondarily with hydrogeomorphic (e.g., surficial geology and topography), climate (e.g., runoff), and land-use (e.g., agriculture) variables, providing insight into potential drivers of freshwater composition and distribution. The connectivity spatial patterns observed suggest that relying solely on freshwater abundance measures in macroscale analyses omits unique information on the structural attributes of freshwater systems that can be critical to key ecological processes. Adopting an integrated freshwater landscape framework to study and manage freshwaters is essential as freshwater systems face broad-scale disturbances that may alter hydrologic connections and subsequently may impact ecosystem processes and services.

Key words: cluster analysis; freshwater abundance; freshwater hydrologic connectivity; glaciation; integrated freshwater landscape; lakes; macroscale; streams; wetlands.

† **E-mail:** emifergus@gmail.com

Introduction

Freshwater ecosystems are active components in the landscape that influence a number of ecological processes and provide invaluable ecosystem services. Freshwater connections that link lakes, wetlands, and streams to one another and the surrounding landscape are major structural components that influence critical freshwater processes such as water, nutrient, and carbon fluxes (Quinlan et al. 2003, Cardille et al. 2007, Acuña and Tockner 2010, Lottig et al. 2011), nutrient and carbon processing (Weller et al. 1996, Kling et al. 2000, Strayer et al. 2003), and the dispersal of organisms influencing population and community dynamics (Pringle 2001, Crump et al. 2007, Bouvier et al. 2009, Nelson et al. 2009). These processes can influence broad-scale ecological phenomena with implications for regional and global biogeochemical budgets and metapopulation dynamics. Quantifying patterns of freshwater abundance and connectivity is essential to inform upscaling of biogeochemical processing and for conservation and management of populations at broad spatial extents. However, we have an incomplete view of the integrated freshwater landscape that includes lakes, wetlands, and streams and their spatial connections to one another at macroscales, defined as regional to continental extents spanning hundreds to thousands of kilometers (Heffernan et al. 2014). While progress has been made on quantifying the abundance of individual ecosystem types across the landscape, an integrative picture of the combined freshwater landscape that explicitly includes spatial connections among these systems has not been developed.

Regional and global abundance has been estimated for lakes (Downing et al. 2006, McDonald et al. 2012, Verpoorter et al. 2014), wetlands (Aselmann and Crutzen 1989, Lehner and Döll 2004), and streams (Downing et al. 2012). These studies were the first to explicitly estimate the distribution of freshwater systems across broad spatial extents and examine underlying geospatial features related to these patterns (i.e., glacial and tectonic activity, geomorphology, and regional climate; Meybeck 1995, Smith et al. 2002, Downing and Duarte 2009). Most of these studies, however, were restricted to one freshwater system type, thus providing a limited view of the diversity of the freshwater landscape. Exceptions to this were studies on macroscale carbon cycles that integrated inland waters by estimating surface area of lakes, reservoirs, and streams (Raymond et al. 2013, Butman et al. 2016). However, both single- and integrated-system studies did not distinguish freshwater systems by their hydrologic connectivity attributes (e.g., drainage, seepage system types), and thus provide general abundance estimates that lump all system types together, ignoring the diversity of hydrologic types across the landscape.

Unlike abundance patterns, freshwater connectivity patterns have not been quantified at regional to global scales, in part because freshwater connectivity can be challenging to define (Ali and Roy 2009), is scale dependent, and fine-scale data on hydrologic flow paths are lacking at macroscale extents. Hence, it is challenging to get precise measures of hydrologic connections among systems at broad spatial extents. However, landscape metrics are useful tools that can address this limitation for broad-scale analyses by capturing aspects of freshwater connectivity within the landscape, which we define as the surface connections linking lakes, wetlands, and streams. Landscape position of freshwater systems is one useful class of metrics that is inherently associated with differences in connectivity characteristics moving along longitudinal gradients of geomorphic and hydrologic settings (Kratz et al. 1997, Soranno et al. 1999, Riera et al. 2000, Martin and Soranno 2006, Müller et al. 2013). For example, lake hydrologic position metrics incorporate lake connections with stream inflows and outflows (Martin and Soranno 2006), and lake order differentiates systems based on their position within stream networks (Riera et al. 2000). While landscape position measures do not provide explicit measures of all possible hydrologic flow paths, they do group systems with similar spatial and landscape characteristics, which can be useful to characterize surface connectivity attributes.

Lake, wetland, and stream landscape position metrics are related to biogeochemical variables (Lohse et al. 2009, Humborg et al. 2010, Racchetti et al. 2010, Sadro et al. 2011), hydrogeomorphic and limnological characteristics (Martin and Soranno 2006, Butman and Raymond 2011, Read et al. 2015), and responses to land-use disturbances

(Detenbeck et al. 1993, Freeman et al. 2007, Soranno et al. 2015*a*), demonstrating that they capture spatial connectivity characteristics that are relevant to ecological processes. For example, within a hydrologic flow system, lakes at high positions tend to be hydrologically dominated by precipitation and, thus, have lower base cation concentrations compared to lakes at lower positions that have greater contributions from groundwater and surface water sources to their water budgets (Kratz et al. 1997, Soranno et al. 1999). These landscape position measures are related to ecosystem structure (i.e., surface connectivity attributes) and function and can provide insight on populations of freshwater systems at macroscales.

There is a need to examine freshwater connectivity attributes at broad scales because they are vulnerable to disturbances such as land-use conversion and climate change that can alter hydrologic properties at multiple spatial and temporal scales and subsequently threaten freshwater integrity and function in the landscape (Carpenter et al. 2011, Steele and Heffernan 2014). The effects of broad-scale disturbances on freshwater systems are likely to vary across the landscape depending on underlying hydrogeomorphic characteristics (Webster et al. 2008), the spatial configuration of freshwater features (Vörösmarty et al. 2010), and interactions with natural features and human hydrological modifications (Jones et al. 2012). Thus, it is challenging to predict the impacts of these potential disturbances without understanding freshwater systems within the context of their geographic setting and at the broad spatial scales that are aligned with the above scales of disturbance (Jones 2011, Moore 2015). Together, these issues create a need to develop broad, macroscale estimates of the abundance and connectivity of freshwater systems in the landscape.

Until recently, it has been challenging to incorporate freshwater connectivity characteristics into broad-scale frameworks because of computational limitations and a lack of integrated lake, wetland, and stream datasets. Technological advances and national-scale geographic data resources (e.g., the high-resolution U.S. National Hydrography Dataset; NHD) have ameliorated some of these constraints, and we are better positioned to estimate freshwater abundance and connectivity at macroscales.

In this study, we characterize the macroscale patterns of the integrated freshwater landscape that includes lakes, wetlands, and streams, across the Midwest and Northeast United States. We quantify lake, wetland, and stream abundance and density and group systems based on their surface connections using landscape position metrics. For the scope of this study, we define freshwater connectivity as the permanent surface hydrologic connections that link lakes, wetlands, and streams, and we measure connectivity as the landscape position of systems within stream networks. Our objectives were to (1) quantify macroscale patterns of surface freshwater characteristics that integrate measures of both abundance and connectivity at a subcontinental spatial extent and (2) relate these macroscale patterns to hydrogeomorphic, land-use, and climatic attributes that are hypothesized to be associated with the distribution and spatial characteristics of surface freshwaters. Within our macroscale extent (~1,800,000 km^2), we found distinct spatial patterns in the abundance and connectivity of freshwaters that can help scientists and managers develop an improved broad-scale understanding of the composition and structure of the freshwater landscape.

Methods

Overview of analytical conceptual approach

No single measure can capture the diversity and spatial complexity of surface freshwater systems, and detailed metrics requiring site-specific data inputs are impractical across broad, regional settings. Therefore, we adopted a multi-step approach to build a synthetic view of the freshwater landscape and investigate broad-scale controls on macroscale patterns. We first mapped freshwater abundance and density of lakes, wetlands, and streams separately across the study extent to determine the proportion of the landscape that is composed of surface freshwater bodies. Second, we examined spatial distributions of lake, wetland, and stream connectivity by characterizing systems based on their stream network landscape position and quantified the relative proportion area for each freshwater system type. We then combined data on abundance and connectivity for lakes, wetlands, and streams to develop an integrated view of the freshwater landscape.

Finally, we quantified associations between integrated freshwater clusters and hydrogeomorphic, land-use, and climate variables to explore underlying controls on spatial patterns at macroscales.

Study extent

The study area spanned a subcontinental extent (~1,800,000 km^2) in the temperate Midwest and Northeast regions of the United States (Fig. 1). This spatial extent presents a distinctive opportunity to capture the diversity of freshwater systems and connectivity characteristics that make up the freshwater landscape because it is rich in lakes, wetlands, and streams and spans regional settings with diverse hydrogeomorphic, land-use, past glacial regime, and climate conditions (Appendix S1: Table S1). To illustrate this, at the Hydrologic Unit 8 scale (HU8), we found a range of values for geologic composition (e.g., alluvial geology 0–89%), hydrologic attributes (e.g., mean runoff 44–762 mm/yr), land-use activities (e.g., agriculture 0–89%), and climate variables (e.g., mean precipitation 566–1376 mm/yr). In addition, the study extent has different glaciation regimes from the most recent glacial period (Wisconsin period ~2.6 million to 11,000 yr ago), with 58% of the study extent having been glaciated, 37% unglaciated, and 3% partially glaciated (i.e., area along glaciation boundary that includes glaciated and unglaciated terrain).

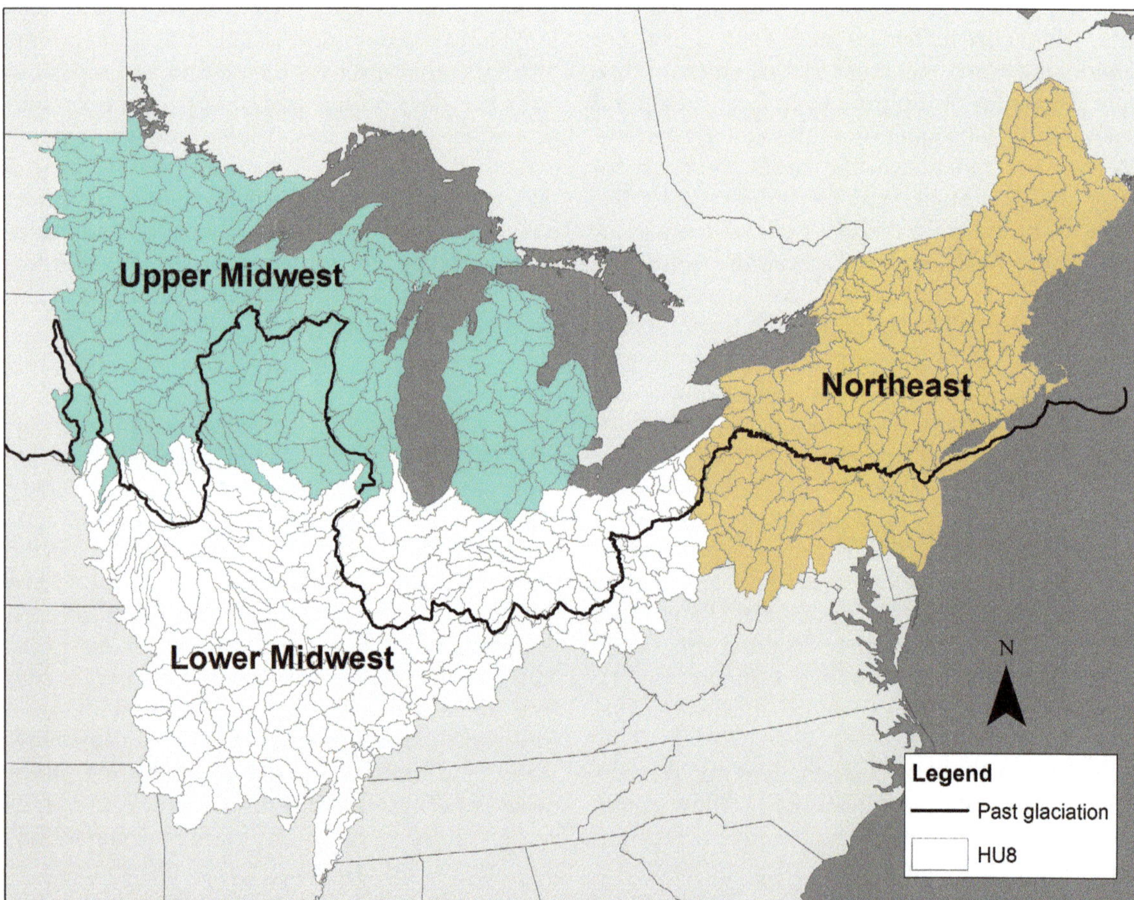

Fig. 1. Study extent regions. The study extent includes three regions in the United States: the Upper Midwest, Lower Midwest, and the Northeast and in total spans ~1,800,000 km^2 in area. Hydrologic Unit 8s (HU8, light gray polygons) are depicted on the map and have a median size of 2880 km^2. The solid black line represents the estimated boundary of the Wisconsin glacial period—north of the line is glaciated area and south of the line is unglaciated area.

Description of datasets

Our analyses used data in the LAGOS-NE$_{GEO}$ database v. 1.03 (LAke multi-scaled GeOSpatial and temporal database), an integrated, multi-thematic geographical database spanning the Midwest and Northeast United States (Soranno et al. 2015*b*). LAGOS-NE$_{GEO}$ includes metrics derived from national-scale geographic data on geology, topography, hydrology, climate, and land use/land cover from multiple sources (Appendix S1: Table S2). Lake and stream data originally came from the National Hydrography Dataset (NHD version 9.3; 1:24,000 resolution; United States Geological Survey [USGS], USGS Headquarters, Reston, Virginia, USA), and wetland data came from the National Wetlands Inventory (NWI version 2 Surface Waters and Wetlands Inventory; U.S. Fish and Wildlife Service, Northwest, Washington, D.C., USA). Other geospatial data in LAGOS-NE$_{GEO}$ came from separate data sources: hydrology (USGS National Water Information System portal), topography (USGS National Elevation Dataset), geology and glaciation limits (USGS surficial materials map database), climate data as 30-yr normal averages (PRISM Climate Group), and land use/land cover (2011 National Land Cover Database, Multi-Resolution Land Characteristic Consortium). See Soranno et al. (2015*b*) for more information on the database.

Definitions of lakes, wetlands, and streams

LAGOS-NE$_{GEO}$ included lake, wetland, and stream features from the original NHD and NWI geographic layers that captured features that were considered to be part of the surface freshwater landscape and to account for limitations in the original geographic data layers (Appendix S1: Table S3). For example, both the NHD and NWI comprehensively map freshwater feature types including artificial systems (e.g., sewage treatment ponds) and artificial flowlines (e.g., lines connecting NHD features) and these highly modified, man-made systems were not incorporated into analyses. Our freshwater system definitions are provided below.

Lake features came from NHD v. 9.3 and were defined as perennial water bodies (as denoted within the NHD) with surface area $\geq$0.04 km^2, including both natural lakes and reservoirs (i.e., impounded streams or rivers), with a total of 50,726 lakes matching our definition within the study extent. Lakes smaller than 0.04 km^2 in size were excluded from the analyses because these smaller water bodies were associated with high identification and digitization error rates compared to lakes that were 0.04 km^2 and larger in the NHD data layer (Appendix S1: Table S3; Soranno et al. 2015*b*). Omitting these small lakes (<0.04 km^2 and $\geq$0.01 km^2) in the NHD did not greatly affect lake area summaries accounting for only 0.73% of total lake area in the study extent; but, they made up over half the number of inland lake bodies (63%). However, there is evidence that lakes smaller than 0.04 km^2 have large digitization error rates (~58%) compared to lakes larger than 0.04 km^2 (~15%; Soranno et al. 2015*b*). We did not distinguish between natural lakes and reservoirs because not all systems in the NHD are properly categorized and many artificial water bodies have been mislabeled as natural lakes (McDonald et al. 2012).

Wetland features came from NWI v.2 and included NWI classified *Palustrine* systems, that is, non-tidal wetlands dominated by trees, shrubs, and persistent emergent vegetation (Cowardin et al. 1979). We removed deep-water systems (lacustrine and riverine) from the dataset and erased lake area from wetland polygons we retained. There were over 4 million wetland polygons remaining in the study extent (n = 4,411,383) following the preprocessing steps. We did not set size restrictions on wetland data because NWI reliably captures wetlands that are 0.002 km^2 in size or larger with a 98% identification accuracy rate (Wetland Mapping Standard; Federal Geographic Data Committee). We kept the NWI wetland class geographic delineations based on vegetation and substrate composition even for polygons that were adjacent to one another. While it can be argued that these adjacent polygons are hydrologically connected to one another, we chose to retain these boundaries rather than to merge them into single wetland polygons because different vegetation and substrate composition is associated with water regime and other hydrologic characteristics (Cowardin et al. 1979), which is important information we wanted to retain. However, future studies should consider how merging wetland polygons influences connectivity estimates.

Finally, stream data included perennial streams classified as *Stream/River*, *Canal/Ditch*, *Connector*, *Pipeline*, *Underground Conduit*, and *Artificial Path* feature types in the NHDFlowline data layer and

NHDArea polygons that captured wide stream and river features. We excluded *Coastlines* linear features and *Artificial Path* features that did not spatially intersect NHDArea polygons with *Stream/River* features. We did not include intermittent stream types in this analysis, although we recognize the importance of intermittent and headwater streams in aquatic ecosystem function and dynamics (Nadeau and Rains 2007). We only included perennial stream features because they likely represent more consistent hydrologic connections and because these headwater and intermittent streams are often underrepresented in the NHD in comparison with field surveys (Fritz et al. 2013). In total, stream length exceeded 1.7 million km in the study extent. Freshwater feature definitions and detailed descriptions of preprocessing steps for the data in LAGOS-NE_{GEO} are provided in Soranno et al. (2015*b*).

There were limitations in the lake, wetland, and stream geographic data that we were not able to address in our analyses (Appendix S1: Table S3). For example, both the NHD and NWI are national datasets but are produced and regularly updated at regional (often state-level) scales by local entities. As a consequence, there are differences in data resolution and digitization intensity of water bodies across the spatial extent. Freshwater features may not be evenly digitized across the study extent, which could affect freshwater abundance and connectivity estimates.

Measuring freshwater features in the landscape at macroscales

Freshwater abundance and connectivity metrics were measured at two HU spatial scales: HU12 and HU8 (Seaber et al. 1987). The HUs are hierarchically nested stream watershed spatial units that are based on USGS 1:24,000-scale topographic maps. The HU12 scale is the smallest nested spatial unit in LAGOS-NE_{GEO} and was used to examine fine-scale heterogeneity in freshwater attributes (n = 18,870; median size = 78.10 km^2). The HU8 scale is an intermediate-sized spatial unit and was used to examine regional-scale heterogeneity in freshwater attributes (n = 445; median size = 2880 km^2).

Freshwater abundance.—Freshwater abundance for lakes and wetlands was quantified as the total proportional area and density within HU12 spatial units. Streams were measured as the total stream density (length per unit area) within HU12 spatial units. We represented freshwater abundance by mapping binned metric values for lakes, wetlands, and streams separately.

Freshwater connectivity.—We characterized freshwater connectivity within the spatial extent as follows: (1) We quantified multiple freshwater connectivity metrics for lakes, streams, and wetlands separately, (2) we performed principal components analysis (PCA) on the connectivity metric values for each freshwater type to reduce collinearity, (3) we performed *k*-means cluster analysis to group spatial units with similar freshwater connectivity characteristics, and (4) we mapped the cluster assignments to visualize spatial patterns.

Freshwater connectivity metrics.—Freshwater connectivity metrics were calculated by first assigning lake, wetland, and stream geographic features into connectivity types based on their spatial arrangement with other freshwater features (Fig. 2). Lake connectivity types were based on landscape position, defined by the surface hydrologic connections of the focal lake with inflowing and outflowing streams and upstream lakes. Lakes were grouped into four hydrologic classes: Isolated, Headwater, Drainage, and Drainage-UPLK (Fig. 2a). Isolated lakes are defined as having no stream inlets or outlets. Headwater lakes have no stream inlets and at least one outlet. Drainage lakes have inlets and outlets and no upstream lakes (≥0.10 km^2 in size). Drainage-UPLK lakes have inlets and outlets and at least one upstream lake (≥0.10 km^2 in size).

Wetland connectivity types also were defined in relation to connections with surface stream networks; however, streamflow directionality was not incorporated into the definition due to computation limitations of processing millions of wetland polygons. Wetlands were grouped into three classes: Upland-Embedded, Headwater, and Drainage (Fig. 2b). We considered wetlands to be stream-connected if located at any point within a 30-m buffer surrounding the stream segment center line or double-banked lines for large rivers (Soranno et al. 2015*b*); this buffer approach was necessary to accommodate spatial data resolution limitations and misalignment between different data layers (NHD and NWI). We called wetlands with no permanent surface stream connections Upland-Embedded. These systems are thought to

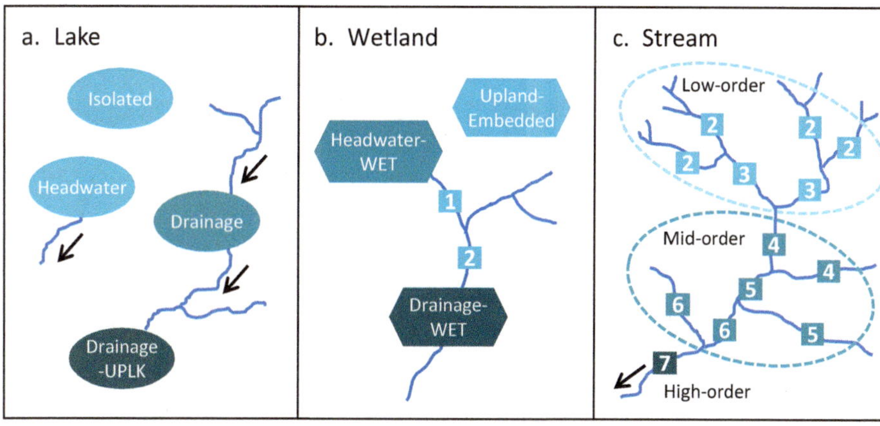

Fig. 2. Diagram of freshwater connectivity metrics for (a) lakes, (b) wetlands, and (c) streams. (a) Lakes (oval) were assigned connectivity types based on spatial relationships with streams and upstream lakes. Lake connectivity types include Isolated, no stream inlets or outlets; Headwater, only stream outlets; Drainage, stream inlets and outlets; and Drainage-UPLK, stream inlets and outlets and at least one upstream lake $\geq$0.10 km^2 in size. (b) Wetlands (hexagon) were assigned connectivity types based on spatial relationships with streams. Wetland connectivity types include Upland-Embedded, no intersecting stream segments; Headwater (Headwater-WET), intersected by one first-order stream segment; and Drainage (Drainage-WET), intersected by a stream segment > first-order or multiple stream segments. (c) Stream segments were assigned Strahler stream order (squares) based on location in the stream network and grouped into stream connectivity classes: Low-order (first–third), Mid-order (fourth–sixth), and High-order (>sixth).

have low connectivity with no persistent surface hydrologic connections but potentially intermittent and/or subsurface hydrologic connections in the landscape (Tiner 2003, Mushet et al. 2015). Because flow directionality of streams intersecting wetlands was not examined and inlets and outlets could not be identified, we defined Headwater wetlands to be systems that are intersected by a Strahler first-order stream segment. And finally Drainage wetlands were defined as systems intersected by either a single second-order or higher stream segment or by multiple stream segments.

Stream features in our study extent were classified by Strahler stream order, a measure of the position of the stream reach in relation to the stream network with order increasing as one moves from the headwater streams to the terminal point (Strahler 1957). We considered stream order to capture differences in stream connectivity characteristics based on the assumption that low-order streams are in headwater watersheds that potentially could have more intermittent flows as compared to high-order streams that may have more persistent connections. Stream order classes were grouped into three categories: Low-order, Mid-order, and High-order (Fig. 2c). Low-order streams include first- to third-order stream reaches, Mid-order streams include fourth- to sixth-order reaches, and High-order streams include greater than sixth-order reaches.

We calculated the relative proportion of each connectivity type to the total area or length of the respective freshwater system in a spatial unit (e.g., Isolated lake area divided by total lake area within the HU12). Relative proportion metrics are better suited to our connectivity analyses as compared to total areal proportion metrics (e.g., Isolated lake area divided by HU12 area) because freshwater features cover only a small fraction of area across our macroscale extent, resulting in many observations having zero or very low values. More information on geoprocessing steps and the GIS toolbox developed to calculate the connectivity metrics is described in Soranno et al. (2015*b*), and the toolbox is available at https://github.com/soranno/LAGOS_GIS_Toolbox.

Analysis

Principal components and k*-means cluster analyses for freshwater connectivity.*—Prior to analyses, we transformed the relative proportion (logit) and stream density (natural log) metric values to meet assumptions of normality and homoscedasticity. We performed PCAs on the transformed connectivity metrics that were quantified at the HU12 scale for lakes, wetlands, and streams separately to reduce codependence among values. Principal components analyses were performed using JMP 10 software (SAS Institute, Cary, North Carolina, USA). Due to co-variation among the variables, the top two axes explained between 70% and 98% of the variation in the different PCAs performed; we only used the scores of these two first axes for the subsequent cluster analysis. The first two axes of the PCA are shown in Appendix S1: Fig. S1.

K-means analyses were performed to group HU12s that had similar freshwater connectivity characteristics, based on PCA scores, using JMP10. Prior to analyses, we used the "Cascade *k*-means" approach (vegan R package; Oksanen et al. 2017) to identify the number of clusters that optimizes the within- vs. among-cluster residual mean square error. There were some spatial units that were not assigned cluster groups because there were no freshwater systems present (e.g., HU12s without lakes present n = 7922 and comprised ~39% of study extent area) or because stream segments were not assigned Strahler stream order class due to digitization errors in streamflow directionality in the original NHD data layer (n = 123 HU12). These sites were assigned their own cluster, named "zero lake" and "no stream order." Finally, we mapped the cluster groups across the study extent to visualize the patterns of freshwater surface connectivity at macroscales.

Patterns of abundance and connectivity integrated across lakes, wetlands, and streams.—To visualize regional patterns of abundance and connectivity integrated across lakes, wetlands, and streams, we performed a similar set of analyses as the HU12 freshwater connectivity analysis (i.e., PCA, cluster analysis, and mapping the cluster groups) at the HU8 scale. We included abundance and connectivity metrics across the three freshwater systems in the PCA because it is expected that these freshwater features may co-vary with one another and be influenced by similar geophysical and climatic drivers. We used the HU8 scale because lakes were present in only 58% of the HU12s; and 60% of HU12s with lakes only had 1–3 lakes, which could result in lakes (or the lack of lakes) playing a disproportionate role in determining broad-scale freshwater patterns. Performing the integrated analyses at the HU8 scale allowed all spatial units to include lakes, providing a more balanced weight to lakes, wetlands, and streams. Eight HU8 spatial units were not assigned cluster groups because they contained streams that were not assigned Strahler stream order class. In total, 447 HU8 spatial units (out of 455) were retained to create integrated freshwater clusters across the study extent.

We examined the geospatial composition of the resulting integrated freshwater clusters by creating boxplots of freshwater abundance and connectivity metrics and other landscape features grouped by cluster. We tested for differences in geospatial composition among cluster groups using non-parametric Kruskal–Wallis tests and using multiple comparison tests to examine differences between individual clusters ($\alpha < 0.05$).

Hydrogeomorphic, climate, and land-use variables associated with freshwater abundance and connectivity.—We performed random forest analyses using randomForest package in R (Liaw and Wiener 2002) to examine associations between the integrated freshwater cluster groups (at HU8 scale) and underlying landscape and climatic characteristics (Appendix S1: Table S1). Random forest is a machine-learning technique based on classification regression trees and combines multiple classification trees to improve classification accuracy (Cutler et al. 2007). The algorithm uses bootstrap samples of the original observations and randomized subsets of predictor variables to build individual trees. Model accuracy and variable importance were estimated from the hold-out observations (about one-third of the observations in the training set; Breiman 2001). We used the VSURF package to inform variable selection in the final random forest model (Genuer et al. 2016). The package provides an automated method to select variables based on importance scores and to minimize redundancy among predictor variables and thus allowed us to identify the geographic metrics that were most closely associated with freshwater abundance and connectivity clusters. The number of bootstrap samples was fixed to the smallest cluster group size to improve prediction accuracy among unbalanced cluster groups and to reduce bias of minimizing error for large cluster groups.

Freshwater clusters, PCA scores, and freshwater metric and landscape composition data at the HU12 and HU8 scales are provided in the Environmental Data Initiative (https://environmentaldatainitiative.org/; Fergus et al. 2017*a*, *b*).

Results

Freshwater abundance

Lake, wetland, and stream abundance measured as areal proportions or stream density within the HU12 exhibited contrasting macroscale patterns within the study extent (Table 1, Fig. 3a–c). Lakes (≥0.04 km^2 in size) were only present in over half of the study extent (61%) in contrast to wetlands and streams that were found throughout the study extent (~99%). In total, lake area made up 1.9% of the study extent area (with 2.2% lake area in glaciated areas and 0.5% in non-glaciated areas), and lake density was 0.03 km^{-2}. In contrast to lakes, wetlands made up 7.8% of the total study extent area and were found at a higher density of 2.8 km^{-2}. Total stream length was 1,794,044 km with an overall density of 1.13 km/km^2 in the study extent. In general, lake and wetland abundance exhibited opposite spatial patterns along a north-to-south gradient compared to stream density. Lake and wetland abundance was higher in

Table 1. Summary statistics of freshwater metrics.

Freshwater metric type	Metric	Median (SD)	Range
HU8 scale			
Lake abundance	Lake proportion	0.01 (0.04)	0–0.41
Lake size	Lake size (km^2)	0.48 (803)	0–1,173,449
Lake–stream connectivity	Isolated	0.08 (0.18)	0–1.00
Relative proportion	Headwater	0.05 (0.11)	0–0.95
	Drainage	0.30 (0.26)	0–1.00
	Drainage-UPLK	0.41 (0.31)	0–1.00
Wetland abundance	Wetland proportion	0.05 (0.11)	0–0.95
Wetland size	Wetland size (km^2)	0.02 (0.04)	0–44.35
Wetland–stream connectivity	Upland-Embedded	0.27 (0.14)	0.02–0.68
Relative proportion	Headwater-WET	0.11 (0.05)	0–0.27
	Drainage-WET	0.59 (0.17)	0.22–0.97
Stream abundance	Stream density (m/km^2)	0.11 (0.05)	1.95–45.12
Stream connectivity	Low-order	0.87 (0.04)	0.71–1.00
Relative proportion	Mid-order	0.12 (0.04)	0–0.29
	High-order	0.001 (0.02)	0–0.13
HU12 scale			
Lake abundance	Lake proportion	0.001 (0.05)	0–0.99
Lake size	Lake size (km^2)	0.18 (5665)	0–82,139
Lake–stream connectivity	Isolated	0 (0.35)	0–1.00
Relative proportion	Headwater	0 (0.26)	0–1.00
	Drainage	0.29 (0.40)	0–1.00
	Drainage-UPLK	0 (0.35)	0–1.00
Wetland abundance	Wetland proportion	0.03 (0.11)	0–0.99
Wetland size	Wetland size (km^2)	0.02 (0.11)	0–425
Wetland–stream connectivity	Upland-Embedded	0.26 (0.23)	0–1.00
Relative proportion	Headwater-WET	0.09 (0.10)	0–1.00
	Drainage-WET	0.59 (0.26)	0–1.00
Stream abundance	Stream density (m/km^2)	0.11 (0.05)	0–70.27
Stream connectivity	Low-order	0.90 (0.13)	0–1.00
Relative proportion	Mid-order	0.09 (0.12)	0–1.00
	High-order	0 (0.05)	0–0.99

Notes: SD, standard deviation. Lake, wetland, and stream abundance and connectivity metrics were quantified at two Hydrologic Unit (HU) spatial scales: HU8 (n = 447) and HU12 (n = 18,876). Lake and wetland connectivity metrics were calculated as relative proportions—the area of each connectivity type divided by the total lake or wetland area within a spatial unit. Stream connectivity was calculated as the relative density of stream connectivity type out of the total stream length within the spatial unit.

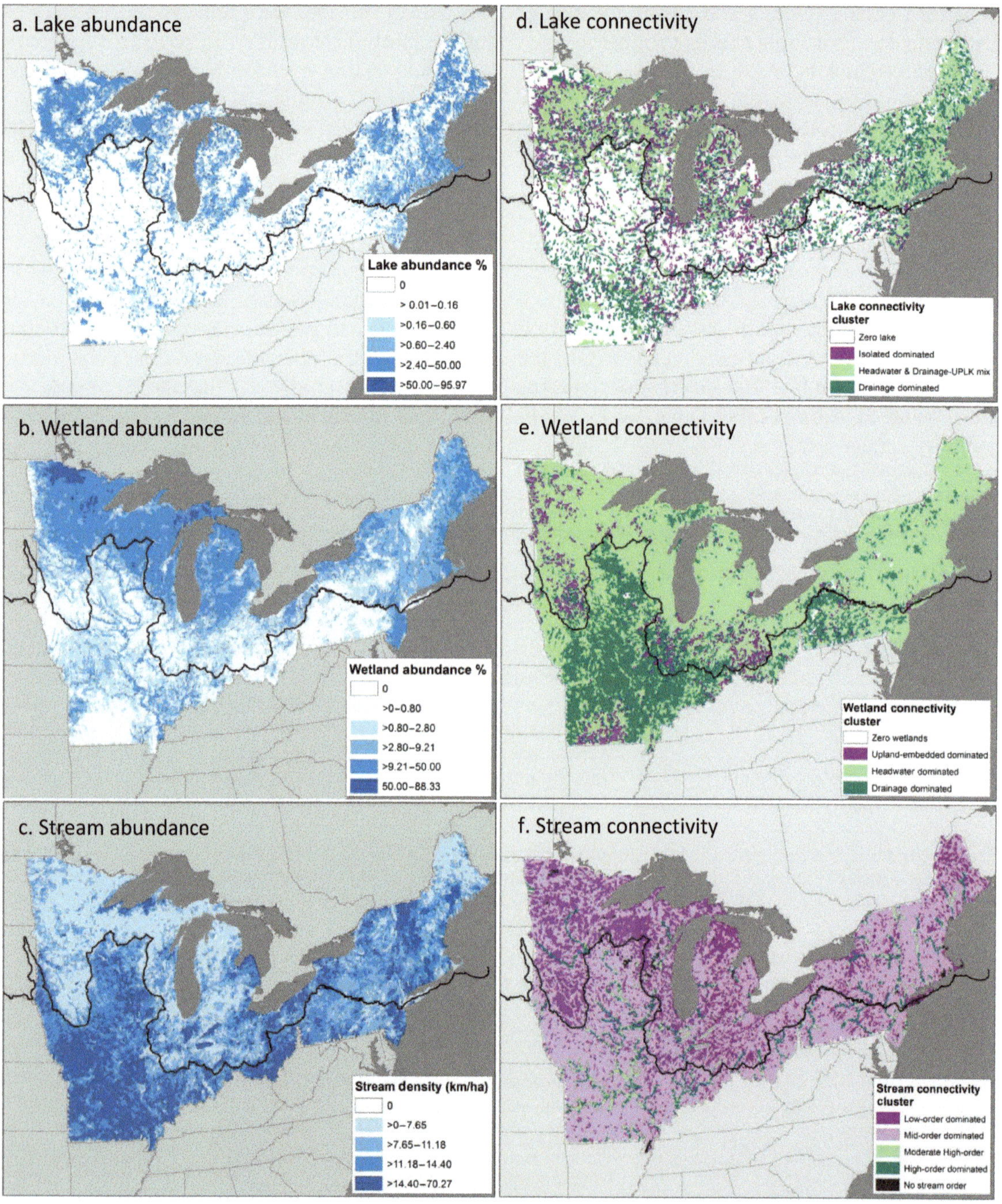

Fig. 3. Freshwater abundance and connectivity maps by system type. Freshwater abundance is quantified as the total proportion area or stream density within the Hydrologic Unit (HU) 12 spatial unit for lakes (a), wetlands (b), and streams (c). Abundance values are binned as quantiles. Freshwater connectivity for lakes (d), wetlands (e), and streams (f) is represented by connectivity cluster groups determined by *k*-means cluster group using principal components analysis (PCA) scores from lake, wetland, and stream connectivity metrics at the HU12 spatial scale. Dominated is in reference to where spatial units plotted on the PCA axes using relative proportion connectivity metric values. The solid black line represents the estimated boundary of the Wisconsin glacial period—north of the line is glaciated area and south of the line is unglaciated area.

the northern area of the study extent compared to the southern areas, whereas stream density was higher in the southern area of the study extent compared to northern areas. However, there were exceptions to this general trend with glaciation extent. Lakes were not present and wetlands were in low abundance in parts of the Upper Midwest (Fig. 3a, b), and stream density in the Northeast did not show as strong a north–south trend (Fig. 3c).

Freshwater connectivity

Lake, wetland, and stream connectivity clusters assigned at the HU12 scale were derived separately from PCA scores. For lakes, two principal component axes explained ~70% of variation in lake connectivity metrics and resulted in three lake connectivity clusters. For wetlands, ~96% of variation in wetland connectivity metrics was explained by two PC axes, and PCA scores resulted in three wetland connectivity clusters. For streams, ~99% of variation in stream order density metrics was explained by two PC axes, and four stream connectivity clusters were retained (Appendix S1: Fig. S1).

Spatial patterns in freshwater connectivity clusters (Fig. 3d–f) differed from corresponding freshwater abundance patterns. Across the study extent, lake connectivity clusters exhibited non-contiguous spatial patterns (Fig. 3d). For example, in the Midwest where lake abundance was high, connectivity clusters included Headwater and Drainage-UPLK lake clusters and Isolated lake clusters. In contrast, wetland connectivity clusters were contiguous and similar to wetland abundance patterns (Fig. 3e). Clusters characterized by Headwater wetlands were the most prominent cluster group (66% of HU12s) and tended to be located in the northern extent of the study area where wetland abundance was high. Areas where wetland abundance was low tended to be dominated by Drainage or Upland-Embedded wetlands. For streams, Low-order streams were prominent in the Upper Midwest HU12s and tended to be in areas where stream density was high.

The freshwater landscape: integrating lake, wetland, and stream abundance and connectivity

We combined freshwater abundance and connectivity measures for lakes, wetlands, and streams that were quantified at the HU8 spatial unit (Table 1) to evaluate the integrated freshwater landscape. Due to the higher number of variables included and the weaker co-variation when different variable types are included, the first two axes of the PCA performed on all the abundance and connectivity metrics captured ~43% of variation in the freshwater metrics (Appendix S1: Fig. S2), lower than the variation captured in the individual HU12 connectivity analyses on lake, wetland, and stream systems (Appendix S1: Fig. S1). Nevertheless, the first two axes of the PCA performed on 15 different input variables captured nearly half of the variability, showing that there are consistent patterns across connectivity, abundance, and system types that can be summed into synthetic PCA axes. Thus, based on multivariate associations among these metrics (PCA scores), we identified five integrated freshwater cluster groups with distinct freshwater abundance and connectivity attributes (Table 2; Appendix S1: Fig. S2). There were significant

Table 2. Composition of integrated freshwater abundance and connectivity cluster groups at the HU8 scale.

				Connectivity									
	Abundance			Lake				Wetland			Stream order		
Cluster assignment	Lake	Wet	Stream	I	HW	DR	UPLK	U	HW	DR	Low	Mid	High
A	−	−	+			+				+	+		+
B	−	−	+			+				+			+
C	+	+	−	+	+			+	+			+	
D	+	+	−				+		+		+		
E	+	+	−				+	+			+		

Notes: Integrated freshwater cluster groups (A–E) are composed of different freshwater abundance and connectivity characteristics. The proportion of lake and wetland area and stream (density) in clusters ranges from high (+) and low (−) abundance. Freshwater connectivity types that are dominant in clusters are represented with (+). Freshwater connectivity types include lakes and wetlands: I, Isolated; U, Upland-Embedded; HW, Headwater; DR, Drainage; UPLK, Drainage-UPLK; and streams Low = first- to third-order stream segments; Mid = fourth- to sixth-order stream segments; and High = >sixth-order stream segments.

differences ($\alpha < 0.05$) in freshwater abundance and connectivity metrics among the clusters based on Kruskal–Wallis tests (Appendix S1: Table S4). However, not all cluster groups were significantly different from one another for a given metric (Appendix S1: Table S5). For example, Clusters C and D were not significantly different in lake, wetland, and stream abundance composition from one another, but they were significantly different from the lake, wetland, and stream abundance composition in Clusters A and B.

These integrated freshwater clusters exhibited macroscale spatial patterns that were largely distinguished by glaciation extent boundaries (Fig. 4). In general, we found that Clusters C, D, and E mostly located north of the glaciation boundary had significantly greater lake and wetland abundance compared to the other clusters and were composed of a variety of both isolated, upland-embedded, and stream-connected lake and wetland connectivity types, whereas Clusters A and B mostly located south of the glaciation boundary had significantly greater stream density compared to glaciated areas and were dominated by Drainage connectivity types for both lakes and wetlands (Figs. 4, 5; Appendix S1: Table S5).

The most common integrated freshwater cluster group in the study extent was Cluster C (149

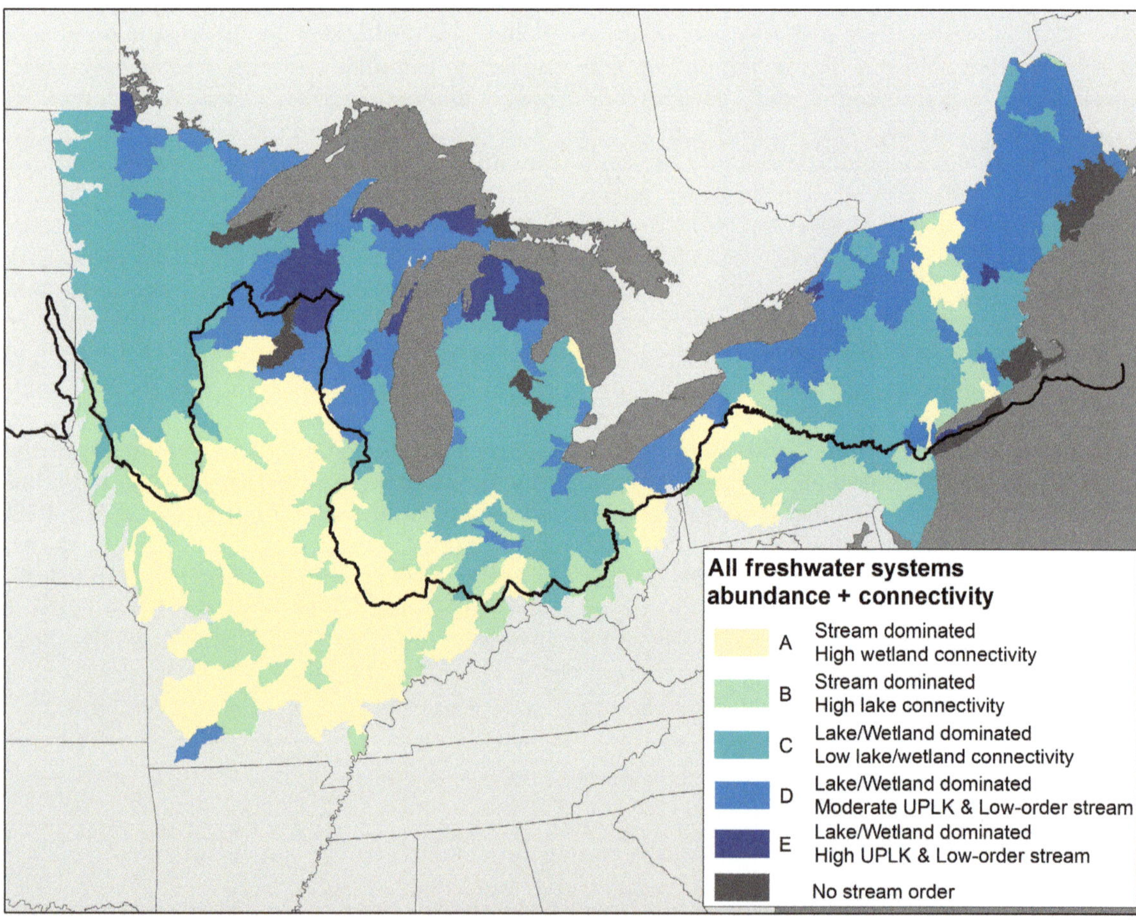

Fig. 4. Integrated freshwater abundance and connectivity map. Integrated freshwater abundance and connectivity clusters were assigned using *k*-means cluster analysis based on principal components analysis scores from lake, wetland, and stream abundance and connectivity metrics quantified at the Hydrologic Unit (HU) 8 scale (HU8 = 447). Spatial units not assigned a cluster group were missing stream order data for a portion of the area. Interpretation of cluster groups is provided in Table 1. The solid black line represents the estimated boundary of the Wisconsin glacial period—north of the line is glaciated area and south of the line is unglaciated area.

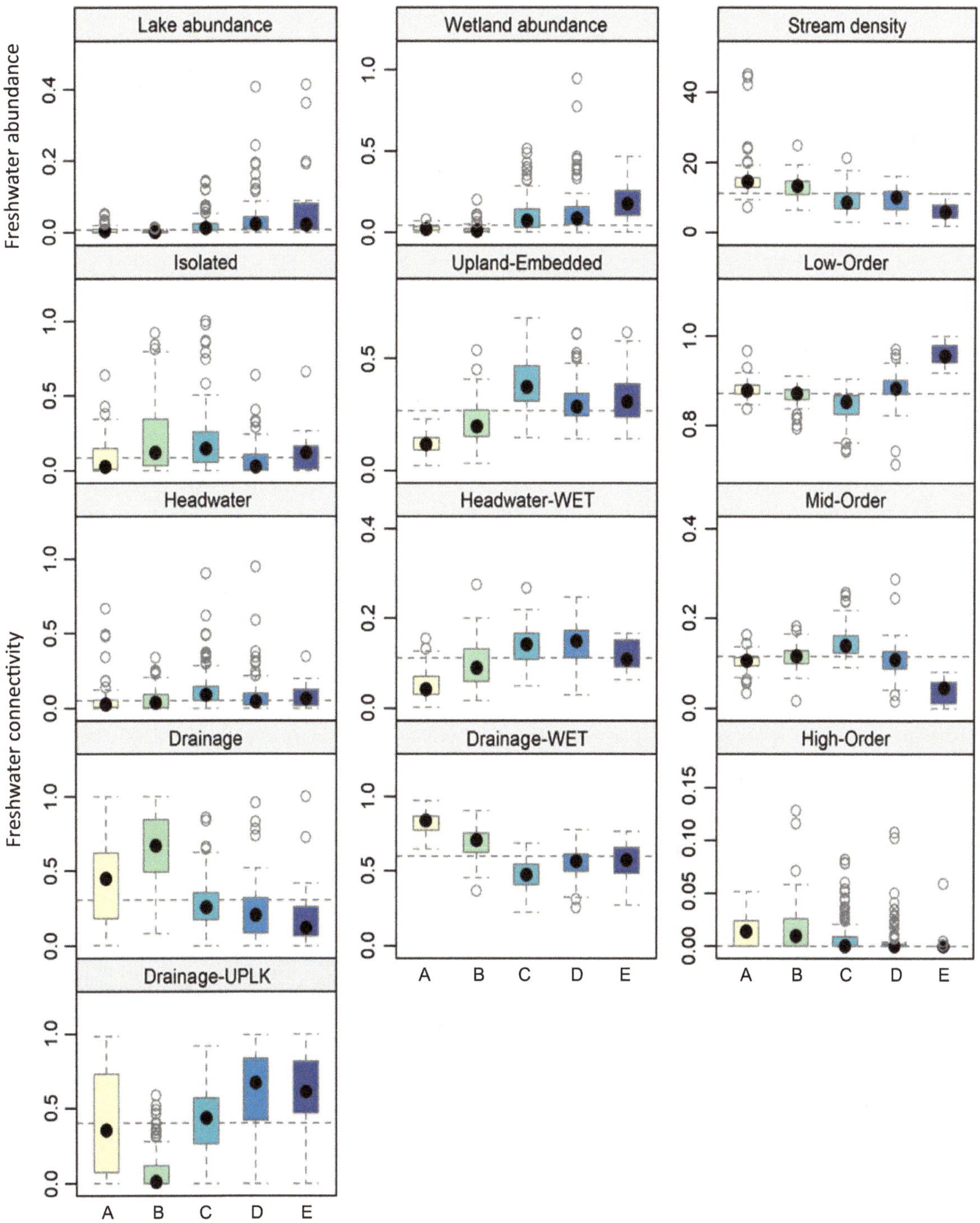

Fig. 5. Boxplots of freshwater abundance and connectivity metrics by the integrated freshwater cluster groups. Boxplots represent the distributions of lake and wetland abundance (proportion area in Hydrologic Unit [HU] 8 unit), stream density (m/km^2), and freshwater connectivity measures (relative proportion of freshwater connectivity type among respective freshwater system total area or density) among the integrated freshwater HU8 clusters (A–E). The dashed gray line represents the median metric value. Integrated freshwater clusters captured significant differences in freshwater abundance and connectivity metrics based on Kruskal–Wallis tests ($\alpha < 0.05$).

out of a total of 447 HU8s). This cluster group was distributed across the Upper Midwest and Northeast (Fig. 4) and was characterized as having significantly greater relative proportions of Isolated, Upland-Embedded, and Headwater lakes and wetlands and lower stream density that was dominated by Mid-order streams compared to the other cluster groups—except that Clusters C and B were not significantly different for Isolated lake metrics (Fig. 5; Appendix S1: Tables S4, S5). The least common integrated freshwater cluster group, Cluster E (23 out of 447 HU8s), was located primarily in the Upper Midwest. This cluster group was characterized as having significantly higher relative proportions of Low-order stream length compared to other cluster groups (Appendix S1: Table S5). However, Cluster E included HU8s with large variation in freshwater composition, and this cluster group did not significantly differ in lake and wetland connectivity metrics from other clusters (Appendix S1: Table S5).

The integrated freshwater landscape exhibited distinct, non-random spatial patterns at broad extents which provide a framework for investigating macroscale relationships that influence freshwater distributions and their connections with the surrounding landscape.

Hydrogeomorphic, climate, and land-use variables associated with broad-scale freshwater landscape attributes

Freshwater abundance patterns aligned with past glacial activity, but freshwater connectivity patterns did not completely follow this spatial gradient, suggesting that there are additional underlying drivers shaping the freshwater landscape. Of the 21 predictor variables tested in random forest models analyzing integrated freshwater abundance and connectivity clusters (HU8 scale), glaciation regime was the top predictor. In fact, there were significant differences in lake, wetland, and stream abundance metrics across glaciation regimes (Kruskal–Wallis tests: lake abundance, $H = 93.69$, df = 2, $P < 0.0001$; wetland abundance, $H = 133.91$, df = 2, $P < 0.0001$; stream density, $H = 138.79$, df = 2, $P < 0.0001$). But Isolated lake metrics were not significantly different among glaciation regimes ($H = 4.28$, df = 2, $P = 0.12$), and Low-order stream metrics were only marginally significantly different among glaciation regimes ($H = 8.50$, df = 2, $P = 0.01$). The associations suggest that past geological activity (i.e., glacial processes) is likely a key driver that affects the presence and abundance of freshwater systems, while specific hydrologic, geologic, and human land-use activity may influence freshwater connectivity within regions in different ways.

The top performing random forest model accurately predicted 57% of integrated freshwater cluster assignment (HU8 scale) based on out-of-bag samples, meaning that the predictor variables did better than random at predicting the overall cluster membership. The most important variables associated with the integrated clusters were glaciation regime, hydrology (mean runoff and baseflow), geology (glacial fluvial outwash and alluvial deposits), mean precipitation, and human land use (pasture and agriculture; Appendix S1: Fig. S3). These variables were significantly different among the five integrated cluster groups based on Kruskal–Wallis tests (Appendix S1: Table S6). Prediction accuracy varied among individual integrated freshwater clusters with Cluster A having the highest classification accuracy (65%) and Cluster B having the lowest (45%; Appendix S1: Table S7).

The integrated freshwater clusters differed in hydrogeomorphic features, climate, and land use (Fig. 6). In general, areas that were lake and wetland rich (Clusters C, D, and E) tended to have high runoff, baseflow, and glacial fluvial outwash geology. In contrast, areas that were rich in streams (Clusters A and B) tended to have high precipitation, topographic slope, alluvial geology, and pasture and agricultural land-use activities. The location of integrated freshwater clusters tended to vary with glaciation regime with Clusters A and B observed in higher frequency in unglaciated areas and Clusters C, D, and E observed in higher frequency in glaciated areas (Appendix S1: Fig. S4). Although glaciation regime co-varied with other geospatial predictors such as mean baseflow, mean runoff, percent agriculture, and mean slope based on Kruskal–Wallis tests (Appendix S1: Fig. S5), redundancy analyses in the variable selection step (VSURF package) retained these variables, indicating that they contained independent information from glaciation regime.

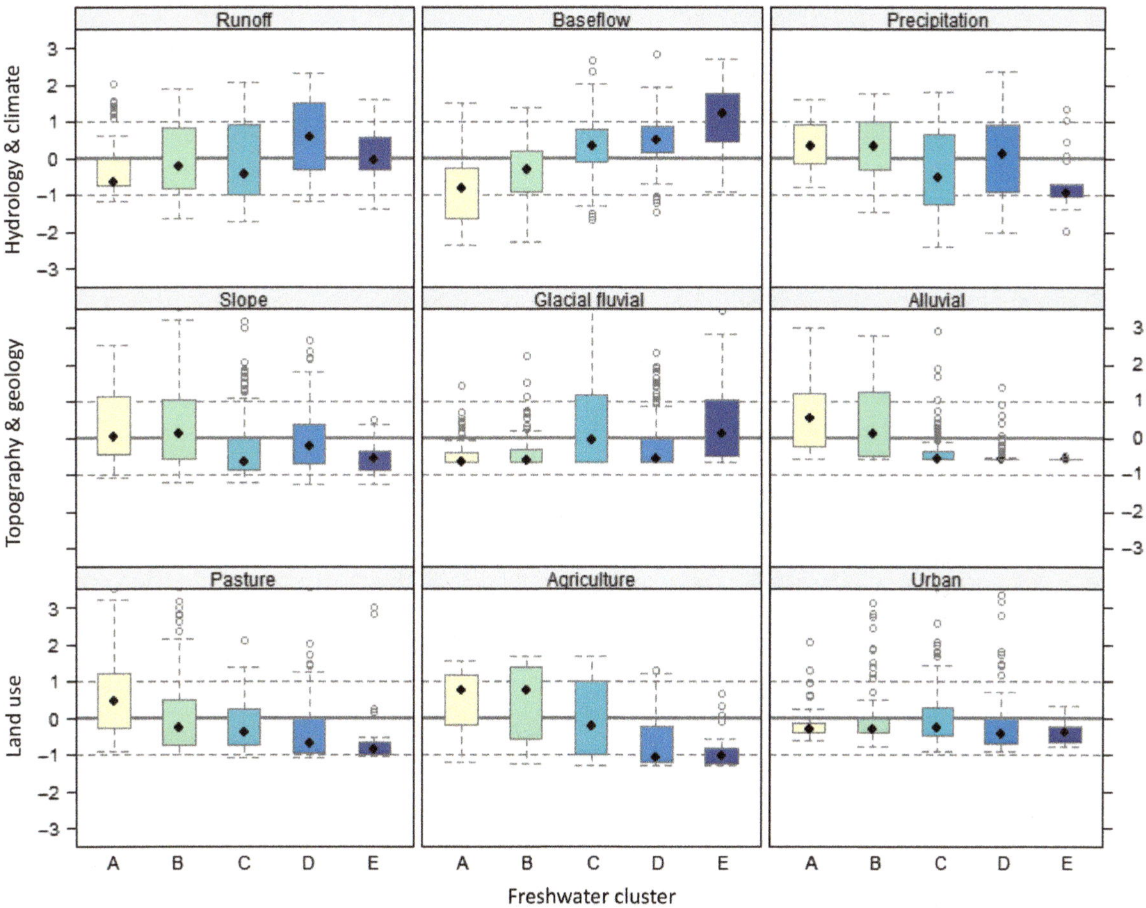

Fig. 6. Boxplots of the geophysical predictors of the integrated freshwater clusters. Boxplots represent the standardized distributions of geophysical variables among the integrated freshwater Hydrologic Unit 8 clusters (A–E). Geophysical variables were the top predictors of the integrated freshwater clusters in random forest analyses and are grouped as hydrology and climate, topography and geology, and land use. Values were standardized by subtracting the mean and dividing by the standard deviation. The solid gray line represents the mean, and the dashed lines represent ± 1 standard deviation above or below the mean. There were overall significant differences in hydrology, climate, topography, geology and land-use values among the clusters based on Kruskal–Wallis tests ($\alpha < 0.05$).

Discussion

We present a synthetic view of the integrated freshwater landscape. Our analyses of lakes, wetlands, and streams illustrate patterns and diversity of freshwater systems and surface connectivity at macroscales. We found four important results. (1) Abundance of lakes and wetlands exhibit spatial patterns that are opposite from stream density and that mostly follow glaciation extent boundaries. Lake and wetland abundance is higher in glaciated areas compared to unglaciated areas, whereas stream density is lower in glaciated areas compared to unglaciated areas. (2) We found distinct, broad-scale patterns among lake, wetland, and stream connectivity that reveal a layer of complexity that abundance measures alone did not capture, suggesting that freshwater abundance and connectivity may be influenced by different underlying processes. (3) There were spatially contiguous patterns in abundance and connectivity measures across all three freshwater system types (i.e., lakes, wetlands, and streams). (4) These patterns in abundance and connectivity were associated with underlying hydrogeomorphic, climate, and land-use variables. We have illustrated a robust approach to quantitatively measure freshwater abundance and connectivity that can be incorporated into models at macroscales. These results can inform both fundamental and applied scientific questions to better understand and manage freshwater systems.

The freshwater landscape: lake, wetland, and stream abundance and connectivity

The freshwater abundance and connectivity analyses presented here provide a broad-scale, integrated picture of the freshwater landscape. The patterns and distributions of the integrated freshwater landscape (Figs. 3, 4) can inform empirical and applied sciences alike. While it is common for freshwater systems to be studied in isolation within disciplinary boundaries, it is becoming widely recognized that systems need to be studied together and that their hydrologic connections are important to ecosystem attributes and processes. Our analyses further our understanding of the freshwater landscape by incorporating freshwater surface connections into a macroscale framework. By ignoring freshwater connectivity, we lose information on structural attributes of freshwater systems that can be critical to key ecological processes. These connections are important to advance understanding in biogeochemistry (Strayer et al. 2003, Fisher et al. 2004, Racchetti et al. 2010), aquatic population and community dynamics (Crump et al. 2007), and to more accurately incorporate freshwater systems in regional and global carbon and nutrient budgets (Cardille et al. 2007, Butman et al. 2016).

Freshwater abundance estimates at regional and global extents have been conducted on lakes, wetlands, and streams separately through inventories (Lehner and Döll 2004, McDonald et al. 2012, Verpoorter et al. 2014) or extrapolation methods using size-distribution scaling laws (Meybeck 1995, Downing et al. 2006, 2012). Our results were in agreement with the spatial patterns and range of values reported in the freshwater abundance literature. Mean lake areal percentages estimated at the HU12 spatial unit were 2.2% in glaciated areas and 0.5% in unglaciated areas in comparison with estimates for the entire United States described by Meybeck (1995; 2.8% in glaciated areas and 0.09% in unglaciated areas). Mean wetland areal percent estimated at the HU12 was 7.8%, slightly larger than total wetland area (including freshwater, estuarine, and marine systems) estimate for the conterminous United States of 5.5% (Dahl 2006); but our study extent did not include arid and mountainous regions, which would likely lower wetland abundance estimates. We also found that freshwater abundance patterns generally followed the Wisconsin glaciation boundary but with inverse patterns for lakes and wetlands compared to streams. Other studies support that lake abundance (for lakes ≥1 km^2) is greater in glaciated areas compared to unglaciated areas (Meybeck 1995, Lehner and Döll 2004), but distributions in relation to glaciation regime have not been explicitly examined for wetlands and streams.

In addition to abundance estimates, we observed general freshwater connectivity patterns that are supported by previous studies conducted at other scales. Isolated and Upland-Embedded lakes and wetlands tended to be smaller in size (median size 0.001 and 0.0001 km^2, respectively) compared to Drainage lakes and wetlands (median size 0.002 and 0.0004 km^2, respectively) in the spatial extent. In fact, there is an increasing probability of a wetland being geographically isolated from surface connections with decreasing patch size (Cohen et al. 2016). The association between freshwater connectivity characteristics and water body size suggests that freshwater connectivity types may follow size-frequency scaling distributions, similar to lake abundance trends (Downing et al. 2006), to some degree, which could aid in extrapolation methods to estimate connectivity at macroscales.

In addition, we found that Low-order (first–third) streams dominated total stream length (88%) in the study extent, and the proportion of stream length declined with increasing stream order with Mid-order streams making up 11% and High-order streams making 1% of the total stream length. These trends support global and regional stream density estimates that indicate that lower-order streams have the greatest number of stream segments and make up the most total stream length (Rodriguez-Iturbe et al. 1994, Butman and Raymond 2011, Downing et al. 2012). Understanding stream network patterns are critical to upscale freshwater processes to broad extents because stream network structure can influence biogeochemical processes (Fisher et al. 2004) such as CO_2 dynamics (Humborg et al. 2010).

Other studies have characterized hydrologic settings at broad spatial extents as we have done here, but they did not explicitly use freshwater surface features, nor connectivity characteristics as data inputs. Hydrologic regions delineated within the United States were based on well-understood topography, geology, and climate characteristics and were designed to identify similar hydrologic settings for surface waters and groundwaters in the landscape (Winter 2001, Wolock et al. 2004). The objective of these hydrologic regions was to identify combinations of topographic, geologic, and climatic conditions that interact to affect hydrologic systems in the landscape; the predictors best describing connectivity reflect those used to develop hydrologic regions. Our study differed from the deductive approach used by Wolock et al. (2004) in that we used an inductive approach based on freshwater metrics rather than land- and climate-based predictors to examine spatial patterns in connectivity. Our approach complements these other hydrologic studies in that we provide a framework to study the macroscale patterns of freshwater systems and their connectivity by directly using data on lakes, wetlands, and streams as inputs in the analyses. With this framework, we can visualize where freshwater systems and connectivity attributes are located on the landscape and how they may interact with other geospatial features to affect ecological attributes and processes.

Geospatial associations with the freshwater landscape

Freshwater composition and connectivity characteristics in the landscape are shaped by a suite of drivers that influence landform (past geologic activity, topographic relief), water source (climate and hydrology), permeability of substrate to hold water (surficial geology), and by human modifications (e.g., land-use activities) that can affect all of the above. Our analysis identified hydrogeomorphic, climate, and land-use variables that may be related to these different drivers. It is expected that several of these landscape features are spatially correlated with one another, which makes it challenging to identify individual drivers of freshwater abundance and connectivity patterns. However, the variables have unique information from one another, support hypothesized relationships from the literature, and provide insight into the broad-scale variables that affect the freshwater landscape. Glaciation regime was one of the top predictors of the integrated freshwater clusters. The most recent glacial period (Wisconsin stage) scoured the landscape to create depressions and sedimentary deposits, and glaciated areas are associated with increased lake and wetland abundance (Meybeck 1995, Winter 2000, Lehner and Döll 2004). With over half of our study extent being glaciated (59%), it was expected that glaciation regime would be significantly related to patterns in the freshwater landscape and associated with increased lake and wetland abundance.

However, glaciation was not an exclusive driver of the freshwater landscape. Within glaciated areas, there can be a great deal of variability in the abundance and types of freshwater bodies present due to differences in rock type and relief (Meybeck 1995, Winter 2000). In our study extent, surficial geology was important to predict freshwater clusters and indicated past geologic activity and current hydrologic conditions. High glacial fluvial outwash deposits were associated with lake- and wetland-rich areas. This substrate is composed of sediment, sand, and gravel deposits from past glacial activity and is associated with modern groundwater aquifers. Stream-rich areas had high alluvial deposits (clay, silt, sand, and gravel) associated with past geomorphic processes, and the permeability of these geologic materials can influence surface and subsurface water exchange (Winter 2000, Fisher et al. 2004) and shape the freshwater landscape. We also found that hydrologic variables were associated with the integrated freshwater clusters—with lake- and wetland-rich areas tending to have higher runoff and baseflow compared to stream-rich areas. These variables can be indicators of surface and groundwater sources to freshwater systems. Climate variables were also associated with the freshwater landscape: Mean precipitation was significantly higher in areas with high stream densities (Clusters A and B) compared to areas with low stream densities (Clusters C and E; Fig. 6; Appendix S1: Table S6). This trend was also observed at a global scale where a positive correlation between stream abundance and precipitation was observed (Raymond et al. 2013).

Finally, we found an association between the human land use in the HU8 and the integrated freshwater cluster groups. Stream-rich clusters were associated with higher proportions of mean agriculture compared to lake- and wetland-rich areas (Fig. 6; Appendix S1: Table S6). Agricultural activity can directly modify the freshwater landscape by diverting or extracting water, creating impoundments, and draining wetlands (Smith et al. 2002, Wright and Wimberly 2013) and may preferentially remove geographically isolated (e.g., Upland-Embedded) systems (Cohen et al. 2016, Rains et al. 2016). However, agriculture may be correlated with other variables such as topography and soil composition that may influence freshwater abundance and connectivity attributes.

The hydrogeomorphic, climate, and land-use variables identified in the analyses may be indicators of the diverse drivers that shape freshwater composition and connectivity characteristics across the landscape. At macroscales, past geological activity (i.e., glacial and fluvial processes) is a key broad-scale driver that affects the presence of freshwater systems with hydrologic, geologic, and human land-use activity influencing smaller-scale variability in freshwater characteristics including connectivity across regions. It should be noted that our objectives were to identify potential geographic variables that may be associated with freshwater abundance and connectivity characteristics. The associations do not imply causative relationships but rather highlight potential variables to examine in greater depth, at finer spatial scales in relation to the freshwater landscape, particularly in the face of global change in the coming decades and centuries.

Future needs for freshwater metric developments

Our freshwater connectivity metrics are simplified representations of the spatial configuration of freshwater systems in the landscape and thus do not capture all aspects of freshwater connectivity. In particular, data are lacking on groundwater, reservoirs, and dams at broad geographic extents. Also, the metrics are only as good as the resolution and accuracy of the original data layers, and our metrics do not capture temporal changes in freshwater connectivity. Surface hydrologic connectivity is a spatial characteristic, but the magnitude and presence of these connections are dynamic through time based on seasonal changes in climate and hydrology. This is especially true for upland-embedded wetlands and intermittent and ephemeral streams. These systems lack persistent surface water connections producing conditions that support unique biogeochemical and biological functions (Larned et al. 2010, Datry et al. 2014, Cohen et al. 2016, Rains et al. 2016). But isolated and ephemeral systems are being modified and lost at high rates due to land-use activities and climate change, and thus, there is a clear need to assess what is currently present (Larned et al. 2010, Rains et al. 2016). Our metrics are based on snapshots of surface water features represented in the NHD and NWI that are based on remotely sensed imagery, and thus lack information on connectivity changes over time. Temporal changes in connectivity could be assessed by incorporating land-cover data measured at other time periods (Pekel et al. 2016) or perhaps modeling hydrologic flow dynamics using topographic and other geographic data layers, but such analysis is beyond the scope of the current study.

Our freshwater connectivity metrics represent the complex surface hydrologic network that links lakes, wetlands, and streams and moderates the flow of water, materials, nutrients, and organisms across the landscape. These metrics are relatively easy to calculate using widely available geospatial data and can be applied at spatial scales aligned with disturbance assessment and management. This study is one of the first attempts at measuring freshwater connectivity at broad scales and we expect our ability to do so will improve in the future as the underlying data improve (e.g., spatial and temporal resolution) and as new methods are developed to measure abundance and connectivity.

Implications for an integrated freshwater perspective

While the objectives of our analyses were to examine the distribution and diversity of freshwater abundance and connectivity and how landscape setting was related to these patterns, our freshwater clusters provide a framework in which connectivity attributes can be integrated into macroscale studies. Freshwater connectivity characteristics can have important implications for management and conservation because they play an integral role in physical, chemical, and biological integrity of freshwater ecosystems (Pringle 2001, 2003). Because of this, connectivity characteristics have been used as criteria to determine which water bodies are given protection under the U.S. Clean Water Act (Leibowitz et al. 2008). These connections can sustain hydrologic and biogeochemical conditions and support metapopulation dynamics. But, they also can impair ecosystems by transporting nutrients, contaminants, and spreading non-native species (Pringle 2001, Rahel 2007), which can have broad-scale regional consequences to downstream receiving water bodies (Freeman et al. 2007). Our integrated freshwater landscape clusters capture differences in lake, wetland, and stream connectivity attributes across a macroextent, and these clusters may also be related to differences in freshwater response such as nutrient transport processes or aquatic community composition. For example, landscape clusters with high proportions of lakes connected to upstream lakes (i.e., lake chains) may show a stronger spatial coherence in indices of water quality to regional changes in the landscape, or may have different species richness and diversity characteristics compared to clusters with less connected lake systems. This information could inform conservation and management strategies conducted at broad spatial scales. However, there is a need to evaluate the ecological relevancy of the integrated clusters by applying them to macroscale studies.

Further, there is growing recognition that upland-embedded and non-perennial systems, not just connected systems, can affect ecological integrity of downstream waters (U.S. EPA 2015). These systems may be considered isolated from persistent surface waters but can be connected to hydrologic networks through subsurface or overland flow and thus provide many of the ecosystem services that are associated with connected systems (Cohen et al. 2016, Rains et al. 2016). Without characterization of the distribution of geographically isolated systems at macroscales, we have poor understanding of the regional landscape context in which these systems operate, which can impede conservation and management actions to protect these vulnerable systems.

A macroscale perspective is warranted to study the patterns of the freshwater landscape at scales that are aligned with some of the leading disturbances to freshwater systems. Broad-scale disturbances such as land-use conversion and climate change threaten the integrity and function of the freshwater landscape by altering freshwater structural attributes. Urbanization and agricultural land use physically alter the size, shape, and connectivity characteristics of freshwater systems through water extraction and diversion, stream channelization, impoundment and burial of headwater bodies, wetland drainage, and altered flow regimes through dam and reservoir construction (Zedler and Kercher 2005, Freeman et al. 2007, Vörösmarty et al. 2010, Carpenter et al. 2011, Steele and Heffernan 2014, Van Meter and Basu 2015). In addition, changes in temperature and precipitation patterns associated with climate change are likely to impact freshwater systems and connectivity characteristics. Because the integrated freshwater landscape framework we developed here accounts for spatial variation in connectivity as well as density, it has the potential to contribute toward improved assessment of ecosystem integrity from disturbances and can help inform appropriate management actions necessary to maintain ecosystem services provided by our lakes, wetlands, and streams.

Acknowledgments

This project was supported by the National Science Foundation MacroSystems Biology Program in the Emerging Frontiers Division of Biological Sciences Directorate (EF-1065786). We thank Ed Bissell, Nicole Smith, and Scott Stopyak for their work in developing the LAGOS-NE$_{GEO}$ database, the CSI-Limnology research group and Mary Bremigan for providing feedback throughout the development of this project, and the reviewers who provided insightful and constructive feedback. We thank Corinna Gries for helping upload data to the EDI depository and reviewing metadata documents. The manuscript idea was conceived of by CEF and refined and further developed through group meetings with the coauthors. CEF was the lead author drafting the majority of the text and creating tables, figures, and maps in the manuscript. All coauthors contributed to the conceptual development of the project, interpretation of results, and provided critical reviews and edits to the manuscript drafts. JFL substantially shaped the conceptual structure and approaches adopted in the study and performed PCA and *k*-means cluster analyses to quantify freshwater connectivity at landscape scales. SKO provided R code to perform random forest analyses and assisted with interpretation of model output. NKS assisted with wetland data acquisition, metadata development, and interpretation of wetland results. KSC provided constructive feedback and edits on conceptual structure of the manuscript. KEW contributed to the conceptual structure of the manuscript and assisted with developing R code for preprocessing steps and to create figures. PAS supervised and assisted in all stages of the development of the study by editing multiple drafts and providing constructive feedback on analyses and interpretation of results.

Literature Cited

Acuña, V., and K. Tockner. 2010. The effects of alterations in temperature and flow regime on organic carbon dynamics in Mediterranean river networks. Global Change Biology 16:2638–2650.

Ali, G. A., and A. G. Roy. 2009. Revisiting hydrologic sampling strategies for an accurate assessment of hydrologic connectivity in humid temperate systems. Geography Compass 3:350–374.

Aselmann, I., and P. J. Crutzen. 1989. Global distribution of natural freshwater wetlands and rice paddies, their net primary productivity, seasonality and possible methane emissions. Journal of Atmospheric Chemistry 8:307–358.

Bouvier, L. D., K. Cottenie, and S. E. Doka. 2009. Aquatic connectivity and fish metacommunities in wetlands of the lower Great Lakes. Canadian Journal of Fisheries and Aquatic Sciences 66:933–948.

Breiman, L. 2001. Random forests. Machine Learning 45:5–32.

Butman, D., and P. A. Raymond. 2011. Significant efflux of carbon dioxide from streams and rivers in the United States. Nature Geoscience 4:839–842.

Butman, D., S. Stackpoole, E. Stets, C. P. McDonald, D. W. Clow, and R. G. Striegl. 2016. Aquatic carbon cycling in the conterminous United States and implications for terrestrial carbon accounting. Proceedings of the National Academy of Sciences USA 113:58–63.

Cardille, J. A., S. R. Carpenter, M. T. Coe, J. A. Foley, P. C. Hanson, M. G. Turner, and J. A. Vano. 2007. Carbon and water cycling in lake-rich landscapes: landscape connections, lake hydrology, and biogeochemistry. Journal of Geophysical Research: Biogeosciences 112:G02031.

Carpenter, S. R., E. H. Stanley, and M. J. V. Zanden. 2011. State of the world's freshwater ecosystems: physical, chemical, and biological changes. Annual Review of Environment and Resources 36:75–99.

Cohen, M. J., et al. 2016. Do geographically isolated wetlands influence landscape functions? Proceedings of the National Academy of Sciences USA 1978–1986:201512650. https://doi.org/10.1073/pnas.1512650113

Cowardin, L. M., V. Carter, F. C. Golet, and E. T. LaRoe. 1979. Classification of wetlands and deepwater habitats of the United States. U.S. Department of the Interior, Fish and Wildlife Service, Washington, D.C., USA.

Crump, B. C., H. E. Adams, J. E. Hobbie, and G. W. Kling. 2007. Biogeography of bacterioplankton in lakes and streams of an Arctic tundra catchment. Ecology 88:1365–1378.

Cutler, D. R., T. C. Edwards, K. H. Beard, A. Cutler, K. T. Hess, J. Gibson, and J. J. Lawler. 2007. Random forests for classification in ecology. Ecology 88:2783–2792.

Dahl, T. E. 2006. Status and trends of wetlands in the conterminous United States 1998 to 2004. U.S. Department of the Interior; Fish and Wildlife Service, Washington, D.C., USA.

Datry, T., S. T. Larned, and K. Tockner. 2014. Intermittent rivers: a challenge for freshwater ecology. BioScience 64:bit027.

Detenbeck, N. E., C. A. Johnston, and G. J. Niemi. 1993. Wetland effects on lake water quality in the Minneapolis/St. Paul metropolitan area. Landscape Ecology 8:39–61.

Downing, J. A., J. J. Cole, C. A. Duarte, J. J. Middelburg, J. M. Melack, Y. T. Prairie, P. Kortelainen, R. G. Striegl, W. H. McDowell, and L. J. Tranvik. 2012. Global abundance and size distribution of streams and rivers. Inland Waters 2:229–236.

Downing, J. A., and C. M. Duarte. 2009. Abundance and size distribution of lakes, ponds, and impoundments. Elsevier, Oxford, UK.

Downing, J. A., et al. 2006. The global abundance and size distribution of lakes, ponds, and impoundments. Limnology and Oceanography 51:2388–2397.

Fergus, C., J. Lapierre, S. Oliver, N. Skaff, K. Cheruvelil, K. Webster, C. Scott, and P. Soranno. 2017*a*. Freshwater connectivity clusters for lakes, wetlands, and streams at the Hydrologic Unit 12 scale in the Midwest and Northeast U.S.A. – freshwater metric variables and k-means cluster assignment. Environmental Data Initiative. https://doi.org/10.6073/pasta/1dc5b024390539f1f7dda70470d35766

Fergus, C., J. Lapierre, S. Oliver, N. Skaff, K. Cheruvelil, K. Webster, C. Scott, and P. Soranno. 2017*b*. Integrated freshwater abundance and connectivity clusters at the Hydrologic Unit 8 scale for the Midwest and Northeast U.S.A. – freshwater metric variables and k-means cluster assignment. Environmental Data Initiative. https://doi.org/10.6073/pasta/e69b7495674e403baa19a22ffbfb40e1

Fisher, S. G., R. A. Sponseller, and J. B. Heffernan. 2004. Horizons in stream biogeochemistry: flowpaths to progress. Ecology 85:2369–2379.

Freeman, M. C., C. M. Pringle, and C. R. Jackson. 2007. Hydrologic connectivity and the contribution of stream headwaters to ecological integrity at regional scales. JAWRA Journal of the American Water Resources Association 43:5–14.

Fritz, K. M., E. Hagenbuch, E. D'Amico, M. Reif, P. J. Wigington, S. G. Leibowitz, R. L. Comeleo, J. L. Ebersole, and T.-L. Nadeau. 2013. Comparing the extent and permanence of headwater streams from two field surveys to values from hydrographic databases and maps. JAWRA Journal of the American Water Resources Association 49:867–882.

Genuer, R., J. M. Poggi, and C. Tuleau-Malot. 2016. VSURF: Variable selection using random forests. R package version 1.0.3. https://CRAN.R-project.org/package=VSURF

Heffernan, J. B., et al. 2014. Macrosystems ecology: understanding ecological patterns and processes at continental scales. Frontiers in Ecology and the Environment 12:5–14.

Humborg, C., C.-M. Mörth, M. Sundbom, H. Borg, T. Blenckner, R. Giesler, and V. Ittekkot. 2010. CO2 supersaturation along the aquatic conduit in Swedish watersheds as constrained by terrestrial respiration, aquatic respiration and weathering. Global Change Biology 16:1966–1978.

Jones, J. A. 2011. Hydrologic responses to climate change: considering geographic context and alternative hypotheses. Hydrological Processes 25: 1996–2000.

Jones, J. A., et al. 2012. Ecosystem processes and human influences regulate streamflow response to climate change at long-term ecological research sites. BioScience 62:390–404.

Kling, G. W., G. W. Kipphut, M. M. Miller, and W. J. O'Brien. 2000. Integration of lakes and streams in a landscape perspective: the importance of material processing on spatial patterns and temporal coherence. Freshwater Biology 43:477–497.

Kratz, T., K. Webster, C. Bowser, J. Maguson, and B. Benson. 1997. The influence of landscape position on lakes in northern Wisconsin. Freshwater Biology 37:209–217.

Larned, S. T., T. Datry, D. B. Arscott, and K. Tockner. 2010. Emerging concepts in temporary-river ecology. Freshwater Biology 55:717–738.

Lehner, B., and P. Döll. 2004. Development and validation of a global database of lakes, reservoirs and wetlands. Journal of Hydrology 296:1–22.

Leibowitz, S. G., P. J. Wigington, M. C. Rains, and D. M. Downing. 2008. Non-navigable streams and adjacent wetlands: addressing science needs following the Supreme Court's Rapanos decision. Frontiers in Ecology and the Environment 6:364–371.

Liaw, A., and M. Wiener. 2002. Classification and regression by randomForest. R News 2:18–22.

Lohse, K. A., P. D. Brooks, J. C. McIntosh, T. Meixner, and T. E. Huxman. 2009. Interactions between biogeochemistry and hydrologic systems. Annual Review of Environment and Resources 34:65–96.

Lottig, N. R., E. H. Stanley, P. C. Hanson, and T. K. Kratz. 2011. Comparison of regional stream and lake chemistry: differences, similarities, and potential drivers. Limnology and Oceanography 56: 1551–1562.

Martin, S. L., and P. A. Soranno. 2006. Lake landscape position: relationships to hydrologic connectivity and landscape features. Limnology and Oceanography 51:801–814.

McDonald, C. P., J. A. Rover, E. G. Stets, and R. G. Striegl. 2012. The regional abundance and size distribution of lakes and reservoirs in the United States and implications for estimates of global lake extent. Limnology and Oceanography 57:597–606.

Meybeck, M. 1995. Global distribution of lakes. Pages 1–35 *in* P. A. Lerman, P. D. D. M. Imboden, and P. J. R. Gat, editors. Physics and chemistry of lakes. Springer, Berlin, Germany; Heidelberg, Germany.

Moore, J. W. 2015. Bidirectional connectivity in rivers and implications for watershed stability and management. Canadian Journal of Fisheries and Aquatic Sciences 72:785–795.

Müller, R. A., M. N. Futter, S. Sobek, J. Nisell, K. Bishop, and G. A. Weyhenmeyer. 2013. Water renewal along the aquatic continuum offsets cumulative retention by lakes: implications for the character of organic carbon in boreal lakes. Aquatic Sciences 75:535–545.

Mushet, D. M., A. J. K. Calhoun, L. C. Alexander, M. J. Cohen, E. S. DeKeyser, L. Fowler, C. R. Lane, M. W. Lang, M. C. Rains, and S. C. Walls. 2015. Geographically isolated wetlands: rethinking a misnomer. Wetlands 35:423–431.

Nadeau, T.-L., and M. C. Rains. 2007. Hydrological connectivity between headwater streams and downstream waters: how science can inform policy. JAWRA Journal of the American Water Resources Association 43:118–133.

Nelson, C. E., S. Sadro, and J. M. Melack. 2009. Contrasting the influences of stream inputs and landscape position on bacterioplankton community structure and dissolved organic matter composition in high-elevation lake chains. Limnology and Oceanography 54:1292–1305.

Oksanen, J., et al. 2017. vegan: community ecology package. https://CRAN.R-project.org/package=vegan

Pekel, J.-F., A. Cottam, N. Gorelick, and A. S. Belward. 2016. High-resolution mapping of global surface water and its long-term changes. Nature 540: 418–422.

Pringle, C. M. 2001. Hydrologic connectivity and the management of biological reserves: a global perspective. Ecological Applications 11:981–998.

Pringle, C. 2003. What is hydrologic connectivity and why is it ecologically important? Hydrological Processes 17:2685–2689.

Quinlan, R., A. M. Paterson, R. I. Hall, P. J. Dillon, A. N. Wilkinson, B. F. Cumming, M. S. V. Douglas, and J. P. Smol. 2003. A landscape approach to examining spatial patterns of limnological variables and long-term environmental change in a southern Canadian lake district. Freshwater Biology 48:1676–1697.

Racchetti, E., M. Bartoli, E. Soana, D. Longhi, R. R. Christian, M. Pinardi, and P. Viaroli. 2010. Influence of hydrological connectivity of riverine wetlands on nitrogen removal via denitrification. Biogeochemistry 103:335–354.

Rahel, F. J. 2007. Biogeographic barriers, connectivity and homogenization of freshwater faunas: It's a small world after all. Freshwater Biology 52:696–710.

Rains, M. C., S. G. Leibowitz, M. J. Cohen, I. F. Creed, H. E. Golden, J. W. Jawitz, P. Kalla, C. R. Lane, M. W. Lang, and D. L. McLaughlin. 2016. Geographically isolated wetlands are part of the hydrological landscape. Hydrological Processes 30:153–160.

Raymond, P. A., et al. 2013. Global carbon dioxide emissions from inland waters. Nature 503:355–359.

Read, E. K., et al. 2015. The importance of lake-specific characteristics for water quality across the continental United States. Ecological Applications 25: 943–955.

Riera, J. L., J. J. Magnuson, T. K. Kratz, and K. E. Webster. 2000. A geomorphic template for the analysis of lake districts applied to the Northern Highland Lake District, Wisconsin, U.S.A. Freshwater Biology 43:301–318.

Rodriguez-Iturbe, I., M. Marani, R. Rigon, and A. Rinaldo. 1994. Self-organized river basin landscapes: fractal and multifractal characteristics. Water Resources Research 30:3531–3539.

Sadro, S., C. E. Nelson, and J. M. Melack. 2011. The influence of landscape position and catchment characteristics on aquatic biogeochemistry in high-elevation lake-chains. Ecosystems 15:363–386.

Seaber, P. R., F. P. Kapinos, and G. L. Knapp. 1987. Hydrologic unit maps. U.S. Government Printing Office, Washington, D.C., USA.

Smith, S. V., W. H. Renwick, J. D. Bartley, and R. W. Buddemeier. 2002. Distribution and significance of small, artificial water bodies across the United States landscape. Science of the Total Environment 299:21–36.

Soranno, P. A., K. S. Cheruvelil, T. Wagner, K. E. Webster, and M. T. Bremigan. 2015*a*. Effects of land use on lake nutrients: the importance of scale, hydrologic connectivity, and region. PLoS ONE 10:e0135454.

Soranno, P. A., et al. 1999. Spatial variation among lakes within landscapes: ecological organization along lake chains. Ecosystems 2:395–410.

Soranno, P. A., et al. 2015*b*. Building a multi-scaled geospatial temporal ecology database from disparate data sources: fostering open science and data reuse. GigaScience 4:28.

Steele, M. K., and J. B. Heffernan. 2014. Morphological characteristics of urban water bodies: mechanisms of change and implications for ecosystem function. Ecological Applications 24:1070–1084.

Strahler, A. 1957. Quantitative analysis of watershed geomorphology. Transactions of the American Geophysical Union 8:913–920.

Strayer, D. L., R. E. Beighley, L. C. Thompson, S. Brooks, C. Nilsson, G. Pinay, and R. J. Naiman. 2003. Effects of land cover on stream ecosystems: roles of empirical models and scaling issues. Ecosystems 6:407–423.

Tiner, R. W. 2003. Geographically isolated wetlands of the United States. Wetlands 23:494–516.

U.S. EPA. 2015. Connectivity of Streams and Wetlands to Downstream Waters: A Review and Synthesis of the Scientific Evidence. Final Report. U.S. Environmental Protection Agency, Washington, D.C., USA.

Van Meter, K. J., and N. B. Basu. 2015. Signatures of human impact: size distributions and spatial organization of wetlands in the Prairie Pothole landscape. Ecological Applications 25:451–465.

Verpoorter, C., T. Kutser, D. A. Seekell, and L. J. Tranvik. 2014. A global inventory of lakes based on high-resolution satellite imagery. Geophysical Research Letters 41:2014GL060641.

Vörösmarty, C. J., et al. 2010. Global threats to human water security and river biodiversity. Nature 467:555–561.

Webster, K. E., P. A. Soranno, K. S. Cheruvelil, M. T. Bremigan, J. A. Downing, P. D. Vaux, T. R. Asplund, L. C. Bacon, and J. Connor. 2008. An empirical evaluation of the nutrient-color paradigm for lakes. Limnology and Oceanography 53:1137–1148.

Weller, C. M., M. C. Watzin, and D. Wang. 1996. Role of wetlands in reducing phosphorus loading to surface water in eight watersheds in the Lake

Champlain Basin. Environmental Management 20: 731–739.
Winter, T. C. 2000. The vulnerability of wetlands to climate change: a hydrologic landscape perspective. JAWRA Journal of the American Water Resources Association 36:305–311.
Winter, T. C. 2001. The concept of hydrologic landscapes. JAWRA Journal of the American Water Resources Association 37:335–349.
Wolock, D. M., T. C. Winter, and G. McMahon. 2004. Delineation and evaluation of hydrologic-landscape regions in the United States using geographic information system tools and multivariate statistical analyses. Environmental Management 34:S71–S88.
Wright, C. K., and M. C. Wimberly. 2013. Recent land use change in the Western Corn Belt threatens grasslands and wetlands. Proceedings of the National Academy of Sciences USA 110:4134–4139.
Zedler, J. B., and S. Kercher. 2005. Wetland resources: status, trends, ecosystem services, and restorability. Annual Review of Environment and Resources 30:39–74.

12

Linear and nonlinear effects of temperature and precipitation on ecosystem properties in tidal saline wetlands

LAURA C. FEHER,[1,†] MICHAEL J. OSLAND,[1] KEREEN T. GRIFFITH,[2] JAMES B. GRACE,[1] REBECCA J. HOWARD,[1] CAMILLE L. STAGG,[1] NICHOLAS M. ENWRIGHT,[1] KEN W. KRAUSS,[1] CHRISTOPHER A. GABLER,[3] RICHARD H. DAY,[1] AND KERRYLEE ROGERS[4]

[1]*U.S. Geological Survey, Wetland and Aquatic Research Center, Lafayette, Louisiana 70506 USA*
[2]*Griffith Consulting Services at U.S. Geological Survey, Lafayette, Louisiana 70506 USA*
[3]*School of Earth, Environmental, and Marine Sciences, University of Texas Rio Grande Valley, Brownsville, Texas 78520 USA*
[4]*School of Earth and Environmental Sciences, University of Wollongong, Wollongong, New South Wales 2522 Australia*

Abstract. Climate greatly influences the structure and functioning of tidal saline wetland ecosystems. However, there is a need to better quantify the effects of climatic drivers on ecosystem properties, particularly near climate-sensitive ecological transition zones. Here, we used climate- and literature-derived ecological data from tidal saline wetlands to test hypotheses regarding the influence of climatic drivers (i.e., temperature and precipitation regimes) on the following six ecosystem properties: canopy height, biomass, productivity, decomposition, soil carbon density, and soil carbon accumulation. Our analyses quantify and elucidate linear and nonlinear effects of climatic drivers. We quantified positive linear relationships between temperature and above-ground productivity and strong positive nonlinear (sigmoidal) relationships between (1) temperature and above-ground biomass and canopy height and (2) precipitation and canopy height. Near temperature-controlled mangrove range limits, small changes in temperature are expected to trigger comparatively large changes in biomass and canopy height, as mangrove forests grow, expand, and, in some cases, replace salt marshes. However, within these same transition zones, temperature-induced changes in productivity are expected to be comparatively small. Interestingly, despite the significant above-ground height, biomass, and productivity relationships across the tropical–temperate mangrove–marsh transition zone, the relationships between temperature and soil carbon density or soil carbon accumulation were not significant. Our literature review identifies several ecosystem properties and many regions of the world for which there are insufficient data to fully evaluate the influence of climatic drivers, and the identified data gaps can be used by scientists to guide future research. Our analyses indicate that near precipitation-controlled transition zones, small changes in precipitation are expected to trigger comparatively large changes in canopy height. However, there are scant data to evaluate the influence of precipitation on other ecosystem properties. There is a need for more decomposition data across climatic gradients, and to advance understanding of the influence of changes in precipitation and freshwater availability, additional ecological data are needed from tidal saline wetlands in arid climates. Collectively, our results can help scientists and managers better anticipate the linear and nonlinear ecological consequences of climate change for coastal wetlands.

Key words: biomass; canopy height; climate change; coastal wetland; mangrove forest; nonlinear; precipitation; productivity; salt marsh; soil carbon; temperature; threshold.

† **E-mail:** lfeher@usgs.gov

Introduction

Across the globe, climatic drivers (i.e., temperature and precipitation regimes) greatly influence the structure and functioning of ecosystems (Jenny 1941, Holdridge 1967, Whittaker 1970, Woodward 1987, Jobbágy and Jackson 2000, Chapin et al. 2011). Within the context of climate change, understanding the ecological influence of climatic drivers is important, because it provides a foundation for anticipating and preparing for the effects of future change (Glick et al. 2011). Knowledge of ecological thresholds, nonlinear relationships, and climate-sensitive ecosystem properties can be particularly useful for predicting where and when small changes in temperature or precipitation regimes may trigger comparatively large and abrupt changes in ecosystem structure and function (i.e., regime shifts sensu Scheffer et al. 2001). Here, we used a combination of climate- and literature-derived ecological data to investigate the influence of temperature and precipitation regimes upon ecosystem structure and function in tidal saline wetlands.

Mangrove forests, salt marshes, and tidal salt flats are all tidal saline wetland ecosystems. Despite vegetation differences, all three of these ecosystem types occupy similar hydrogeomorphic positions within the landscape (i.e., sheltered tidally inundated saline wetland habitats). In addition to providing fish and wildlife habitat, these highly productive ecosystems improve water quality, protect coastlines, store carbon, and provide recreational opportunities (Barbier et al. 2011). Hence, the conservation and maintenance of these systems in the face of climate change are a high priority for coastal environmental managers (Alongi 2008, McKee et al. 2012, Kirwan and Megonigal 2013).

Climatic drivers govern the distribution and abundance of mangrove forests, salt marshes, and salt flats (Davis 1940, West 1977, Twilley 1982, Semeniuk 1983, Woodroffe and Grindrod 1991, Saenger 2002). Mangroves are woody plants that are sensitive to freezing and chilling temperatures; hence, mangrove forests are most common in the tropics and subtropics (Tomlinson 1986, Ross et al. 2001, Stuart et al. 2007). In contrast, salt marshes are freeze tolerant and most common in temperate, sub-arctic, and arctic climes (Pennings and Bertness 2001, Ibáñez et al. 2012). Salt flats are hypersaline ecosystems that lack vascular plants and are most abundant along arid and semi-arid coasts (Zedler 1983, Adam 1990, Semeniuk 2013).

Climate-focused research on tidal saline wetlands has historically treated mangrove forests, salt marshes, and salt flats as different systems that occur independently. Mangrove ecologists have often examined climatic controls on ecosystem structure and function solely within mangrove forests, salt marsh ecologists have typically focused solely on salt marshes, and very little work has been conducted in salt flats partly due to lack of data (Table 1). Within a given estuary, local abiotic and biotic factors (e.g., hydrology, geomorphology, salinity, competition, facilitation) greatly influence the distribution and abundance of salt marshes, mangrove forests, and salt flats (Lugo and Snedaker 1974, Pennings and Callaway 1992, Roy et al. 2001, Krauss et al. 2008, Friess et al. 2012, Jones et al. 2016*b*, Gabler et al. 2017, Hayes et al. 2017); however, climatic factors greatly influence those local abiotic and biotic conditions. Hence, at the global scale, climatic factors affect the relative distribution and abundance of these different kinds of tidal saline wetland ecosystems. Several studies have used regional- or global-scale remotely sensed mangrove, salt marsh, and/or salt flat data to examine the influence of climate on plant functional group abundance (Deegan et al. 1986, Bucher and Saenger 1994, Longley 1994, Montagna et al. 2011, Osland et al. 2013, 2014*b*, 2017*a*, Cavanaugh et al. 2014, 2015, Giri and Long 2016, Rodriguez et al. 2016). However, to our knowledge, the work of Chmura et al. (2003), Gabler et al. (2017), Ouyang et al. (2017), and Simpson et al. (2017) are the only regional- or global-scale studies that have used field-based data from both mangrove forests and salt marshes to evaluate the influence of climatic drivers on a particular coastal wetland ecosystem property. These studies show that there is much to gain from examining climatic controls on these tidal saline wetland ecosystems together, as they occupy the same broad hydrogeomorphic niche (i.e., the tidal saline wetland niche) across wide regional climatic gradients; studies that just examine one system (e.g., mangrove forest or salt marsh) only examine a portion of the tidal saline wetland niche. Ecological regime shifts and oscillations are common in tidal saline wetlands, and

Table 1. Literature used to develop and test hypotheses regarding the effects of climatic drivers on ecosystem properties in tidal saline wetlands.

Property	Literature sources†,‡
Canopy height	Lot-Helgueras et al. (1975)[m,3], Anderson and Treshow (1980)[s,1], Saenger and Snedaker (1993)[m,4], Morrisey et al. (2007)[m,2], Komiyama et al. (2008)[m,4], Méndez-Alonzo et al. (2008)[m,3], McCall and Pennings (2012)[s,1,3], Soares et al. (2012)[m,4], Lovelock et al. (2016)[m,4], Rovai et al. (2016)[m,1,3], Yando et al. (2016)[s,m,3], Gabler et al. (2017)[s,m,3], Guo et al. (2017)[s,m,3], Simpson et al. (2017)[s,m,1]
Above-ground biomass	Turner (1976)[s,1], Twilley et al. (1992)[m,4], Saenger and Snedaker (1993)[m,4], Bucher and Saenger (1994)[m,2], Saenger (2002)[m,1,2], Morrisey et al. (2007, 2010)[m,2], Komiyama et al. (2008)[m,4], Alongi (2009)[m,4], Kirwan et al. (2009)[s,1], Twilley and Day (2012)[m,4], Asbridge et al. (2015)[m,2], Lovelock et al. (2016)[m,4], Rovai et al. (2016)[m,1,3], Sanders et al. (2016)[m,2], Yando et al. (2016)[s,m,3], Gabler et al. (2017)[s,m,3], Schile et al. (2017)[s,m,4]
Above-ground primary productivity	Turner (1976)[s,1], Twilley et al. (1992)[m,4], Saenger and Snedaker (1993)[m,4], Saenger (2002)[m,1,2], Komiyama et al. (2008)[m,4], Alongi (2009)[m,4], González Trilla et al. (2009)[s,4], Kirwan et al. (2009)[s,1], Morrisey et al. (2007, 2010)[m,2], Twilley and Day (2012)[m,4], Day et al. (2013)[s,1], Osland et al. (2014*a*, *b*)[s,m,3], Asbridge et al. (2015)[m,2], Lovelock et al. (2016)[m,4]
Decomposition	Kirwan and Blum (2011)[s,1], Kirwan et al. (2014)[s,1], Crosby et al. (2017)[s,1], Ouyang et al. (2017)[s,m,1]
Soil carbon density and accumulation	Chmura et al. (2003)[s,m,4], Craft (2007)[s,1], Kristensen et al. (2008)[m,4], Perry and Mendelssohn (2009)[s,m,3], Comeaux et al. (2012)[s,m,3], Bianchi et al. (2013)[s,m,3], Henry and Twilley (2013)[s,m,3], Jardine and Siikamäki (2014)[m,4], Ouyang and Lee (2014)[s,4], Doughty et al. (2016)[m,1], Kelleway et al. (2016)[s,m,2], Sanders et al. (2016)[m,2], Yando et al. (2016)[s,m,3], Guo et al. (2017)[s,m,3], Schile et al. (2017)[s,m,4]

† Habitat: s, salt marsh; m, mangrove.
‡ Transition zone: 1, eastern North America; 2, eastern Australia; 3, western Gulf of Mexico; 4, global.

the identification of global- and regional-scale ecological thresholds, transition zones, and climate–ecological linkages can be achieved via holistic analyses that span the entire tidal saline wetland niche and include ecological data from mangrove forests, salt marshes, and salt flats (Osland et al. 2016, Gabler et al. 2017).

In this study, our overarching aim was to quantify the influence of temperature and precipitation regimes upon the following six ecosystem properties in tidal saline wetlands: canopy height, above-ground biomass, above-ground productivity, decomposition, soil carbon density, and soil carbon accumulation. These properties greatly influence the many ecosystem goods and services that wetlands provide including storm protection, erosion prevention, nutrient cycling, trophic linkages, fish and wildlife habitat, and carbon storage (Barbier et al. 2011, Kelleway et al. 2017). Due in part to high productivity and anaerobic conditions that limit decomposition, tidal saline wetlands have the potential for high rates of soil carbon storage; soil carbon stocks and carbon accumulation rates in tidal saline wetlands are often higher than their terrestrial counterparts (Chmura et al. 2003, Donato et al. 2011, McLeod et al. 2011, Atwood et al. 2017, Macreadie et al. 2017). Hence, carbon cycling and storage have been widely studied and emphasized in the tidal saline wetland literature, and data for the six selected ecosystem properties are available and consolidated in reviews (Table 1). Our objectives were to use literature-derived data from all tidal saline wetland types to determine the influence of climatic drivers (i.e., temperature and precipitation) on the selected ecosystem properties. Based upon a review of the coastal wetland and terrestrial ecological literature (Table 1), we developed hypotheses regarding the influence of temperature and precipitation upon each of the six targeted ecosystem properties (Fig. 1). Given space constraints in this introduction, our description of and rationale for each of the selected hypotheses is explained in detail in tandem with the results in the discussion section of this manuscript. For each of the hypothesized relationships, we sought regional-scale literature-derived ecological data that could be used to quantify climate–ecological linkages. These literature-derived data were obtained from prior reviews as well as individual studies that examined global patterns of ecosystem structure and function in tidal saline wetlands (Table 1). Note that most of these reviews focus on salt marshes or mangrove forests separately (see symbols in Table 1), and Chmura et al. (2003) and Ouyang et al. (2017) are the only two reviews that have incorporated data from both mangrove forests and salt marshes. We

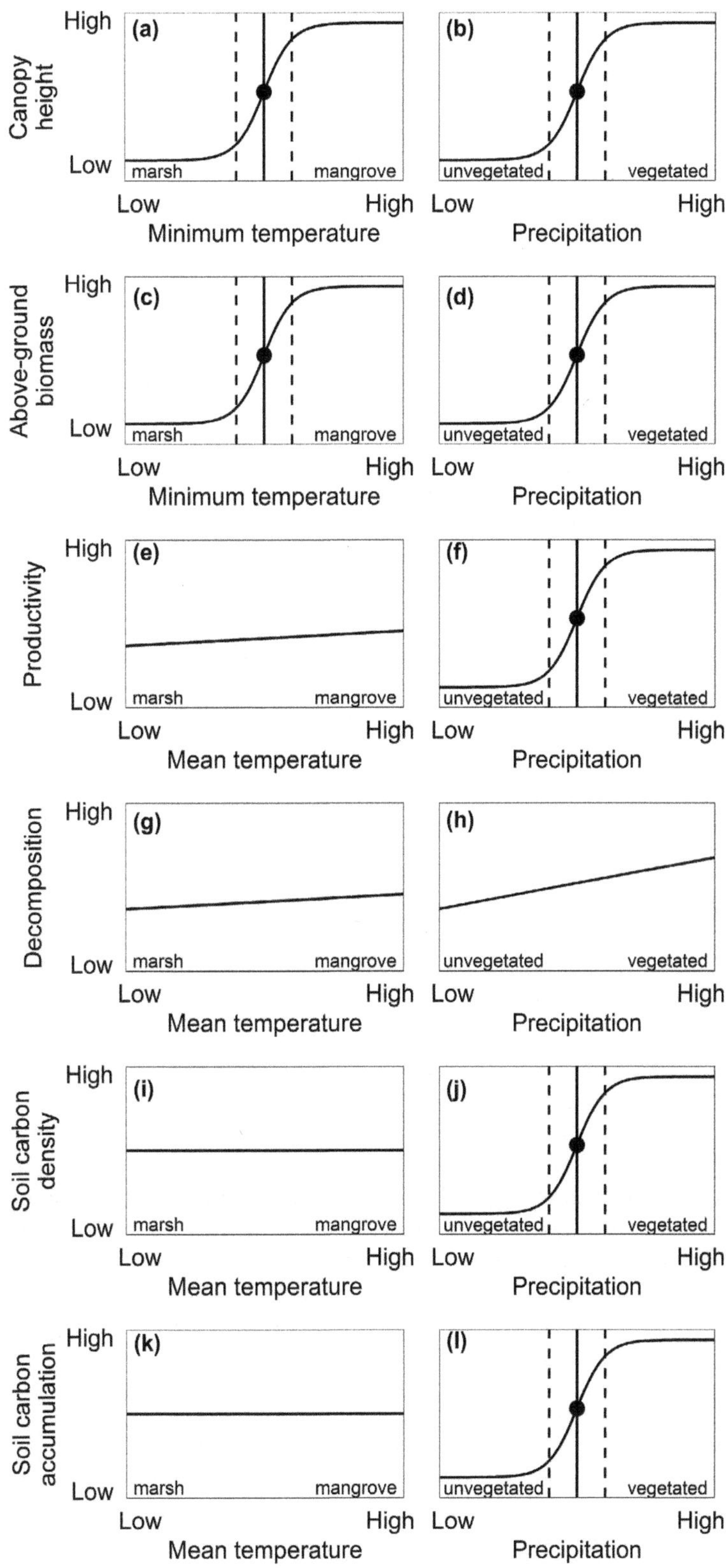

Fig. 1. Hypothesized relationships between climatic drivers and ecological properties of tidal saline wetlands:

(Fig. 1. *Continued*)
vegetation canopy height (a, b), above-ground biomass (c, d), above-ground productivity (e, f), decomposition (g, h), soil carbon density (i, j), and soil carbon accumulation (k, l). For sigmoidal relationships, the inflection point (black dot) represents a discrete threshold and the dashed lines represent a threshold zone. The rationale for these hypotheses is provided in the discussion section.

used our climate- and literature-derived data to address the following questions: (1) "What is the influence of temperature and precipitation on canopy height, above-ground biomass, above-ground primary productivity, decomposition, soil carbon density, and soil carbon accumulation in tidal saline wetlands?" (2) "Are there nonlinear relationships and temperature- and precipitation-based thresholds for these six ecological properties?" (3) "How do the identified climate–ecological relationships for tidal saline wetlands compare to those from terrestrial ecosystems?" and (4) "What data limitations affect our understanding of the influence of climatic drivers and climate change on ecosystem structure and function in tidal saline wetlands?"

Methods

Study area

Our study area initially encompassed the entire globe, and we attempted to obtain literature-derived ecological data within each of the 14 regional-scale temperature- and precipitation-controlled ecological transition zones described by Osland et al. (2016, 2017*b*; see also Saenger 2002, Spalding et al. 2010). These transitions include (1) areas where winter air temperature gradients control mangrove forest structure and/or produce a regional-scale mangrove forest-to-salt marsh transition and (2) areas where precipitation gradients produce a regional-scale vegetated-to-unvegetated transition (i.e., from mangrove forests and/or salt marshes to hypersaline salt flats without vascular plants). To quantify climate–ecological linkages within a transition zone, we sought ecological data that were spatially distributed across the relevant climatic gradients. We found, however, that adequate ecological data were only readily available for the following three regions of the world: eastern North America and the Caribbean Islands (hereafter called eastern North America), eastern Australia, and western Gulf of Mexico (Fig. 2). Therefore, our analyses focus solely on these three regions. Whereas the western Gulf of Mexico transition zone contains a precipitation-driven vegetated-to-unvegetated transition (Fig. 2b), the eastern North America and eastern Australia transition zones contain temperature-controlled salt marsh-to-mangrove transitions (Fig. 2c, d). We used climate data to identify the boundaries for each transition zone (Fig. 2b–d). In eastern North America and eastern Australia, the region boundaries spanned gradients in minimum air temperature from −40° to 20°C and −10° to 15°C, respectively. In the western Gulf of Mexico, the region boundaries spanned a mean annual precipitation gradient that ranged from 650 to 2000 mm. Within each region, we created a grid of cells with a resolution of 0.5° (i.e., ~55 km cell size at the equator). We converted all data (i.e., the gridded global climate data as well as the literature-derived ecological data) to this 0.5-degree resolution.

Climate data

We selected the following three climatic variables for our analyses: minimum air temperature, mean annual air temperature, and mean annual precipitation. Minimum air temperature was selected as a variable due to its critical role in controlling the distribution and abundance of mangrove forests relative to salt marshes (Woodroffe and Grindrod 1991, Osland et al. 2013, Cavanaugh et al. 2014, Gabler et al. 2017); freezing and/or chilling temperatures affect plant performance and result in mangrove damage or mortality (Markley et al. 1982, McMillan and Sherrod 1986, Pickens and Hester 2011, Osland et al. 2015, 2017*a*, Lovelock et al. 2016). Minimum air temperature is a particularly important variable for quantifying the ecological thresholds associated with the mangrove–marsh transition zone (Stuart et al. 2007, Osland et al. 2013, 2017*b*). Hence, we used minimum air temperature for our temperature-focused analyses of canopy height and above-ground biomass. Mean annual air temperature was also selected as a

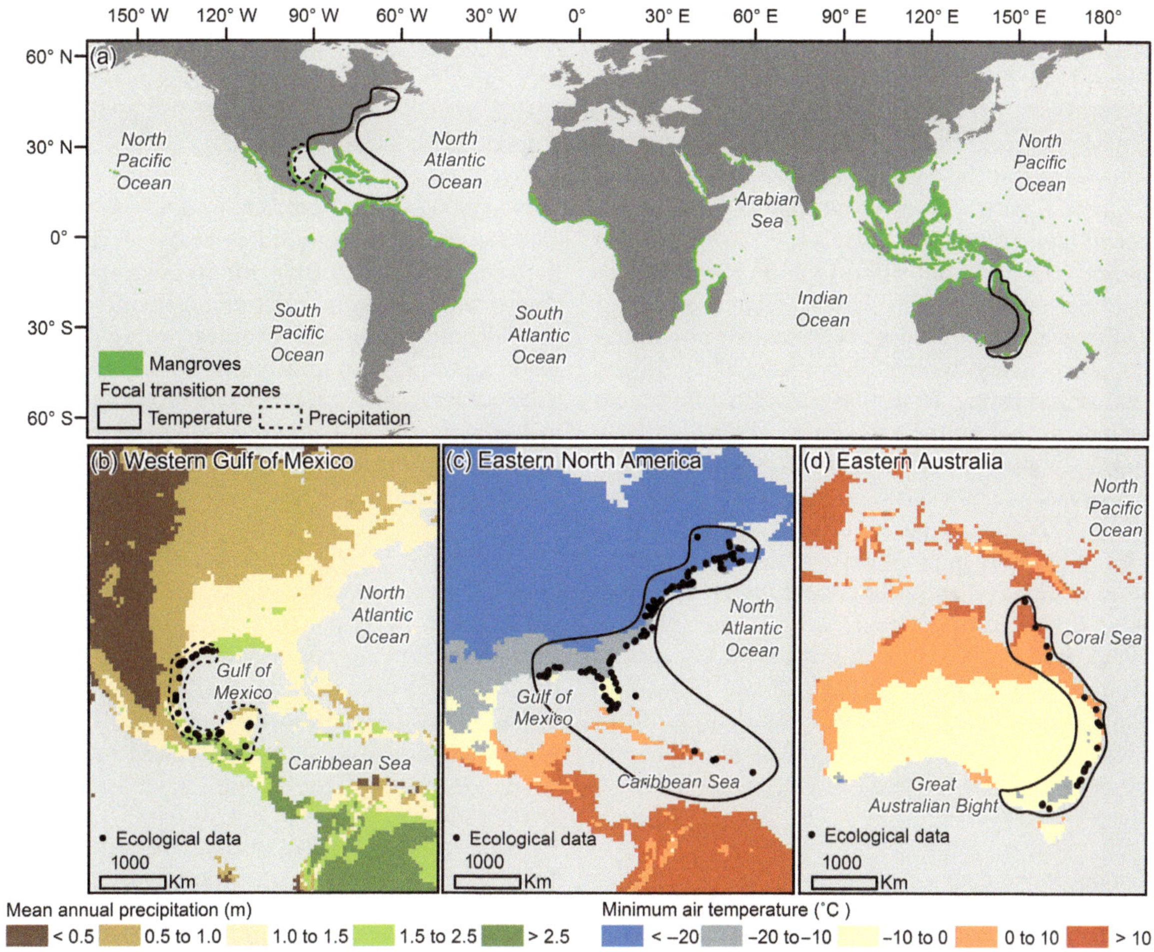

Fig. 2. Maps of global mangrove distribution (a) and mean annual precipitation (b) or minimum air temperatures (c, d) in the focal transition zones. The polygons denote the climatic transition zones evaluated in this study. Polygon line styles indicate the climatic driver hypothesized to be controlling tidal saline wetlands structure and function in each transition zone. Black dots in b–d indicate the locations of literature-derived ecological data used in our analyses.

variable due to its influence on plant and microbial physiology; mean annual temperature is directly correlated with the number of growing degree days (Appendix S1: Fig. S1) and has been shown to regulate primary productivity (Kirwan et al. 2009) and decomposition (Kirwan et al. 2014, Ouyang et al. 2017) in tidal saline wetlands. We used mean annual air temperature for our temperature-focused analyses of aboveground productivity, soil carbon density, and soil carbon accumulation. Mean annual precipitation was selected as a variable for consideration due to its critical role in controlling the abundance of tidal saline wetland plants (Bucher and Saenger 1994, Osland et al. 2014*b*, Gabler et al. 2017). Low precipitation can lead to hypersaline conditions as oceanic salts accumulate, which in turn can limit vegetation establishment and performance (Zedler 1983, Pennings and Bertness 1999, Semeniuk 2013, Reef and Lovelock 2014). We used mean annual precipitation for all of our precipitation-focused analyses.

We obtained air temperature and precipitation data from gridded climate datasets. For minimum air temperature ($T_{\min}$), we used multiple data sources to characterize the absolute coldest air temperature that occurred within each 0.5-degree resolution cell across a multi-decadal period (for context, see Osland et al. 2013, Cavanaugh et al. 2014). For cells within the conterminous United States, we used 2.5-arc-min resolution data (~5 km cell size at the equator) created by the PRISM Climate Group (Oregon State University; http://prism.oregonstate.edu; Daly et al. 2008), for the period extending from 1981 to 2010. For continental cells outside of the United States, we used 1-degree resolution data (~111 km cell size at the equator) created by Sheffield et al. (2006), for the period extending from 1981 to 2010. For most islands outside of the conterminous United States (e.g., Caribbean and Pacific Islands), we used 0.5-degree resolution data created by Maurer et al. (2009), for the period extending from 1971 to 2000. For mean annual temperature and precipitation, we used 30-arc-second resolution (~1 km cell size at the equator) mean annual precipitation (P_{ma}) and mean annual temperature (T_{ma}) data from the WorldClim Global Climate Data (Hijmans et al. 2005), for the period extending from 1950 to 2000. Where necessary, climate data were resampled to match the 0.5-degree resolution of the study grids.

Canopy height data

Mangrove forest and/or salt marsh vegetation canopy height data were obtained for all three of the transition zones. While some of these data were acquired from individual studies, most of these data were obtained from the following reviews: Anderson and Treshow (1980), Saenger and Snedaker (1993), Komiyama et al. (2008), Soares et al. (2012), Rovai et al. (2016; Appendix S2: Tables S1 and S2).

Above-ground biomass data

Sufficient total above-ground biomass data were only available for the eastern North America and eastern Australia transition zones. The methods used to determine above-ground biomass data varied, and categories included in our analyses for these methods are (1) end-of-season live biomass; (2) live and dead biomass averaged over multiple months; (3) end-of-season live and dead biomass; (4) end-of-season biomass (live or dead not specified); (5) biomass derived from allometric equations; or (6) method not reported. Whereas some of these data were acquired from individual studies, most of the data were obtained from the following reviews: Saenger and Snedaker (1993), Morrisey et al. (2007, 2010), Komiyama et al. (2008), Kirwan et al. (2009), Rovai et al. (2016; Appendix S3: Tables S1 and S2).

Above-ground primary productivity data

Adequate above-ground primary productivity data were only available for the eastern North America and eastern Australia transition zones. In addition to productivity measurements using peak standing crop and litterfall methods, other productivity measurement methods are also included in the dataset, including the methods described in Smalley (1959), Wiegert and Evans (1964), Lomnicki et al. (1968), Milner and Hughes (1968), Valiela et al. (1975), Wiegert and McGinnis (1975), and Shew et al. (1981). While some of these data were acquired from individual studies, most of these data were obtained from the following reviews: Saenger and Snedaker (1993), Komiyama et al. (2008), González Trilla et al. (2009), Morrisey et al. (2010), and Day et al. (2013; Appendix S4: Tables S1 and S2).

Decomposition data

Sufficient decomposition data were not available for any of the three transition zones.

Soil carbon density and carbon accumulation data

Soil carbon density and accumulation data were only available for the eastern North America transition zone. Soil carbon data were limited to samples collected from the upper 30 cm of the soil profile so as to capture the effects of current plant communities on soil carbon distribution and to minimize any potential influence of prior habitat shifts (Owers et al. 2016). Carbon density data were obtained from the following reviews: Chmura et al. (2003), Kristensen et al. (2008), and Ouyang and Lee (2014; Appendix S5: Tables S1 and S2). Carbon accumulation data were acquired from some individual studies, but most were obtained from the following reviews: Chmura et al. (2003) and Ouyang and Lee (2014; Appendix S6: Tables S1 and S2).

Data analyses

Cell-level means were used to convert the ecological data to the 0.5-degree resolution of the study grids. We used linear and sigmoidal regression analyses to examine climate–ecological relationships within each transition zone. Minimum air temperature was the independent variable for analyses of canopy height and above-ground biomass in eastern North America and eastern Australia because extreme minimum temperature is the major climatic control on the distribution of mangroves relative to salt marshes in these areas. Mean annual air temperature was the independent variable for analyses of productivity, decomposition, soil carbon density, and soil carbon accumulation in eastern North America and eastern Australia because these properties have been shown to be strongly related to growing season characteristics. In the western Gulf of Mexico, mean annual precipitation was the independent variable. For sigmoidal regression analyses, the following equation was used:

$$y(x) = \frac{a}{1 + e^{-\left(\frac{x-c}{b}\right)}}$$

where y = the ecological property variable, x = the climatic variable of interest (T_{min}, T_{ma}, or P_{ma}), a = the asymptote, b = the function growth rate, and c = the function midpoint. In order to identify thresholds and transition zones, we calculated the local maxima of the first derivative (T) of each sigmoidal model, which represents the point of the maximum rate of change (i.e., inflection point). The area between the local maximum and minimum peaks of the second derivative of each sigmoidal equation was calculated in order to identify the area of the maximum rate of change (AMRC) for each transition zone (Scheffer et al. 2001, Osland et al. 2014*b*). From an ecological perspective, the AMRC represents a threshold zone and T represents a discrete threshold. Both values represent areas where relatively small changes in a climatic driver can trigger large ecological responses. Linear and sigmoidal regression analyses were conducted in R (R Core Team 2016). Where linear and sigmoidal results were both significant (i.e., for the canopy height and above-ground biomass analyses), the r^2 values were used to select the most appropriate model (Appendix S7: Tables S1 and S2). All mapping and geospatial analyses were performed using Esri ArcGIS 10.3.1 (Environmental Systems Research Institute, Redlands, California, USA).

Results

There were significant positive sigmoidal relationships between (1) minimum air temperature and canopy height in eastern North America (Fig. 3a) and in eastern Australia (Fig. 3b); (2) mean annual precipitation and canopy height in the western Gulf of Mexico (Fig. 3c); and (3) minimum air temperature and above-ground biomass in eastern North America (Fig. 3d) and in eastern Australia (Fig. 3e; Appendix S7: Table S2). There were significant positive linear relationships between mean annual air temperature and above-ground productivity in eastern North America (Fig. 3g) and in eastern Australia (Fig. 3h; Appendix S7: Table S1). There were no significant relationships between mean annual air temperature and soil carbon density (Fig. 3m) or soil carbon accumulation (Fig. 3p) in eastern North America (Appendix S7: Tables S1 and S2). Equations for all of these relationships are provided in Tables S1 and S2 of Appendix S7.

In eastern North America, the minimum air temperature threshold zones (the AMRC for this metric, as previously defined) for canopy height and above-ground biomass were −6.9° to 4.1°C (Fig. 3a) and −8.1° to 2.9°C (Fig. 3d), respectively (Appendix S7: Table S2). In eastern Australia, the minimum air temperature threshold zones for canopy height and above-ground biomass were −4.6° to 2.2°C (Fig. 3b) and 0.2° to 4.8°C (Fig. 3e), respectively (Appendix S7: Table S2). In the western Gulf of Mexico, the mean annual precipitation threshold zone for canopy height was 1308–1780 mm (Fig. 3c; Appendix S7: Table S2).

Discussion

Ecologists have long known that climatic drivers influence ecosystem structure and function in mangrove forests, salt marshes, and salt flats, and the results from this study help to quantify those climate–coastal wetland relationships. In the subsequent sections, we discuss our hypotheses and results concerning the influence of climatic drivers on each ecosystem property. We also identify critical data limitations and discuss the broader relevance of our findings, particularly in the context of climate change.

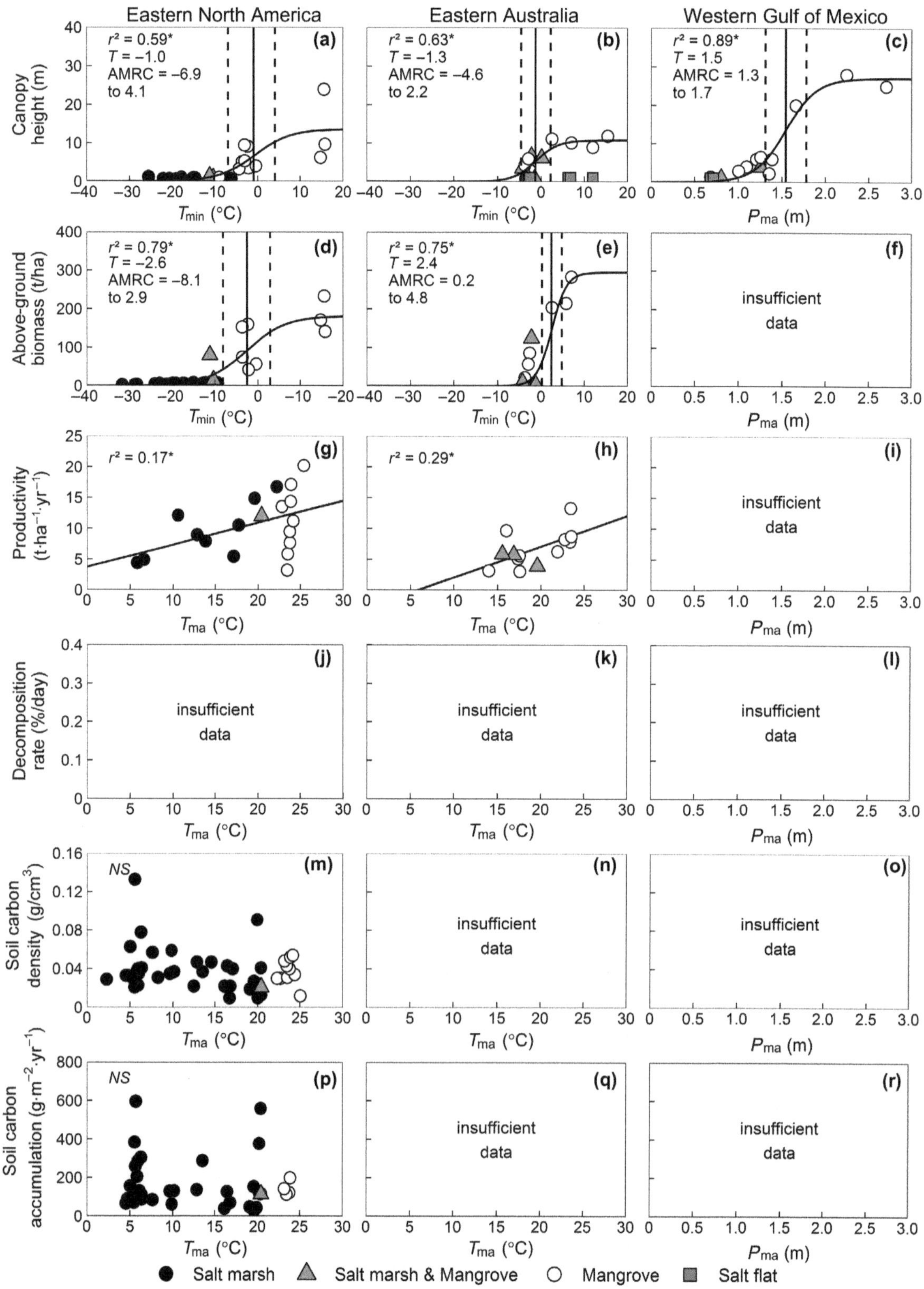

Fig. 3. The relationships between minimum air temperature (T_{min}), mean annual air temperature (T_{ma}), or mean

(Fig. 3. *Continued*)
annual precipitation (P_{ma}) and canopy height (a–c), above-ground biomass (d–f), productivity (g–i), decomposition (j–l), soil carbon density (m–o), and soil carbon accumulation (p–r) across the three transition zones. In panels a–e, the discrete climate threshold (T) is represented by a solid line and the dashed lines represent the threshold zone boundaries (i.e., the area of the maximum rate of change [AMRC]). $^*P < 0.05$; NS, not significant.

Canopy height and above-ground biomass

Studies focused on the mangrove–marsh ecotone in eastern North America and the temperature-driven mangrove transition zones in eastern Australia and New Zealand have shown that air temperature regimes can have a large effect on tidal saline vegetation canopy height and above-ground biomass (Lugo and Patterson-Zucca 1977, Osunkoya and Creese 1997, Morrisey et al. 2010, Osland et al. 2014*a*, 2015, Doughty et al. 2016, Yando et al. 2016, Gabler et al. 2017). Twilley et al. (1992), Saenger and Snedaker (1993), Alongi (2009), Twilley and Day (2012), and Sanders et al. (2016) characterized negative relationships between latitude (a temperature proxy) and mangrove forest above-ground biomass. Rovai et al. (2016) recently developed a continental-scale model for mangrove forest above-ground biomass, which includes temperature as a critical variable. In salt marshes, there are also negative latitude–biomass and latitude–height relationships (Turner 1976, Kirwan et al. 2009, McCall and Pennings 2012). Despite significant independent progress in the fields of mangrove forest and salt marsh ecology and recent studies that used field-based data from the northern Gulf of Mexico (Gabler et al. 2017) and the Atlantic coast of North America (Simpson et al. 2017), no study, to our knowledge, has evaluated temperature–height or temperature–biomass relationships using salt marsh and mangrove forest literature review-based data together. Here, we hypothesized that there are positive sigmoidal relationships between minimum air temperature and canopy height and minimum air temperature and above-ground biomass (Fig. 1a, c, respectively; Gabler et al. 2017). Our results from eastern North America and eastern Australia support these hypotheses; near poleward mangrove transition zones, there are regional, minimum temperature-controlled thresholds for canopy height and above-ground biomass in tidal saline wetlands. In terrestrial ecosystems, minimum air temperature gradients in the transition zone between tropical and temperate climates do not produce grassland-to-forest transitions or decreases in forest height and biomass; similar climatic gradients merely produce terrestrial forest-to-forest transitions without comparable changes in height or biomass (Stegen et al. 2011, Pan et al. 2013, Liu et al. 2014). Hence, the temperature–height and temperature–biomass relationships present in tidal saline wetlands differ greatly from those relationships present in terrestrial ecosystems. These results imply that, near the tropical-to-temperate climatic transition zone, the effects of future changes in temperature regimes upon canopy height and above-ground biomass may be greater in tidal saline wetlands than in terrestrial ecosystems. The implications of these structural changes for the provision of certain ecosystem goods and services are important. For example, changes in above-ground structure will greatly affect avian habitat (Chavez-Ramirez and Wehtje 2012, Guo et al. 2017), above-ground carbon stocks (Doughty et al. 2016, Yando et al. 2016), and potentially, coastal protection processes related to wave attenuation and erosion prevention (Gedan et al. 2011, Shepard et al. 2011, Marois and Mitsch 2015, Sutton-Grier et al. 2015, Doughty et al. 2017, Kelleway et al. 2017).

Our precipitation-based hypotheses for height and biomass are similar to our temperature-based hypotheses. Across precipitation gradients, we hypothesized that there are positive sigmoidal relationships between mean annual precipitation and canopy height and mean annual precipitation and above-ground biomass (Fig. 1b, d, respectively). The hypersaline and physiologically demanding conditions present in arid and semi-arid estuaries can lead to mangrove mortality (Duke et al. 2017, Lovelock et al. 2017) and result in the development of expansive salt flat ecosystems that lack vegetation (Adam 1990, Bucher and Saenger 1994, Saenger 2002, Semeniuk 2013). Reef and Lovelock (2014), Asbridge et al. (2015), and Lovelock et al. (2016) provide valuable recent reviews of the effects of salinity and precipitation regimes on mangrove forest structure and function. Previous studies have shown that there is a positive sigmoidal relationship between precipitation and vegetation abundance in tidal saline wetlands (Longley 1995, Montagna et al. 2011, Osland et al. 2014*b*, Gabler et al. 2017). Our hypothesis of a sigmoidal relationship between precipitation and canopy height was supported by our analysis of literature-derived data from the western Gulf of Mexico. We were not able to obtain sufficient biomass data across a relevant precipitation gradient to test our precipitation–biomass hypothesis. However, since biomass is correlated with height at this scale (Rovai et al. 2016), we expect that our precipitation–biomass hypothesis is also valid. In a recent study, Gabler et al. (2017) used field-collected canopy height and biomass proxy data across climatic gradients along the northern Gulf of Mexico coast to quantify sigmoidal precipitation–height and precipitation–biomass relationships. These findings reinforce the valuable precipitation gradient-focused data and discussions contained within Lot-Helgueras et al. (1975), Lugo and Patterson-Zucca (1977), Bucher and Saenger (1994), Saenger (2002), Méndez-Alonzo et al. (2008), and Yando et al. (2016). Collectively, these results are also complemented by temporal analyses showing that precipitation fluctuations, along with concomitant fluctuations in sea level in certain areas, can result in the expansion and contraction of vegetated tidal saline wetlands (Diop et al. 1997, Eslami-Andargoli et al. 2009, Rogers et al. 2014, Saintilan et al. 2014, Duke et al. 2017, Lovelock et al. 2017). In terrestrial ecosystems, there are also positive precipitation–biomass and precipitation–height linkages (e.g., across precipitation-induced desert-to-grassland-to-forest transitions; Whittaker 1970, Pan et al. 2013, Liu et al. 2014); however, due to the physically demanding and dynamic abiotic conditions present in tidal saline wetlands, we expect that the rate of ecological change across precipitation gradients may be higher in tidal saline wetlands compared to their terrestrial counterparts. Similar to changes in temperature regimes, precipitation-induced changes in above-ground biomass and canopy height would greatly influence certain ecosystem goods and services provided by tidal saline wetlands.

Above-ground productivity

Temperature regimes greatly influence ecosystem productivity in terrestrial ecosystems (Churkina and Running 1998, Nemani et al. 2003, Chu et al. 2016). Despite large differences in above-ground biomass (i.e., relatively lower biomass in temperate salt marshes compared to tropical mangrove forests), mangrove forests and salt marshes both have the potential to be highly productive ecosystems under optimal physicochemical conditions (Mendelssohn and Morris 2002, Twilley and Day 2012). To our knowledge, our study is the first to use productivity data from both salt marshes and mangrove forests to characterize the influence of climatic drivers on tidal saline wetlands. In mangrove forests, Twilley et al. (1992), Saenger and Snedaker (1993), and Alongi (2009) identified negative linear relationships between latitude and above-ground productivity. In salt marshes, Turner (1976) and Kirwan et al. (2009) identified a negative linear relationship between latitude and above-ground productivity. Kirwan et al. (2009) also identified a negative linear relationship between mean annual temperature and above-ground productivity in salt marshes. Based upon these results, we hypothesized that there would be a positive linear relationship between mean annual air temperatures and above-ground primary productivity in all tidal saline wetlands (Fig. 1e). Our findings from eastern North America and eastern Australia support this hypothesis, as we identified a positive linear temperature–productivity relationship despite sigmoidal temperature–height and temperature–biomass relationships. Since the ability of coastal wetlands to adjust to sea-level rise is partly dependent upon ecogeomorphic feedbacks between inundation, plant growth, and sedimentation (Morris et al. 2002, Kirwan et al. 2010), a linear productivity response (versus a sigmoidal response) has important implications for models of coastal wetland response to sea-level rise. Whereas a sigmoidal relationship might indicate that temperature-induced mangrove forest encroachment into salt marsh may dramatically increase productivity and increase the ability of coastal wetlands to build surface elevations with progressive sea-level rise, a linear relationship might reflect minor effects of temperature-induced mangrove encroachment upon ecosystem productivity and ecogeomorphic feedbacks relative to highly productive salt marshes (Yando et al. 2016). The significant relationships between mean annual temperatures and productivity identified for eastern North America and eastern Australia suggest that productivity will be impacted by changes in mean temperature (e.g., growing season characteristics) as opposed to changes in air temperature extremes.

Across precipitation gradients, a decrease in tidal saline wetland above-ground productivity is expected due partially to physiological stress associated with hypersaline conditions and low freshwater availability (Ibáñez et al. 2012, Reef and Lovelock 2014, Lovelock et al. 2016, Osland et al. 2016). In terrestrial ecosystems, there is a positive relationship between precipitation and productivity (Churkina and Running 1998, Knapp and Smith 2001, Chu et al. 2016). We hypothesized that there is a positive sigmoidal relationship between mean annual precipitation and above-ground productivity as highly productive mangrove forests or salt marshes would be expected to replace tidal flats or low-growing succulent-dominated marshes as the availability of freshwater increases across a rainfall gradient (Fig. 1f). Although we were unable to find sufficient productivity data across precipitation gradients to test this hypothesis, previous studies have found sigmoidal patterns for vegetation abundance across the precipitation gradient in the western Gulf of Mexico (Longley 1995, Montagna et al. 2011, Osland et al. 2014*b*, Gabler et al. 2017). On a physiological level, high salinities and low porewater osmotic potentials affect water uptake, transport, and loss in tidal saline wetland plants, and plant persistence during hypersaline conditions requires water acquisition adaptations that are energetically costly (Ball 1988, Callaway and Sabraw 1994, Field 1995, Naidoo 2006, Stachelek and Dunton 2013, Reef and Lovelock 2014, Nguyen et al. 2017). Whereas wetlands in arid and semi-arid regions may be resistant to minor episodes of drought, long-term changes in freshwater availability and estuarine salinity can result in mortality and significantly alter the structure and functioning of coastal wetlands (Noe and Zedler 2001, Forbes and Dunton 2006, Krauss et al. 2007, Lovelock et al. 2017). Additionally, the ecosystem impacts of future changes in precipitation regimes may be compounded by concurrent changes in sea level and air temperatures and also by human land-use pressures resulting in altered river discharge and tidal exchange patterns (Gilman et al. 2008, McKee et al. 2012, Alongi 2015, Ward et al. 2016).

Decomposition

For terrestrial ecosystems, Zhang et al. (2008) identified positive linear relationships between climatic variables (i.e., temperature, precipitation) and litter decomposition (see also Kirschbaum 1995, Davidson and Janssens 2006, García-Palacios et al. 2013). For salt flats, salt marshes, and mangrove forests, we expected a positive linear relationship between climatic variables (i.e., mean annual air temperature, mean annual precipitation) and decomposition rates (Fig. 1g, h, respectively). However, we did not have sufficient data to fully evaluate the effects of climatic drivers on decomposition. In comparison with the other ecosystem properties examined in this study, decomposition data are labor intensive and difficult to collect. In a global review of the limited root decomposition data available from tidal saline wetlands, Ouyang et al. (2017) identified significant effects of temperature, latitude, and precipitation. Support for our temperature-based hypothesis can also be found in relevant studies from salt marsh ecosystems. Craft (2007) discussed the potential edaphic effects of a positive temperature–decomposition relationship. Within temperate salt marshes that span a temperature gradient in eastern North America, Kirwan et al. (2014) identified a positive relationship between temperature and decomposition rates. In the same general region and also within salt marshes, Crosby et al. (2017) identified a negative relationship between latitude and decomposition. In an experiment within a single salt marsh in eastern North America, Kirwan and Blum (2011) identified a positive relationship between mean daily temperature and decomposition. In a west-central Florida site that contains mangrove forest and salt marsh, Lewis et al. (2014) found that higher temperatures stimulated decomposition.

We hypothesized that decomposition would be positively influenced by precipitation (Fig. 1h). The magnitude and frequency of precipitation events can modulate tidal saline wetland plant community dynamics via direct controls on freshwater availability as well as indirect effects on estuarine salinity gradients, amount and timing of river discharge, nutrient inflows, and soil salinity and moisture (Zedler 1983, Dunton et al. 2001, Alexander and Dunton 2002). In the only study to directly examine the influence of precipitation on decomposition in a tidal saline wetland habitat, Charles and Dukes (2009) demonstrated that decomposition in a New England (USA) salt marsh was accelerated by increased precipitation and slowed by drought. However, decomposition rates in coastal wetlands are complex and influenced by many different factors (Jones et al. 2016*a*, Janousek et al. 2017, Stagg et al. 2017, Wu et al. 2017), and there is a need for decomposition data from coastal wetlands located in arid and semi-arid climates.

Soil carbon density and accumulation

Soil carbon stocks and accumulation rates in tidal saline wetlands are often much larger than in their terrestrial counterpart ecosystems (Donato et al. 2011, McLeod et al. 2011, Duarte et al. 2013). In a seminal contribution and one of the only global-scale studies to include ecosystem property data from both mangrove forests and salt marshes, Chmura et al. (2003) investigated patterns of soil carbon density and soil carbon accumulation; they found that soil carbon accumulation rates were similar in salt marshes and mangrove forests despite differences in soil carbon density. In a recent comparison of salt marshes and mangrove forests near the subtropical mangrove–marsh transition zone in the southeastern USA, Yando et al. (2016) found no soil carbon density differences between mangrove forests and adjacent salt marshes in areas with abundant precipitation (i.e., Louisiana and Florida; see also Perry and Mendelssohn 2009, Henry and Twilley 2013, Doughty et al. 2016). Given these findings and because we expected that productivity and decomposition vary in a positive linear fashion with temperature, we expected to find no relationship between minimum temperature and soil carbon density or soil carbon accumulation. Our results from eastern North America support these hypotheses. We expect that the lack of significant temperature–soil carbon relationships may be attributed to the following hypothesis: Temperature-driven increases in productivity (Fig. 1e) are offset by comparable temperature-driven increases in decomposition (Fig. 1g), resulting in relatively constant carbon density and accumulation values across temperature gradients (Fig. 1i, k).

Mangrove forest expansion into salt marsh is a phenomenon that is occurring on multiple continents in diverse environmental settings in response to various environmental drivers (Saintilan et al. 2014, Saintilan and Rogers 2015). For example, changes in sea level and precipitation regimes can lead to the landward expansion of mangrove forests at the expense of salt marshes or other upslope ecosystems (Saintilan and Williams 1999, Krauss et al. 2011, Enwright et al. 2016). Similarly, changing air temperature regimes can lead to poleward mangrove expansion and salt marsh contraction (Michot et al. 2010, Cavanaugh et al. 2014, Osland et al. 2017*a*). In certain locations (e.g., tidal saline wetlands where marshes are dominated by succulent plants due to elevated salinity), mangrove expansion into salt marsh has resulted in elevated bulk soil carbon properties (e.g., Comeaux et al. 2012, Bianchi et al. 2013, Kelleway et al. 2016, Yando et al. 2016, Guo et al. 2017), and in certain arid coastal wetlands, soil carbon stocks can be higher in mangrove forests than in other wetland types (Schile et al. 2017). However, in wetter climes, where salt marshes are dominated by highly productive graminoids, mangrove expansion into salt marsh has typically resulted in no change in bulk soil carbon properties (Perry and Mendelssohn 2009, Henry and Twilley 2013, Doughty et al. 2016, Yando et al. 2016). In southeast Queensland (Australia), Hayes et al. (2017) found that while soil carbon stocks did not differ between mangroves and salt marshes, soil carbon stocks were strongly controlled by hydrogeomorphic setting. Using data from 84 tidal marsh sites around Australia, Macreadie et al. (2017) also show that geomorphology greatly influences soil organic carbon stocks. We highlight these differences to underscore the importance of considering differences in environmental setting when discussing the ramifications of mangrove expansion into salt marsh. The implications of mangrove expansion for below-ground ecological properties are diverse and context dependent (Howe et al. 2009, Livesley and Andrusiak 2012, Saintilan et al. 2013, Lovelock et al. 2014, Yando et al. 2016, Hayes et al. 2017). While it is tempting to generalize about the ecological effects of mangrove encroachment into salt marsh, it is important to recognize that mangrove encroachment into salt marsh is occurring in many different settings in response to multiple drivers.

The influence of precipitation and hypersaline conditions on soil carbon cycling and storage has not received much attention in the coastal wetland literature (Osland et al. 2016, Yando et al. 2016, Schile et al. 2017). In terrestrial ecosystems, there is often a positive linear relationship between precipitation and soil carbon stocks (Burke et al. 1989, Jobbágy and Jackson 2000). Woody expansion in dryland terrestrial ecosystems can sometimes result in local increases in soil carbon (Jackson et al. 2002, Eldridge et al. 2011, Berthrong et al. 2012). Recent global-scale analyses of data from mangrove forests indicate that there is a positive relationship between precipitation and soil carbon stocks (Jardine and Siikamäki 2014, Sanders et al. 2016). Given the positive sigmoidal relationship between precipitation and vegetation abundance in tidal saline wetlands (Longley 1995, Montagna et al. 2011, Osland et al. 2014*b*), we hypothesized that there are positive sigmoidal relationships between mean annual precipitation and soil carbon density and soil carbon accumulation (Figs. 1j, l and 3). Unfortunately, ecological data from salt flats are scant, and we were not able to identify sufficient soil carbon data across precipitation gradients to test our hypotheses. However, preliminary data and support for this hypothesis are provided in Yando et al. (2016) and Schile et al. (2017), where above- and below-ground carbon stocks were lower in wetlands with lower precipitation compared to wetlands located along coastal reaches with abundant precipitation.

Critical data gaps and limitations

Ecological research in tidal saline wetlands has primarily been concentrated in areas with sufficient precipitation to support abundant salt marshes and/or mangrove forests (i.e., humid and sub-humid regions of the world). Very little research has been conducted in hypersaline tidal salt flats, which are most abundant in arid and semi-arid estuaries (Withers 2002, Semeniuk 2013, Osland et al. 2016, Gabler et al. 2017). Although we found canopy height data distributed across a precipitation gradient, we could not find sufficient ecological data to test our hypotheses regarding the influence of precipitation upon above-ground biomass, productivity, decomposition, soil carbon density, or soil carbon accumulation. In order to advance understanding of the influence of precipitation and freshwater availability on tidal saline wetlands, ecological data are needed from tidal saline wetlands in arid and semi-arid climates (e.g., tidal salt flats, succulent-dominated salt marshes, and arid mangrove forests). Consistent and targeted data collections across regional-scale precipitation gradients would be particularly useful for testing and refining the hypothesized relationships between precipitation and wetland ecosystem properties.

In contrast to precipitation gradients, we found relevant ecological data across minimum temperature gradients in eastern North America and eastern Australia. However, additional data that are strategically collected across these gradients would help refine the identified relationships. Although we tested many of our temperature-focused hypotheses, data from temperature gradients in other regions (e.g., China, New Zealand, southern Africa, western South America) would help validate these results. To test our temperature–decomposition hypothesis, decomposition data distributed across broad temperature gradients (i.e., across tropical-to-temperate climatic transition zones) are needed.

Differences between the eastern North America and eastern Australia transition zones

There are important differences between the temperature-driven coastal wetland transition zones in eastern North America and eastern Australia. In eastern North America, winter air temperature extremes are more intense than in Australia and can cause mangrove mortality and loss of above-ground biomass (Lugo and Patterson-Zucca 1977, Stuart et al. 2007). In contrast, mangrove damage and mortality from extreme minimum temperature events in eastern Australia are very rare; however, mean winter minimum temperatures in the Australian transition zone limit mangrove performance (e.g., plant growth and reproduction; Duke 1990, Saintilan et al. 2014, Boon 2017). Although temperature clearly affects the structure and function of mangrove forests along the eastern Australian coast, the coastline ends before mangrove forests are fully replaced by salt marshes. As a result, *A. marina* individuals in eastern Australia are likely not at their physiological limit, and a temperature-induced mangrove-to-marsh transition zone is not clearly present (Saintilan et al. 2014). The absence of this transition affects our analyses from eastern Australian because the transition zone in eastern Australia spans a narrower and incomplete temperature gradient compared to the eastern North American transition.

In contrast, the coastal wetlands along the Gulf of Mexico and Atlantic coasts of North America provide one of the best natural laboratories in the world for investigating the effects of minimum temperature on the mangrove–marsh ecotone. Across the tropical-to-temperate climatic transition zone in eastern North America, coastal wetlands are abundant and continuously present. Moreover, precipitation regimes and the geomorphic characteristics of estuaries across this transition zone are relatively consistent, and the coastline spans the entire mangrove-to-marsh transition zone without interruption. As a result, analyses that span the temperature gradient in eastern North America can fully characterize the influence of air temperature on tidal saline wetland structure and function. In contrast, coastal wetlands in the eastern Australian transition zone are less ubiquitous and affected by concomitant hydrogeomorphic and precipitation-related factors that affect ecosystem structure and function (Saintilan and Williams 1999, Hayes et al. 2017). For example, the geomorphic characteristics of estuaries that span the temperature gradient in eastern Australia are diverse in terms of their geological setting, sedimentary environments, and hydrologic properties (Roy et al. 2001), which affect the distribution of mangrove forests, salt marshes, and salt flats.

Beyond just climate: the role of geomorphology, tidal position, and other factors

In this communication, we have focused primarily on the influence of climatic drivers. To simplify our message, we have intentionally avoided incorporating the influence of other important factors (e.g., geomorphology, tidal position). However, in coastal wetlands, hydrology and geomorphology are tremendously important variables that greatly influence ecosystem structure and function (Lugo and Snedaker 1974, Adam 1990, Ibáñez et al. 2012, Twilley and Day 2012, Hayes et al. 2017). Within a given wetland or estuary, inundation and salinity regimes greatly determine the position and relative abundance of mangrove forests, salt marshes, and salt flats (Eleuterius and Eleuterius 1979, Odum et al. 1982, Semeniuk 1983, Bird 1986, Rogers et al. 2006, Gabler et al. 2017). Our aim here is merely to provide a foundation for considering the role of climatic drivers in tidal saline wetlands, and we anticipate that future work will be improved via analyses that also incorporate the effects of geomorphology, tidal position, sea-level rise, atmospheric carbon dioxide concentrations, and other factors that also influence tidal saline wetland ecosystem structure and function.

Conclusions

Climatic factors can greatly influence the structure and functioning of tidal wetland ecosystems, but most analyses have examined data from mangrove forests, salt marshes, and salt flats separately (Turner 1976, Twilley 1982, Woodroffe and Grindrod 1991, Saenger 2002). Due to constraints in data quality and access, climatic controls of ecological properties in tidal saline wetland ecosystems have not been fully elucidated. Here, we present hypotheses regarding the influence of temperature and precipitation regimes on various tidal saline wetland ecosystem properties (i.e., canopy height, biomass, productivity, decomposition, soil carbon density, soil carbon accumulation). We tested many of these hypotheses using literature-derived ecological data from mangrove forests, salt marshes, and tidal salt flats. Ecological thresholds and regime shifts are common in coastal wetlands (Scheffer et al. 2001, Folke et al. 2004). Changing winter temperature regimes can result in marsh-to-mangrove conversions (Saintilan et al. 2014), and changing precipitation regimes can result in marsh-to-salt flat or mangrove-to-salt flat conversions (Semeniuk 2013, Osland et al. 2014*b*, Gabler et al. 2017). Our results identify linear and nonlinear ecological effects related to these regime shifts, and advance understanding of climatic controls of ecosystem structure and function in tidal saline wetlands. Our work also highlights critical data gaps and limitations that can be improved upon via future research.

Acknowledgments

This research was partially supported by the USGS Ecosystems Mission Area, the USGS Land Change Science Program, the Department of Interior South Central Climate Science Center, the Department of Interior Southeast Climate Science Center, the EPA Gulf of Mexico Program, the USGS Greater Everglades Priority Ecosystems Science Program, and the USGS Wetland and Aquatic Research Center. We thank Karen McKee and three anonymous reviewers for their comments on a previous version of this manuscript. Any use of trade, firm, or product names is for descriptive purposes only and does not imply endorsement by the U.S. Government. This manuscript is submitted for publication with the understanding that the U.S. Government is authorized to reproduce and distribute reprints for governmental purposes.

Literature Cited

Adam, P. 1990. Saltmarsh ecology. Cambridge University Press, Cambridge, UK.

Alexander, H. D., and K. H. Dunton. 2002. Freshwater inundation effects on emergent vegetation of a hypersaline salt marsh. Estuaries 25:1426–1435.

Alongi, D. M. 2008. Mangrove forests: resilience, protection from tsunamis, and responses to global climate change. Estuarine, Coastal and Shelf Science 76:1–13.

Alongi, D. M. 2009. The energetics of mangrove forests. Springer, New York, New York, USA.

Alongi, D. M. 2015. The impact of climate change on mangrove forests. Current Climate Change Reports 1:30–39.

Anderson, C. M., and M. Treshow. 1980. A review of environmental and genetic factors that affect height in *Spartina alterniflora* Loisel. (salt marsh cord grass). Estuaries 3:168–176.

Asbridge, E., R. Lucas, A. Accad, and R. Dowling. 2015. Mangrove response to environmental changes predicted under varying climates: case studies from Australia. Current Forestry Reports 1:178–194.

Atwood, T. B., et al. 2017. Global patterns in mangrove soil carbon stocks and losses. Nature Climate Change 7:523–528.

Ball, M. C. 1988. Ecophysiology of mangroves. Trees 2:129–142.

Barbier, E. B., S. D. Hacker, C. Kennedy, E. W. Koch, A. C. Stier, and B. R. Silliman. 2011. The value of estuarine and coastal ecosystem services. Ecological Monographs 81:169–193.

Berthrong, S. T., G. Piñeiro, E. G. Jobbágy, and R. B. Jackson. 2012. Soil C and N changes with afforestation of grasslands across gradients of precipitation and plantation age. Ecological Applications 22:76–86.

Bianchi, T. S., M. A. Allison, J. Zhao, X. Li, R. S. Comeaux, R. A. Feagin, and R. W. Kulawardhana. 2013. Historical reconstruction of mangrove expansion in the Gulf of Mexico: linking climate change with carbon sequestration in coastal wetlands. Estuarine, Coastal and Shelf Science 119:7–16.

Bird, E. C. F. 1986. Mangroves and intertidal morphology in Westernport Bay, Victoria, Australia. Marine Geology 69:251–271.

Boon, P. I. 2017. Are mangroves in Victoria (south-eastern Australia) already responding to climate change? Marine and Freshwater Research, *in press*. https://doi.org/10.1071/MF17015

Bucher, D., and P. Saenger. 1994. A classification of tropical and subtropical Australian estuaries. Aquatic Conservation: Marine and Freshwater Ecosystems 4:1–19.

Burke, I. C., C. M. Yonker, W. J. Parton, C. V. Cole, D. S. Schimel, and K. Flach. 1989. Texture, climate, and cultivation effects on soil organic matter content in US grassland soils. Soil Science Society of America Journal 53:800–805.

Callaway, R. M., and C. S. Sabraw. 1994. Effects of variable precipitation on the structure and diversity of a California salt marsh community. Journal of Vegetation Science 5:433–438.

Cavanaugh, K. C., J. R. Kellner, A. J. Forde, D. S. Gruner, J. D. Parker, W. Rodriguez, and I. C. Feller. 2014. Poleward expansion of mangroves is a threshold response to decreased frequency of extreme cold events. Proceedings of the National Academy of Sciences USA 111:723–727.

Cavanaugh, K. C., J. D. Parker, S. C. Cook-Patton, I. C. Feller, A. P. Williams, and J. R. Kellner. 2015. Integrating physiological threshold experiments with climate modeling to project mangrove species' range expansion. Global Change Biology 21:1928–1938.

Chapin, F. S., P. A. Matson, and P. M. Vitousek. 2011. Principles of terrestrial ecosystem ecology. Second edition. Springer, New York, New York, USA.

Charles, H., and J. S. Dukes. 2009. Effects of warming and altered precipitation on plant and nutrient dynamics of a New England salt marsh. Ecological Applications 19:1758–1773.

Chavez-Ramirez, F., and W. Wehtje. 2012. Potential impact of climate change scenarios on whooping crane life history. Wetlands 32:11–20.

Chmura, G. L., S. C. Anisfeld, D. R. Cahoon, and J. C. Lynch. 2003. Global carbon sequestration in tidal, saline wetland soils. Global Biogeochemical Cycles 17:1–12.

Chu, C., M. Bartlett, Y. Wang, F. He, J. Weiner, J. Chave, and L. Sack. 2016. Does climate directly influence NPP globally? Global Change Biology 22:12–24.

Churkina, G., and S. W. Running. 1998. Contrasting climatic controls on the estimated productivity of global terrestrial biomes. Ecosystems 1:206–215.

Comeaux, R. S., M. A. Allison, and T. S. Bianchi. 2012. Mangrove expansion in the Gulf of Mexico with climate change: implications for wetland health and resistance to rising sea levels. Estuarine, Coastal and Shelf Science 96:81–95.

Craft, C. 2007. Freshwater input structures soil properties, vertical accretion, and nutrient accumulation of Georgia and U.S. tidal marshes. Limnology and Oceanography 52:1220–1230.

Crosby, S. C., A. Angermeyer, J. M. Adler, M. D. Bertness, L. A. Deegan, N. Sibinga, and H. M. Leslie. 2017. *Spartina alterniflora* biomass allocation and temperature: implications for salt marsh persistence with sea-level rise. Estuaries and Coasts 40:213–223.

Daly, C., M. Halbleib, J. I. Smith, W. P. Gibson, M. K. Doggett, G. H. Taylor, J. Curtis, and P. P. Pasteris. 2008. Physiographically sensitive mapping of climatological temperature and precipitation across the conterminous United States. International Journal of Climatology 28:2031–2064.

Davidson, E. A., and I. A. Janssens. 2006. Temperature sensitivity of soil carbon decomposition and feedbacks to climate change. Nature 440:165–173.

Davis, J. H. 1940. The ecology and geologic role of mangroves in Florida. Tortugas Laboratory Papers. Carnegie Institute of Washington, Washington, D.C., USA.

Day, J. W., W. M. Kemp, A. Yáñez-Arancibia, and B. C. Crump. 2013. Estuarine ecology. Second edition. Wiley-Blackwell, Hoboken, New Jersey, USA.

Deegan, L. A., J. W. Day, J. G. Gosselink, A. Yáñez-Arancibia, G. S. Chávez, and P. Sánchez-Gil. 1986. Relationships among physical characteristics, vegetation distribution and fisheries yield in Gulf of Mexico estuaries. Pages 83–100 *in* D. A. Wolfe, editor. Estuarine variability. Academic Press, New York, New York, USA.

Diop, E. S., A. Soumare, N. Diallo, and A. Guisse. 1997. Recent changes of the mangroves of the Saloum River estuary, Senegal. Mangroves and Salt Marshes 1:163–172.

Donato, D. C., J. B. Kauffman, D. Murdiyarso, S. Kurnianto, M. Stidham, and M. Kanninen. 2011. Mangroves among the most carbon-rich forests in the tropics. Nature Geoscience 4:293–297.

Doughty, C. L., K. C. Cavanaugh, C. R. Hall, I. C. Feller, and S. K. Chapman. 2017. Impacts of mangrove encroachment and mosquito impoundment management on coastal protection services. Hydrobiologia, *in press*. https://doi.org/10.1007/s10750-017-3225-0

Doughty, C. L., J. A. Langley, W. S. Walker, I. C. Feller, R. Schaub, and S. K. Chapman. 2016. Mangrove range expansion rapidly increases coastal wetland carbon storage. Estuaries and Coasts 39: 385–396.

Duarte, C. M., I. J. Losada, I. E. Hendriks, I. Mazarrasa, and N. Marbá. 2013. The role of coastal plant communities for climate change mitigation and adaptation. Nature Climate Change 3:961–968.

Duke, N. C. 1990. Phenological trends with latitude in the mangrove tree *Avicennia marina*. Journal of Ecology 78:113–133.

Duke, N. C., J. M. Kovacs, A. D. Griffiths, L. Preece, D. J. E. Hill, P. van Oosterzee, J. Mackenzie, H. S. Morning, and D. Burrows. 2017. Large-scale dieback of mangroves in Australia's Gulf of Carpentaria: a severe ecosystem response, coincidental with an unusually extreme weather event. Marine and Freshwater Research, *in press*. https://doi.org/10.1071/MF16322

Dunton, K. H., B. Hardegree, and T. E. Whitledge. 2001. Response of estuarine marsh vegetation to interannual variations in precipitation. Estuaries 24:851–861.

Eldridge, D. J., M. A. Bowker, F. T. Maestre, E. Roger, J. F. Reynolds, and W. G. Whitford. 2011. Impacts of shrub encroachment on ecosystem structure and functioning: towards a global synthesis. Ecology Letters 14:709–722.

Eleuterius, L. N., and C. K. Eleuterius. 1979. Tide levels and salt marsh zonation. Bulletin of Marine Science 29:394–400.

Enwright, N. M., K. T. Griffith, and M. J. Osland. 2016. Barriers to and opportunities for landward migration of coastal wetlands with sea-level rise. Frontiers in Ecology and the Environment 14:307–316.

Eslami-Andargoli, L., P. Dale, N. Sipe, and J. Chaseling. 2009. Mangrove expansion and rainfall patterns in Moreton Bay, Southeast Queensland, Australia. Estuarine, Coastal and Shelf Science 85: 292–298.

Field, C. D. 1995. Impact of expected climate change on mangroves. Hydrobiologia 295:75–81.

Folke, C., S. Carpenter, B. Walker, M. Scheffer, T. Elmqvist, L. Gunderson, and C. S. Holling. 2004. Regime shifts, resilience, and biodiversity in ecosystem management. Annual Review of Ecology, Evolution, and Systematics 35:557–581.

Forbes, M. G., and K. H. Dunton. 2006. Response of a subtropical estuarine marsh to local climatic change in the southwestern Gulf of Mexico. Estuaries and Coasts 29:1242–1254.

Friess, D. A., K. W. Krauss, E. M. Horstman, T. Balke, T. J. Bouma, D. Galli, and E. L. Webb. 2012. Are all intertidal wetlands naturally created equal? Bottlenecks, thresholds and knowledge gaps to mangrove and saltmarsh ecosystems. Biological Reviews 87:346–366.

Gabler, C. A., M. J. Osland, J. B. Grace, C. L. Stagg, R. H. Day, S. B. Hartley, N. M. Enwright, A. S. From, M. L. McCoy, and J. L. McLeod. 2017. Macroclimatic change expected to transform coastal wetland ecosystems this century. Nature Climate Change 7:142–147.

García-Palacios, P., F. T. Maestre, J. Kattge, and D. H. Wall. 2013. Climate and litter quality differentially modulate the effects of soil fauna on litter decomposition across biomes. Ecology Letters 16: 1045–1053.

Gedan, K. B., M. L. Kirwan, E. Wolanski, E. B. Barbier, and B. R. Silliman. 2011. The present and future role of coastal wetland vegetation in protecting shorelines: answering recent challenges to the paradigm. Climatic Change 106:7–29.

Gilman, E. L., J. Ellison, N. C. Duke, and C. Field. 2008. Threats to mangroves from climate change and adaptation options: a review. Aquatic Botany 89:237–250.

Giri, C., and J. Long. 2016. Is the geographic range of mangrove forests in the conterminous United States really expanding? Sensors 16:E2010.

Glick, P., B. A. Stein, and N. A. Edelson. 2011. Scanning the conservation horizon: a guide to climate change vulnerability assessment. National Wildlife Federation, Washington, D.C., USA.

González Trilla, G., P. Kandus, V. Negrin, R. Vicari, and J. Marcovecchio. 2009. Tiller dynamic and production on a SW Atlantic *Spartina alterniflora* marsh. Estuarine, Coastal and Shelf Science 85: 126–133.

Guo, H., C. Weaver, S. P. Charles, A. Whitt, S. Dastidar, P. D'Odorico, J. D. Fuentes, J. S. Kominoski, A. R. Armitage, and S. C. Pennings. 2017. Coastal regimes shifts: rapid responses of coastal wetlands to changes in mangrove cover. Ecology 98:762–772.

Hayes, M. A., A. Jesse, B. Hawke, J. Baldock, B. Tabet, D. Lockington, and C. E. Lovelock. 2017. Dynamics of sediment carbon stocks across intertidal wetland habitats of Moreton Bay, Australia. Global Change Biology 23:4222–4234.

Henry, K. M., and R. R. Twilley. 2013. Soil development in a coastal Louisiana wetland during a climate-induced vegetation shift from salt marsh to mangrove. Journal of Coastal Research 29:1273–1283.

Hijmans, R. J., S. E. Cameron, J. L. Parra, P. G. Jones, and A. Jarvis. 2005. Very high resolution interpolated climate surfaces for global land areas. International Journal of Climatology 25:1965–1978.

Holdridge, L. R. 1967. Life zone ecology. Tropical Science Center, San Jose, Costa Rica.

Howe, A. J., J. F. Rodríguez, and P. M. Saco. 2009. Surface evolution and carbon sequestration in disturbed and undisturbed wetland soils of the Hunter estuary, southeast Australia. Estuarine, Coastal and Shelf Science 84:75–83.

Ibáñez, C., J. T. Morris, I. A. Mendelssohn, and J. W. Day. 2012. Coastal marshes. Pages 129–163 *in* J. W. Day, B. C. Crump, W. M. Kemp, and A. Yáñez-Arancibia, editors. Estuarine ecology. Second edition. John Wiley & Sons, Hoboken, New Jersey, USA.

Jackson, R. B., J. L. Banner, E. G. Jobbágy, W. T. Pockman, and D. H. Wall. 2002. Ecosystem carbon loss with woody plant invasion of grasslands. Nature 418:623–626.

Janousek, C. N., K. J. Buffington, G. R. Guntenspergen, K. M. Thorne, B. D. Dugger, and J. Y. Takekawa. 2017. Inundation, vegetation, and sediment effects on litter decomposition in Pacific coast tidal marshes. Ecosystems, *in press*. https://doi.org/10.1007/s10021-017-0111-6

Jardine, S. L., and J. V. Siikamäki. 2014. A global predictive model of carbon in mangrove soils. Environmental Research Letters 9:1–9.

Jenny, H. 1941. Factors of soil formation. McGraw-Hill, New York, New York, USA.

Jobbágy, E. G., and R. B. Jackson. 2000. The vertical distribution of soil organic carbon and its relation to climate and vegetation. Ecological Applications 10:423–436.

Jones, J. A., J. A. Cherry, and K. L. McKee. 2016*a*. Species and tissue type regulate long-term decomposition of brackish marsh plants grown under elevated CO_2 conditions. Estuarine, Coastal and Shelf Science 169:38–45.

Jones, S. F., C. L. Stagg, K. W. Krauss, and M. W. Hester. 2016*b*. Tidal saline wetland regeneration of sentinel vegetation types in the Northern Gulf of Mexico: an overview. Estuarine, Coastal and Shelf Science 174:A1–A10.

Kelleway, J. J., K. Cavanaugh, K. Rogers, I. C. Feller, E. Ens, C. Doughty, and N. Saintilan. 2017. Review of the ecosystem service implications of mangrove encroachment into salt marshes. Global Change Biology 23:3967–3983.

Kelleway, J. J., N. Saintilan, P. I. Macreadie, C. G. Skilbeck, A. Zawadzki, and P. J. Ralph. 2016. Seventy years of continuous encroachment substantially increases "blue carbon" capacity as mangroves replace intertidal salt marshes. Global Change Biology 22:1097–1109.

Kirschbaum, M. U. F. 1995. The temperature dependence of soil organic matter decomposition, and the effect of global warming on soil organic C storage. Soil Biology and Biochemistry 27:753–760.

Kirwan, M. L., and L. K. Blum. 2011. Enhanced decomposition offsets enhanced productivity and soil carbon accumulation in coastal wetlands responding to climate change. Biogeosciences 8:987–993.

Kirwan, M. L., G. R. Guntenspergen, A. D'Alpaos, J. T. Morris, S. M. Mudd, and S. Temmerman. 2010. Limits on the adaptability of coastal marshes to rising sea level. Geophysical Research Letters 37:1–5.

Kirwan, M. L., G. R. Guntenspergen, and J. A. Langley. 2014. Temperature sensitivity of organic-matter decay in tidal marshes. Biogeosciences 11:4801–4808.

Kirwan, M. L., G. R. Guntenspergen, and J. T. Morris. 2009. Latitudinal trends in *Spartina alterniflora* productivity and the response of coastal marshes to global change. Global Change Biology 15:1982–1989.

Kirwan, M. L., and J. P. Megonigal. 2013. Tidal wetland stability in the face of human impacts and sea-level rise. Nature 504:53–60.

Knapp, A. K., and M. D. Smith. 2001. Variation among biomes in temporal dynamics of aboveground primary production. Science 291:481–484.

Komiyama, A., J. E. Ong, and S. Poungparn. 2008. Allometry, biomass, and productivity of mangrove forests: a review. Aquatic Botany 89:128–137.

Krauss, K. W., A. S. From, T. W. Doyle, T. J. Doyle, and M. J. Barry. 2011. Sea-level rise and landscape change influence mangrove encroachment onto marsh in the Ten Thousand Islands region of Florida, USA. Journal of Coastal Conservation 15:629–638.

Krauss, K. W., B. D. Keeland, J. A. Allen, K. C. Ewel, and D. J. Johnson. 2007. Effects of season, rainfall, and hydrogeomorphic setting on mangrove tree growth in Micronesia. Biotropica 39:161–170.

Krauss, K. W., C. E. Lovelock, K. L. McKee, L. López-Hoffman, S. M. L. Ewe, and W. P. Sousa. 2008. Environmental drivers in mangrove establishment and early development: a review. Aquatic Botany 89:105–127.

Kristensen, E., S. Bouillon, T. Dittmar, and C. Marchand. 2008. Organic carbon dynamics in mangrove ecosystems: a review. Aquatic Botany 89:201–219.

Lewis, D. B., J. A. Brown, and K. L. Jimenez. 2014. Effects of flooding and warming on soil organic matter mineralization in *Avicennia germinans* mangrove forests and *Juncus roemerianus* salt marshes. Estuarine, Coastal and Shelf Science 139:11–19.

Liu, H., H. Ren, D. Hui, W. Wang, B. Liao, and Q. Cao. 2014. Carbon stocks and potential carbon storage in the mangrove forests of China. Journal of Environmental Management 133:86–93.

Livesley, S. J., and S. M. Andrusiak. 2012. Temperate mangrove and salt marsh sediments are a small methane and nitrous oxide source but important carbon store. Estuarine, Coastal and Shelf Science 97:19–27.

Lomnicki, A., E. Bandola, and K. Jankowska. 1968. Modification of the Wiegert-Evans method for estimation of net primary production. Ecology 49:147–149.

Longley, W. L. 1994. Freshwater inflows to Texas bays and estuaries: ecological relationships and methods for determination of needs. Texas Water Development Board and Texas Parks and Wildlife Department, Austin, Texas, USA.

Longley, W. L. 1995. Estuaries. Pages 88–118 *in* G. R. North and J. Schmandt, editors. The impact of global warming on Texas: a report to the task force on climate change in Texas. University of Texas Press, Austin, Texas, USA.

Lot-Helgueras, A., C. Vásquez-Yánes, and F. Menéndez. 1975. Physiognomic and floristic changes near the northern limit of mangroves in the Gulf Coast of Mexico. Pages 52–61 *in* G. E. Walsh, S. C. Snedaker, and H. J. Teas, editors. Proceedings of the International Symposium on Biology and Management of Mangroves, Honolulu, Hawaii, October 8–11, 1974. Institute of Food and Agricultural Sciences, University of Florida, Gainesville, Florida, USA.

Lovelock, C. E., M. F. Adame, V. Bennion, M. Hayes, J. O'Mara, R. Reef, and N. S. Santini. 2014. Contemporary rates of carbon sequestration through vertical accretion of sediments in mangrove forests and saltmarshes of South East Queensland, Australia. Estuaries and Coasts 37:763–771.

Lovelock, C. E., I. C. Feller, R. Reef, S. Hickey, and M. C. Ball. 2017. Mangrove dieback during fluctuating sea levels. Scientific Reports 7: Article 1680.

Lovelock, C. E., K. W. Krauss, M. J. Osland, R. Reef, and M. C. Ball. 2016. The physiology of mangrove trees with changing climate. Pages 149–179 *in* G. Goldstein and L. S. Sanitago, editors. Tropical tree physiology: adaptations and responses in a changing environment. Springer, New York, New York, USA.

Lugo, A. E., and C. Patterson-Zucca. 1977. The impact of low temperature stress on mangrove structure and growth. Tropical Ecology 18:149–161.

Lugo, A. E., and S. C. Snedaker. 1974. The ecology of mangroves. The Annual Review of Ecology and Systematics 5:39–64.

Macreadie, P. I., et al. 2017. Carbon sequestration by Australian tidal marshes. Scientific Reports 7: 44071.

Markley, J. L., C. McMillan, and G. A. Thompson Jr. 1982. Latitudinal differentiation in response to chilling temperatures among populations of three mangroves, *Avicennia germinans*, *Laguncularia racemosa*, and *Rhizophora mangle*, from the western tropical Atlantic and Pacific Panama. Canadian Journal of Botany 60:2704–2715.

Marois, D. E., and W. J. Mitsch. 2015. Coastal protection from tsunamis and cyclones provided by mangrove wetlands: a review. International Journal of Biodiversity Science, Ecosystem Services & Management 11:71–83.

Maurer, E. P., J. C. Adam, and A. W. Wood. 2009. Climate model based consensus on the hydrologic impacts of climate change to the Rio Lempa basin of Central America. Hydrology and Earth System Sciences 13:183–194.

McCall, B. D., and S. C. Pennings. 2012. Geographic variation in salt marsh structure and function. Oecologia 170:777–787.

McKee, K., K. Rogers, and N. Saintilan. 2012. Response of salt marsh and mangrove wetlands to changes in atmospheric CO_2, climate, and sea level. Pages 63–92 *in* B. A. Middleton, editor. Global change and the function and distribution of wetlands. Springer, Dordrecht, The Netherlands.

McLeod, E., G. L. Chmura, S. Bouillon, R. Salm, M. Björk, C. M. Duarte, C. E. Lovelock, W. H. Schlesinger, and B. R. Silliman. 2011. A blueprint for blue carbon: toward an improved understanding of the role of vegetated coastal habitats in sequestering CO_2. Frontiers in Ecology and the Environment 9:552–560.

McMillan, C., and C. L. Sherrod. 1986. The chilling tolerance of black mangrove, *Avicennia germinans*, from the Gulf of Mexico coast of Texas, Louisiana and Florida. Contributions in Marine Science 29: 9–16.

Mendelssohn, I. A., and J. T. Morris. 2002. Eco-physiological controls on the productivity of *Spartina alterniflora* Loisel. Pages 59–80 *in* M. P. Weinstein and D. A. Kreeger, editors. Concepts and controversies in tidal marsh ecology. Springer, Dordrecht, The Netherlands.

Méndez-Alonzo, R., J. López-Portillo, and V. H. Rivera-Monroy. 2008. Latitudinal variation in leaf and tree traits of the mangrove *Avicennia germinans* (Avicenniaceae) in the central region of the Gulf of Mexico. Biotropica 40:449–456.

Michot, T. C., R. H. Day, and C. J. Wells. 2010. Increase in black mangrove abundance in coastal Louisiana. Newsletter of the Louisiana Association of Professional Biologists. http://www.labiologists.org/newsletter.php

Milner, C., and R. E. Hughes. 1968. Methods for the measurement of the primary production of grasslands. IBP Handbook No. 6. Blackwell Scientific, Oxford, UK.

Montagna, P. A., J. Brenner, J. C. Gibeaut, and S. Morehead. 2011. Coastal impacts. Pages 96–123 *in* J. Schmandt, G. R. North, and J. Clarkson, editors. The impact of global warming on Texas. Second edition. University of Texas Press, Austin, Texas, USA.

Morris, J. T., P. V. Sundareshwar, C. T. Nietch, B. Kjerfve, and D. R. Cahoon. 2002. Responses of coastal wetlands to rising sea level. Ecology 83: 2869–2877.

Morrisey, D. J., C. M. Beard, M. A. Morrison, R. Craggs, and M. Lowe. 2007. The New Zealand mangrove: review of the current state of knowledge. Technical Publication Number 325. Auckland Regional Council, Auckland, New Zealand.

Morrisey, D. J., A. Swales, S. Dittmann, M. A. Morrison, C. E. Lovelock, and C. M. Beard. 2010. The ecology and management of temperate mangroves. Pages 43–160 *in* R. N. Gibson, R. J. A. Atkinson, and J. D. M. Gordon, editors. Oceanography and marine biology: an annual review. CRC Press, Boca Raton, Florida, USA.

Naidoo, G. 2006. Factors contributing to dwarfing in the mangrove *Avicennia marina*. Annals of Botany 97:1095–1101.

Nemani, R. R., C. D. Keeling, H. Hashimoto, W. M. Jolly, S. C. Piper, C. J. Tucker, R. B. Myneni, and S. W. Running. 2003. Climate-driven increases in global terrestrial net primary production from 1982 to 1999. Science 300:1560–1563.

Nguyen, H. T., P. Meir, L. Sack, J. R. Evans, R. S. Oliveira, and M. C. Ball. 2017. Leaf water storage increases with salinity and aridity in the mangrove *Avicennia marina*: integration of leaf structure, osmotic adjustment, and access to multiple water sources. Plant, Cell & Environment 40:1576–1591.

Noe, G. B., and J. B. Zedler. 2001. Variable rainfall limits the germination of upper intertidal marsh plants in Southern California. Estuaries 24:30–40.

Odum, W. E., C. C. McIvor, and T. J. Smith III. 1982. The ecology of mangroves of south Florida: a community profile. US Fish and Wildlife Service, Office of Biological Services, Washington, D.C., USA.

Osland, M. J., R. H. Day, A. S. From, M. L. McCoy, J. L. McLeod, and J. J. Kelleway. 2015. Life stage influences the resistance and resilience of black mangrove forests to winter climate extremes. Ecosphere 6: Article 160.

Osland, M. J., R. H. Day, C. T. Hall, M. D. Brumfield, J. L. Dugas, and W. R. Jones. 2017*a*. Mangrove expansion and contraction at a poleward range limit: climate extremes and land-ocean temperature gradients. Ecology 98:125–137.

Osland, M. J., R. H. Day, J. C. Larriviere, and A. S. From. 2014*a*. Aboveground allometric models for freeze-affected black mangroves (*Avicennia germinans*): equations for a climate sensitive mangrove-marsh ecotone. PLoS ONE 9:e99604.

Osland, M. J., N. Enwright, R. H. Day, and T. W. Doyle. 2013. Winter climate change and coastal wetland foundation species: salt marshes vs. mangrove forests in the southeastern United States. Global Change Biology 19:1482–1494.

Osland, M. J., N. M. Enwright, R. H. Day, C. A. Gabler, C. L. Stagg, and J. B. Grace. 2016. Beyond just sea--level rise: considering macroclimatic drivers within coastal wetland vulnerability assessments to climate change. Global Change Biology 22:1–11.

Osland, M. J., N. Enwright, and C. L. Stagg. 2014*b*. Freshwater availability and coastal wetland foundation species: ecological transitions along a rainfall gradient. Ecology 95:2789–2802.

Osland, M. J., et al. 2017*b*. Climatic controls on the global distribution, abundance, and species richness of mangrove forests. Ecological Monographs 87:341–359.

Osunkoya, O., and R. G. Creese. 1997. Population structure, spatial pattern and seedling establishment of the grey mangrove, *Avicennia marina* var. *australasica*, in New Zealand. Australian Journal of Botany 45:707–725.

Ouyang, X., and S. Y. Lee. 2014. Updated estimates of carbon accumulation rates in coastal marsh sediments. Biogeosciences 11:5057–5071.

Ouyang, X., S. Y. Lee, and R. M. Connolly. 2017. The role of root decomposition in global mangrove and saltmarsh carbon budgets. Earth-Science Reviews 166:53–63.

Owers, C. J., K. Rogers, D. Mazumder, and C. D. Woodroffe. 2016. Spatial variation in carbon storage: a case study for Currambene Creek, NSW, Australia. Journal of Coastal Research 75:1297–1301.

Pan, Y., R. A. Birdsey, O. L. Phillips, and R. B. Jackson. 2013. The structure, distribution, and biomass of the world's forests. Annual Review of Ecology, Evolution, and Systematics 44:593–622.

Pennings, S. C., and M. D. Bertness. 1999. Using latitudinal variation to examine effects of climate on coastal salt marsh pattern and process. Current Topics in Wetland Biogeochemistry 3:100–111.

Pennings, S. C., and M. D. Bertness. 2001. Salt marsh communities. Pages 289–316 *in* M. D. Bertness, S. D. Gaines, and M. E. Hay, editors. Marine community ecology. First edition. Sinauer, Sunderland, Massachusetts, USA.

Pennings, S. C., and R. M. Callaway. 1992. Salt marsh plant zonation: the relative importance of competition and physical factors. Ecology 73:681–690.

Perry, C. L., and I. A. Mendelssohn. 2009. Ecosystem effects of expanding populations of *Avicennia germinans* in a Louisiana salt marsh. Wetlands 29:396–406.

Pickens, C. N., and M. W. Hester. 2011. Temperature tolerance of early life history stages of black mangrove *Avicennia germinans*: implications for range expansion. Estuaries and Coasts 34:824–830.

R Core Team. 2016. R: a language and environment for statistical computing. R Foundation for Statistical Computing, Vienna, Austria.

Reef, R., and C. E. Lovelock. 2014. Regulation of water balance in mangroves. Annals of Botany 115:385–395.

Rodriguez, W., I. C. Feller, and K. C. Cavanaugh. 2016. Spatio-temporal changes of a mangrove–saltmarsh ecotone in the northeastern coast of Florida, USA. Global Ecology and Conservation 7:245–261.

Rogers, K., N. Saintilan, and C. D. Woodroffe. 2014. Surface elevation change and vegetation distribution dynamics in a subtropical coastal wetland: implications for coastal wetland response to climate change. Estuarine, Coastal and Shelf Science 149:46–56.

Rogers, K., K. M. Wilton, and N. Saintilan. 2006. Vegetation change and surface elevation dynamics in estuarine wetlands of southeast Australia. Estuarine, Coastal and Shelf Science 66:559–569.

Ross, M. S., P. L. Ruiz, G. J. Telesnicki, and J. F. Meeder. 2001. Estimating above-ground biomass and production in mangrove communities of Biscayne National Park, Florida (USA). Wetlands Ecology and Management 9:27–37.

Rovai, A. S., et al. 2016. Scaling mangrove aboveground biomass from site-level to continental-scale. Global Ecology and Biogeography 25:286–298.

Roy, P. S., R. J. Williams, A. R. Jones, I. Yassini, P. J. Gibbs, B. Coates, R. J. West, P. R. Scanes, J. P. Hudson, and S. Nichol. 2001. Structure and function of south-east Australian estuaries. Estuarine, Coastal and Shelf Science 53:351–384.

Saenger, P. 2002. Mangrove ecology, silviculture and conservation. Kluwer Academic, Dordrecht, The Netherlands.

Saenger, P., and S. C. Snedaker. 1993. Pantropical trends in mangrove above-ground biomass and annual litterfall. Oecologia 96:293–299.

Saintilan, N., and K. Rogers. 2015. Woody plant encroachment of grasslands: a comparison of terrestrial and wetland settings. New Phytologist 205:1062–1070.

Saintilan, N., K. Rogers, D. Mazumder, and C. Woodroffe. 2013. Allochthonous and autochthonous contributions to carbon accumulation and carbon store in southeastern Australian coastal wetlands. Estuarine, Coastal and Shelf Science 128:84–92.

Saintilan, N., and R. J. Williams. 1999. Mangrove transgression into saltmarsh environments in south-east Australia. Global Ecology and Biogeography 8: 117–124.

Saintilan, N., N. C. Wilson, K. Rogers, A. Rajkaran, and K. W. Krauss. 2014. Mangrove expansion and salt marsh decline at mangrove poleward limits. Global Change Biology 20:147–157.

Sanders, C. J., D. T. Maher, D. R. Tait, D. Williams, C. Holloway, J. Z. Sippo, and I. R. Santos. 2016. Are global mangrove carbon stocks driven by rainfall? Journal of Geophysical Research: Biogeosciences 121:2600–2609.

Scheffer, M., S. Carpenter, J. A. Foley, C. Folke, and B. Walker. 2001. Catastrophic shifts in ecosystems. Nature 413:591–596.

Schile, L. M., J. B. Kauffman, S. Crooks, J. W. Fourqurean, J. Glavan, and J. P. Megonigal. 2017. Limits on carbon sequestration in arid blue carbon ecosystems. Ecological Applications 27:859–874.

Semeniuk, V. 1983. Mangrove distribution in Northwestern Australia in relationship to regional and local fresh water seepage. Vegetatio 53:11–31.

Semeniuk, V. 2013. Predicted response of coastal wetlands to climate changes: a western Australian model. Hydrobiologia 708:23–43.

Sheffield, J., G. Goteti, and E. F. Wood. 2006. Development of a 50-year high-resolution global dataset of meteorological forcings for land surface modeling. Journal of Climate 19:3088–3111.

Shepard, C. C., C. M. Crain, and M. W. Beck. 2011. The protective role of coastal marshes: a systematic review and meta-analysis. PLoS ONE 6:e27374.

Shew, D. M., R. A. Linthurst, and E. D. Seneca. 1981. Comparison of production computation methods in a southeastern North Carolina *Spartina alterniflora* salt marsh. Estuaries 4:97–109.

Simpson, L. T., T. Z. Osborne, L. J. Duckett, and I. C. Feller. 2017. Carbon storages along a climate induced coastal wetland gradient. Wetlands, *in press*. https://doi.org/10.1007/s13157-017-0937-x

Smalley, A. E. 1959. The role of two invertebrate populations, *Littorina irrorate* and *Orchelimum fidicinum*, in the energy flow of a salt marsh ecosystem. Dissertation. University of Georgia, Athens, Georgia, USA.

Soares, M. L. G., G. C. D. Estrada, V. Fernandez, and M. M. P. Tognella. 2012. Southern limit of the Western South Atlantic mangroves: assessment of the potential effects of global warming from a biogeographical perspective. Estuarine, Coastal and Shelf Science 101:44–53.

Spalding, M., M. Kainuma, and L. Collins. 2010. World atlas of mangroves. Earthscan, London, UK.

Stachelek, J., and K. H. Dunton. 2013. Freshwater inflow requirements for the Nueces Delta, Texas: *Spartina alterniflora* as an indicator of ecosystem condition. Texas Water Journal 4:62–73.

Stagg, C. L., D. R. Schoolmaster, K. W. Krauss, N. Cormier, and W. H. Conner. 2017. Causal mechanisms of soil organic matter decomposition: deconstructing salinity and flooding impacts in coastal wetlands. Ecology 98:2003–2018.

Stegen, J. C., N. G. Swenson, B. J. Enquist, E. P. White, O. L. Phillips, P. M. Jørgensen, M. D. Weiser, A. Monteagudo Mendoza, and P. Núñez Vargas. 2011. Variation in above-ground forest biomass across broad climatic gradients. Global Ecology and Biogeography 20:744–754.

Stuart, S. A., B. Choat, K. C. Martin, N. M. Holbrook, and M. C. Ball. 2007. The role of freezing in setting the latitudinal limits of mangrove forests. New Phytologist 173:576–583.

Sutton-Grier, A. E., K. Wowk, and H. Bamford. 2015. Future of our coasts: the potential for natural and hybrid infrastructure to enhance the resilience of our coastal communities, economies and ecosystems. Environmental Science & Policy 51:137–148.

Tomlinson, P. B. 1986. The botany of mangroves. Cambridge University Press, New York, New York, USA.

Turner, R. E. 1976. Geographic variations in salt marsh macrophyte production: a review. Contributions in Marine Science 20:47–68.

Twilley, R. R. 1982. Litter dynamics and organic carbon exchange in black mangrove (*Avicennia germinans*) basin forests in a southwest Florida estuary. Dissertation. University of Florida, Gainesville, Florida, USA.

Twilley, R. R., R. H. Chen, and T. Hargis. 1992. Carbon sinks in mangroves and their implications to carbon budget of tropical coastal ecosystems. Water, Air and Soil Pollution 64:265–288.

Twilley, R. R., and J. W. Day. 2012. Mangrove wetlands. Pages 165–202 *in* J. W. Day Jr., B. C. Crump, M. W. Kemp, and A. Yáñez-Arancibia, editors. Estuarine ecology. Second edition. John Wiley & Sons, Hoboken, New Jersey, USA.

Valiela, I., J. M. Teal, and W. J. Sass. 1975. Production and dynamics of salt marsh vegetation and the effects of experimental treatment with sewage sludge: biomass, production and species composition. Journal of Applied Ecology 12:973–981.

Ward, R. D., D. A. Friess, R. H. Day, and R. A. MacKenzie. 2016. Impacts of climate change on mangrove ecosystems: a region by region overview. Ecosystem Health and Sustainability 2:e01211.

West, R. C. 1977. Tidal salt-marsh and mangal formations of Middle and South America. Pages 193–213 *in* V. J. Chapman, editor. Wet coastal ecosystems. Elsevier, Amsterdam, The Netherlands.

Whittaker, R. H. 1970. Communities and ecosystems. Second edition. Macmillan USA, New York, New York, USA.

Wiegert, R. G., and F. C. Evans. 1964. Primary production and the disappearance of dead vegetation on an old field in southeastern Michigan. Ecology 45:49–63.

Wiegert, R. G., and J. T. McGinnis. 1975. Annual production and disappearance of detritus on three South Carolina old fields. Ecology 56:129–140.

Withers, K. 2002. Wind-tidal flats. Pages 114–126 *in* J. W. Tunnel and F. W. Judd, editors. The Laguna Madre of Texas and Tamaulipas. Texas A&M University Press, College Station, Texas, USA.

Woodroffe, C. D., and J. Grindrod. 1991. Mangrove biogeography: the role of quaternary environmental and sea-level change. Journal of Biogeography 18:479–492.

Woodward, F. I. 1987. Climate and plant distribution. Cambridge University Press, Cambridge, UK.

Wu, W., H. Huang, P. Biber, and M. Bethel. 2017. Litter decomposition of *Spartina alterniflora* and *Juncus roemerianus*: implications of climate change in salt marshes. Journal of Coastal Research 332:372–384.

Yando, E. S., M. J. Osland, J. M. Willis, R. H. Day, K. W. Krauss, and M. W. Hester. 2016. Salt marsh-mangrove ecotones: using structural gradients to investigate the effects of woody plant encroachment on plant-soil interactions and ecosystem carbon pools. Journal of Ecology 104:1020–1031.

Zedler, J. B. 1983. Freshwater impacts in normally hypersaline marshes. Estuaries 6:346–355.

Zhang, D., D. Hui, Y. Luo, and G. Zhou. 2008. Rates of litter decomposition in terrestrial ecosystems: global patterns and controlling factors. Journal of Plant Ecology 1:85–93.

PERMISSIONS

 Every chapter published in this book has been scrutinized by our experts. Their significance has been extensively debated. The topics covered herein carry significant findings which will fuel the growth of the discipline. They may even be implemented as practical applications or may be referred to as a beginning point for another development.

The contributors of this book come from diverse backgrounds, making this book a truly international effort. This book will bring forth new frontiers with its revolutionizing research information and detailed analysis of the nascent developments around the world.

We would like to thank all the contributing authors for lending their expertise to make the book truly unique. They have played a crucial role in the development of this book. Without their invaluable contributions this book wouldn't have been possible. They have made vital efforts to compile up to date information on the varied aspects of this subject to make this book a valuable addition to the collection of many professionals and students.

This book was conceptualized with the vision of imparting up-to-date information and advanced data in this field. To ensure the same, a matchless editorial board was set up. Every individual on the board went through rigorous rounds of assessment to prove their worth. After which they invested a large part of their time researching and compiling the most relevant data for our readers.

The editorial board has been involved in producing this book since its inception. They have spent rigorous hours researching and exploring the diverse topics which have resulted in the successful publishing of this book. They have passed on their knowledge of decades through this book. To expedite this challenging task, the publisher supported the team at every step. A small team of assistant editors was also appointed to further simplify the editing procedure and attain best results for the readers.

Apart from the editorial board, the designing team has also invested a significant amount of their time in understanding the subject and creating the most relevant covers. They scrutinized every image to scout for the most suitable representation of the subject and create an appropriate cover for the book.

The publishing team has been an ardent support to the editorial, designing and production team. Their endless efforts to recruit the best for this project, has resulted in the accomplishment of this book. They are a veteran in the field of academics and their pool of knowledge is as vast as their experience in printing. Their expertise and guidance has proved useful at every step. Their uncompromising quality standards have made this book an exceptional effort. Their encouragement from time to time has been an inspiration for everyone.

The publisher and the editorial board hope that this book will prove to be a valuable piece of knowledge for researchers, students, practitioners and scholars across the globe.

LIST OF CONTRIBUTORS

Erin F. Abernethy
Odum School of Ecology, University of Georgia, Athens, Georgia 30602 USA
Savannah River Ecology Lab, University of Georgia, Aiken, South Carolina 29802 USA

Kelsey L. Turner and James C. Beasley
Savannah River Ecology Lab, University of Georgia, Aiken, South Carolina 29802 USA
Warnell School of Forestry and Natural Resources, University of Georgia, Athens, Geor

Olin E. Rhodes Jr.
Savannah River Ecology Lab, University of Georgia, Aiken, South Carolina 29802 USA

Nicolas L. Gutierrez, Ray Hilborn and Andre E. Punt
School of Aquatic and Fishery Sciences, University of Washington, Box 355020, Seattle, Washington 98195-5020 USA

Peter Halmay
San Diego Watermen's Association, 11103 Highway 67, Lakeside, California 92040 USA

Stephen Schroeter
Marine Science Institute, University of California Santa Barbara, Isla Vista, California 93106 USA

Emily L. A. Kelly, Yoan Eynaud, Stuart A. Sandin and Jennifer E. Smith
Center for Marine Biodiversity and Conservation, Scripps Institution of Oceanography, University of California San Diego, La Jolla, California 92093 USA

Ivor D. Williams
Ecosystem Science Division (ESD), Pacific Islands Fisheries Science Center (PIFSC), National Marine Fisheries Service, NOAA IRC, Honolulu, Hawaii 96818 USA

Russell T. Sparks
Division of Aquatic Resources, Maui Office, Depart-ment of Land and Natural Resources, Wailuku, Hawaii 96768 USA

Meghan L. Dailer
Department of Botany, University of Hawai'i, 3190 Maile Way, Honolulu, Hawaii 96822 USA

David L. Kimbro and Owen Stokes-Cawley
Department of Marine and Environmental Sciences, Northeastern University, Marine Science Center, Nahant, Massachusetts 01908 USA

J. Wilson White
Department of Fisheries and Wildlife, Oregon State University, Corvallis, Oregon 97331 USA

Hanna Tillotson, Nikkie Cox, Megan Christopher and Samanthayuan
Apalachicola National Estuarine Research Reserve, East Point, Florida 32328 USA

Timothy J. Pusack and Christopher D. Stallings
College of Marine Science, University of South Florida, St. Petersburg, Florida 33701 USA

Jonah Piovia-Scott
School of Biological Sciences, Washington State University, 14204 NE Salmon Creek Avenue, Vancouver, Washington 98686 USA

Louie H. Yang
Department of Entomology and Nematology, University of California, One Shields Avenue, Davis, California 95616 USA

Amber N. Wright
Department of Biology, University of Hawaii, Manoa, 2500 Campus Road, Honolulu, Hawaii 96822 USA

David A. Spiller and Thomas W. Schoener
Department of Evolution and Ecology, University of California, One Shields Avenue, Davis, California 95616 USA

Thomas Pool and Gordon Holtgrieve
School of Aquatic and Fishery Sciences (SAFS), University of Washington, Seattle, Washington 98105 USA

Vittoria Elliott, Chheng Phen, Thach Phanara and Samol Chhuoy
Inland Fisheries Research and Development Institute (IFReDI), Fisheries Administration, Phnom Penh, Cambodia

Kevin Mccann
Department of Integrative Biology, University of Guelph, Guelph, Ontario, Canada

Bailey Mcmeans
Department of Biology, University of Toronto, Mississauga, Ontario, Canada

Neil Rooney
School of Environmental Sciences, University of Guelph, Guelph, Ontario, Canada

Adrianne Smits
Department of Biology, University of Nevada, Reno, Nevada 89557 USA

Michael Cooperman
Gordon and Betty Moore Center for Science, Conservation International, Arlington, Virginia 22202 USA

Sydney Clark
Earth, Environmental and Planetary Sciences, Brown University, Providence, Rhode Island 02912 USA

Hongqing Wang, Brady R. Couvillion and Holly J. Beck
Wetland and Aquatic Research Center, U.S. Geological Survey, Baton Rouge, Louisiana 70803 USA

Gregory D. Steyer
Southeastern Region, U.S. Geological Survey, Baton Rouge, Louisiana 70803 USA

John M. Rybczyk
Department of Environmental Science, Western Washington University, Bellingham, Washington 98225 USA

Victor H. Rivera-Monroy
Department of Oceanography and Coastal Sciences, Louisiana State University, Baton Rouge, Louisiana 70803 USA

Ken W. Krauss
Wetland and Aquatic Research Center, U.S. Geological Survey, Lafayette, Louisiana 70506 USA

Jenneke M. Visser
Institute for Coastal and Water Research, School of Geosciences, University of Louisiana at Lafayette, Lafayette, Louisiana 70506 USA

Fidelis A. Atuo and Timothy J. O'connell
Department of Natural Resource Ecology and Management, Oklahoma State University, Stillwater, Oklahoma 74078 USA

D. E. P. Bryant
The Global Change Institute, ARC Centre for Excellence in Coral Reef Studies and School of Biological Sciences, The University of Queensland, St Lucia, Queensland 4072 Australia
Remote Sensing Research Centre, School of Geography Planning and Environmental Management, The University of Queensland, St Lucia, Queensland 4072 Australia

A. Rodriguez-Ramirez, M. González-Rivero, O. Hoegh-Guldberg, K. T. Brown and S. Dove
The Global Change Institute, ARC Centre for Excellence in Coral Reef Studies and School of Biological Sciences, The University of Queensland, St Lucia, Queensland 4072 Australia

S. Phinn
Remote Sensing Research Centre, School of Geography Planning and Environmental Management, The University of Queensland, St Lucia, Queensland 4072 Australia

B. P. Neal
The Global Change Institute, ARC Centre for Excellence in Coral Reef Studies and School of Biological Sciences, The University of Queensland, St Lucia, Queensland 4072 Australia
Bigelow Laboratory for Ocean Sciences, 60 Bigelow Drive, East Boothbay, Maine 04544 USA

Jeff W. Atkins
Department of Environmental Sciences, University of Virginia, 291 McCormick Ave, Charlottesville, Virginia 22902 USA
Department of Biology, Virginia Commonwealth University, 1000 West Cary Street, Richmond, Virginia 23284 USA

Howard E. Epstein
Department of Environmental Sciences, University of Virginia, 291 McCormick Ave, Charlottesville, Virginia 22902 USA

Daniel L. Welsch
American Public University System, 111 W Congress St, Charles Town, West Virginia 25414 USA

Jean-François Lapierre
Département de Sciences Biologiques, Universite de Montreal, Pavillon Marie-Victorin, 90, Avenue Vincent-d'Indy, Montreal, Québec H2V 2S9 Canada
GRIL Université de Montréal, Pavillon Marie-Victorin, Département de Sciences Biologiques, C.P. 6128, succ. Centre-ville, Montreal, Québec H3C 3J7 Canada

Samantha K. Oliver
Center for Limnology, University of Wisconsin Madison, Madison, Wisconsin 53706 USA

Carol Emi Fergus, Nicholas K. Skaff and Patricia Soranno
Department of Fisheries and Wildlife, Michigan State University, East Lansing, Michigan 48824 USA

Kendra S. Cheruvelil
Department of Fisheries and Wildlife, Michigan State University, East Lansing, Michigan 48824 USA
Lyman Briggs College, Michigan State University, East Lansing, Michigan 48825 USA

Katherine Webster
Department of Fisheries and Wildlife, Michigan State University, East Lansing, Michigan 48824 USA
School of Natural Sciences, Trinity College Dublin, The University of Dublin, College Green, Dublin 2, Ireland

Caren Scott
NEON, 1685, 38th Street, Suite 100, Boulder, Colorado 80301 USA

Laura C. Feher, Michael J. Osland, James B. Grace, Rebecca J. Howard, Camille L. Stagg, Nicholas M. Enwright, Ken W. Krauss and Richard H. Day
U.S. Geological Survey, Wetland and Aquatic Research Center, Lafayette, Louisiana 70506 USA

Kereen T. Griffith
Griffith Consulting Services at U.S. Geological Survey, Lafayette, Louisiana 70506 USA

Christopher A. Gabler
School of Earth, Environmental and Marine Sciences, University of Texas Rio Grande Valley, Brownsville, Texas 78520 USA

Kerrylee Rogers
School of Earth and Environmental Sciences, University of Wollongong, Wollongong, New South Wales 2522 Australia

Index